Practices in Education for Climate Emergency

Júlio J. Conde · Antonio García-Vinuesa ·
Pablo Ángel Meira Cartea
Editors

Practices in Education for Climate Emergency

Disruptive Approaches Beyond the Anglosphere

Springer

Editors
Júlio J. Conde
Facultade de Ciencias da Educación
Universidade de Santiago de Compostela
Santiago de Compostela, Spain

Antonio García-Vinuesa
Facultade de Ciencias da Educación
Universidade de Santiago de Compostela
Santiago de Compostela, Spain

Pablo Ángel Meira Cartea
Facultade de Ciencias da Educación
Universidade de Santiago de Compostela
Santiago de Compostela, Spain

ISBN 978-981-92-2303-9 ISBN 978-981-92-2304-6 (eBook)
https://doi.org/10.1007/978-981-92-2304-6

This Springer imprint is published by the registered company Springer Nature Singapore Pte Ltd.
The registered company address is: 152 Beach Road, #21-01/04 Gateway East, Singapore 189721, Singapore

A Disruptive Foreword

The Existential Threat of Environmental Collapse and the Rejection of Reproductivist Climate Education

In the early 1970s, Hans Liebmann (1976) observed that "the number of biologists who predict an impending ecological catastrophe—whose effects will be incomparably worse than those of all nuclear bombs combined—is steadily increasing". From this diagnosis, he argued that, although "the planet on which we live is on the verge of becoming uninhabitable", human beings had to "live today with this grim prospect [because] there is only one way to avert an impending ecological catastrophe: to recognize the options that still remain for human survival" (p. 11).

The preceding passages capture the continuing relevance of environmental collapse, yet it was written a long time ago, in 1973, and not in a prophetic or sadistic vein. The idea of habitability and survival on a planet genuinely threatened by an anthropogenic ecocatastrophe is not recent. Rather, fifty years ago, in the context of the Stockholm Conference, it became established with profound apprehension in the social imaginary of environmentalism, when the warning message of those who came to be known in the environmental community as the "Prophets of the Apocalypse" began to trouble the promising progress of consumer society in its pursuit of economic prosperity.

That moment was marked not only by the publication of the Meadows report (Meadows et al., 1972) for the Club of Rome, with its dramatic warning about the existence of planetary natural limits that could not be exceeded, as a consequence of environmental deterioration resulting from economic growthism, at the risk of the total destruction of the basis supporting life on Earth. It was understood that the very habitability of this planet, and even human survival itself, were in danger, threatened by the uncontrolled advance of developmentalism now directed by neoliberal doctrine.

Fear of environmental imbalance on a planetary scale entailed a warning about the existential threat that a global environmental crisis could represent. In the year before and the year after the Stockholm Conference, Richard Falk (1971) published

This Endangered Planet; Paul Ehrlich (1971) published *How to Be a Survivor* Laurence Pringle (1971) published *Ecology, Science of Survival*; Barbara Ward and René Dubos (1972) wrote *Only One Earth*, in which they present, in its conclusion, "strategies for survival"; Sicco Mansholt (1973), together with Edgar Morin, Herbert Marcuse, and other colleagues, published *Ecologia: caso de vida ou de morte*; Hans Liebmann (1976) wrote in 1973 the work *Terra, um planeta habitável?*; René Dumont (1973) published *L'utopie ou la mort!*; and later Ernest Snyder (1978) wrote *Please Stop Killing Me!*

The common thread among these works—despite their terrifying titles and their influence both from the new perspective of the metaphorical image of the Earth as a small spaceship wandering through the immensity of the cosmos and from Neo-Malthusianism—was the scientific recognition that anthropogenic environmental impacts could indeed reach the global biospheric scale.

Accompanied by the warning that the course of developmentalism was drawing dangerously close to an imminent planetary environmental crisis, the concern with the existential question of survival was an explicit message recommending prudence in the face of that sign of material abundance and economic prosperity promised by the idea of Progress. After all, resolving the contradiction between a voracious growth-oriented economy and an ecologically fragile and naturally finite planet represented the essential question if humanity was to avoid the path toward biospheric ecocatastrophe and guarantee the rights of future generations.

The existential meaning of Gaia's survival is to take into account the vital conditions of Earth's habitability, and the civilizational mission comes to revolve around guaranteeing the functioning of ecosystem services as fundamental infrastructure. For Bruno Latour (2023), this is defined by bringing together, within the same space fractured by industrial production, the world in which one lives and the world on which one depends for life, because the world from which production is extracted has overburdened the world in which one dwells.

At that initial moment of collapse warning by the Prophets of the Apocalypse, ecology came to be viewed as the science of survival, and an attempt was made to develop public awareness of human dependence on its natural base in order to achieve political debate capable of limiting economic production within ecologically safe margins so that planetary habitability might thrive in full. Guided by the precautionary principle, it was believed that the soundness of scientific forecasting could guide a balance between industrial production and environmental protection, and that it would be enough to popularize awareness of existential risk for the political conditions for future prevention to emerge naturally.

Not by chance, this is one of the structuring roots of Environmental Education, which for some time overlapped with the teaching of ecology, but renewed by reflection on the harmful impact of anthropogenic activities on the environment. Yet that very root was already bifurcating in two distinct directions: while the Prophets of the Apocalypse operated by accentuating a negative affective approach in an emotionally painful pedagogical experience—based on understanding the dramatic nature of ecocatastrophe in order to generate a response commensurate with the risk that had to be avoided at any cost—, the fundamental identity of Environmental Education was forged in parallel, strongly anchored in a positive affective approach, with the desire to cultivate deep biophilic affection for nature. This was based on the assumption "that a constant diet of bad news would probably be unhealthy for young people" (Tanner, 1974), because speaking about a future threatened by a global environmental crisis could result in pessimism and paralysis.

For this reason, it is unusual to regard the messages of the Prophets of the Apocalypse as authentic and legitimate voices within the field of Environmental Education, because they do not correspond to the hegemonic idea that it would be inappropriate to adopt an excessively somber tone in order to trigger an ecological consciousness frightened by the warning of an existential threat. Yet, one way or another, with the pessimism of reason and the optimism of the will, with collapsology on the one hand and mainstream Environmental Education on the other, humanity would undergo a process of ecological conscientization, sensitized and mobilized either by fear of ecocatastrophe or by love of nature.

The warning of the Prophets of the Apocalypse began to sound particularly from the 1970s onward, when the ecological sciences made evident the planetary dimension of the negative environmental effects of the economistic developmentalism that had become hegemonic two decades earlier, immediately after the Second World War. Indeed, environmental impacts began to change abruptly in scale from a specific moment onward: the 1950s, when the US cycle of capital accumulation began and the new regime of production—now mass production—was propelled with the inauguration of consumer society. It is also precisely from this moment, known as the Golden Age of Capitalism, that the American Way of Life initiated the "Metabolic Rift" identified by eco-Marxist authors such as John Bellamy Foster (2005) and Kohei Saito (2021). This, not coincidentally, coincides with what the philosopher István Mészáros (1996) characterized as "Destructive Production," and with what some climate scientists working within the geological framing of the Anthropocene labelled the "Great Acceleration" (McNeill & Engelke, 2016), when capital accumulation began to exceed the boundaries of natural biospheric safety.

Anthropocentric exceptionalism and economic power triumphed over the outcry for caution. There is no shortage of indicators pointing to the critical worsening of the global ecosystemic condition, compromising the planet's habitability. The destination ahead of spaceship Earth was, in the end, the unwanted arrival of ecocatastrophe, which could not be avoided. From the first decade of the twenty-first century onward, marked by the establishment of the "state of climate emergency," a profound inflection began as planetary thresholds were reached (Rockström et al., 2009). The metabolic rift finally materializes. The warnings of the Prophets of the Apocalypse did not produce the expected preventive effect, and the environmental crisis that had been anticipated in advance descended in the form of collapse, marking our historical time with the idea of irreversibility.

This new characterization of climate change, now designated as climate emergency, is emblematic, leaving no doubt about the sense of urgency that must be activated in this decisive decade which Luiz Marques (2023) identifies as an ultimatum to action. This geoclimatic turning point bears witness to the fact that the time has come to consult the "survival manual" in order to rewrite the next lines of history in the face of modernity's breakdown in steering spaceship Earth.

This decisive threshold, which should never have been crossed, must be understood in its deepest sense: existential, because what is now at stake as a societal project is no longer so much the reconciliation between the pursuit of economic prosperity and ecological prudence balanced with social justice, but rather the survival of planetary life. This change of state, from crisis to environmental collapse, once again places the dramatic question of survival and habitability on Earth at the center of political debate. It does so now compounded both by an inescapable sense of urgency and by the frustration of the failure of the early warning about the limits to growth, because the logic agreed upon in the context of the 1992 Rio Conference to save the future, namely to make environmental protection compatible with profit in the market of Sustainable Development, did not prevent the climate emergency.

This situation suggests that it is not reasonable to trust that the responses required to confront the state of climate emergency should continue to be based on reformist solutions within the order of capital, still shaped by the idea of a green economy capable of energizing markets in clean energy and climate resilience, yet without any guarantee that this time it will be able to avert the ultimate environmental collapse within the little time that remains. It is precisely because there is no "Planet B" to rescue us that this distrust is so well represented by the disruptive slogan "change the system, not the climate," so popular in climate demonstrations.

Thus, the problem that should orient climate education becomes existential: it is no longer enough to continue forming the ecological subject who now understands enough about climate change only, in the end, to become an altruistic "friend of the climate," convinced that it is sufficient to do one's individual part to reduce one's carbon footprint, because the cultural industry says that these small but virtuous behavioral changes would already constitute a significant contribution to climate mitigation.

The meaning of the existential threat posed by the climate emergency requires a revision of what Environmental Education has developed as a "normal" tendency of the climate curriculum. To stand at the urgent threshold of the initial phase of environmental collapse impels us to reconsider the functionalist identity of Environmental Education, and this new framing follows a disruptive logic: the reformist norm of climate education must be transformed, clearly situating it in a counter-hegemonic position capable of breaking with the common sense of domesticated collapseist power/knowledge.

For disruptive climate education, entering the Anthropocene means altering the understanding one has of the functionalist societal norm, designed to form uncritical subjects who are politically incapable of contesting the unjust and ecocidal social order. Reformism, as a reproductivist strategy, neutralizes critique of the system and naturalizes the social order, leaving the prevailing societal model unchanged. In this state of climate emergency, it is no longer possible to practice reformist pedagogical processes that preserve the economic imperative intact and contribute to the reproduction of capitalist sociability.

Now, in the Anthropocene, climate education must be rethought, beginning with recognition of the mechanisms of ideological co-optation that are continuously reproduced in "normal" educational practices concerning climate. From the disruptive perspective, what matters fundamentally is not to continue repeating the greatest climatic half-truth that the climate crisis is derived generically from human activities. If that premise is accepted, then once again the solution would lie in the individual responsibilization of the climate-friendly subject who virtuously and obediently does one's part to reduce one's carbon footprint, and nothing more. There is no social conflict whatsoever in this logic, only the personal examination of conscience of guilt; therefore, in a society conceived without conflict, there is no reason at all to rebel, collectively, against those who truly are the perpetrators of the climate crisis.

Climate education must be rethought in the real context of the climate emergency, because it is contaminated by a reformist educational project that insists on erasing the structural cause, always masked by the guiding idea of the "guilt" of the aggregate of human activities for the environmental crisis. The naturalization of this tacit premise of domesticated power/knowledge implies dispensing with critique of the structural causes of the crisis, which go unnoticed or are underestimated in "normal" pedagogical mediation.

Within this curricular configuration of erasing the structural dimension of the climate emergency, it is not possible to activate disruptive mechanisms that dare to change the system rather than the climate, such as the controversial recourse to civil disobedience in the climate action of radical environmentalism, which vehemently demands from political-economic elites the immediate end of fossil fuels and the transition to ecosocialism. Perhaps an enlightened understanding of the structural causes of climate instability is what allows the just anger of the oppressed, in the Freirean pedagogy of indignation, to emerge against a model of society conceived in inequality, the basis of the suffering of social injustice and environmental degradation, because this is the dynamic that drives capital.

The crucial point is the neoliberal domestication present in Environmental Education, aligned with the capitalist sociability of the Sustainable Development paradigm, which operates in the reproductivist direction of the contemporary social order. The concealment of certain forms of knowledge that are excluded from the curriculum is the central mechanism of domination, flowing discursively through the circulation of the subtle version of denialism, expressed through climate half-truths, that is, information that is correct but imprecise or incomplete. Ideologically, this results in the shaping of servile obedience to the socioeconomic order, because it communicates with the contemporary imaginary of the social norm that—in Foucauldian terms—is civilized, docile, and useful, in the name of the stability of capitalist sociability. It seems unlikely that one could hope to change the system while teaching obedience in relation to what should be done as climate action: perhaps the power/knowledge of the Prophets of the Apocalypse points to a disruptive path.

But why is it important not to dismiss political-pedagogical reflection on the structural causes of the climate emergency, regardless of the thematic focus selected when choosing contents to compose the climate curriculum? Because situating the existential problem of environmental collapse as deriving from fossil capitalism, and not exactly from generic human activities, makes explicit the reason for the concealment of this ecopolitical knowledge that advances anticapitalist critique: the interest in erasing class struggle and the values and powers of the petro-bourgeois elite in the dispute over the direction of the societal project that encompasses us.

The climate emergency affects class struggle but also redefines it through new alliances in the context of the ultimate Metabolic Rift that projected us into the Anthropocene. With its new sources of socio-environmental suffering and new fronts of resistance articulated around the cry for climate justice, the existential meaning of survival is the common horizon that makes it possible to update ecopolitical class struggle in a contemporary configuration. This configuration essentially antagonizes the powers and interests of those who want to change the system so as not to change the climate and therefore adopt the disruptive perspective; and those representatives of capital, linked to the petro-bourgeoisie, who, on the contrary, want to change the climate, but never the system. They adopt reproductivist ideological strategies of social control in order to keep the order intact, so that there may be no obstacles to capital accumulation.

Even in the ultimate state of climate emergency in which we now find ourselves, capitalism continues to respond to the existential dilemma with the same structural formula based on a growth-oriented free-market economy, because what matters to keep alive is not the planet, but the system that continuously and indefinitely expands capital accumulation through the successive creation of new markets, even at the cost of climate change. At the 30th United Nations Climate Change Conference, held in the Amazon in 2025, the petro-bourgeoisie chose its path when it decided to focus its investments on the economy of climate adaptation, once again rejecting structural mitigation through the urgent energy transition away from fossil fuels. Those at the top abandoned the habitability of the Earth; their political responses remain out of step with the state of climate emergency, the outcome of a vision concerned solely with the possibility of reforming capitalism, even when the biophysical limits of planetary ecosystemic support have been dangerously exceeded.

This implies a reorganization of what had previously been the consciousness of the popular classes, which now becomes ecopolitical, capable of generating a sense of collective belonging around the same ideal and with clarity about who the rivals are in the state of climate emergency: it is also in this sense that the disruptive educational perspective is framed, through the formation of consciousness of belonging to the ecopolitical class that wants to change the system, not the climate. It is not enough, for example, to form an ecological subject who is a "friend of the climate" if that subject is not also an ecopolitical subject who is an enemy of oil.

Finally, disruptive climate education avoids the concealment of reality as an ideological strategy of social reproductivism that keeps the social order unchanged. Counter-hegemonic and anti-systemic climate education is that whose political project does not servilely follow the masking of the structural causes of the climate crisis derived from fossil capitalism, whether through the repetition of vague and imprecise discourses, through shifting the focus of attention from mitigation to adaptation—supposedly now a priority—, or through the repetition of the neoliberal discourse of the altruistic ecological subject, which tells such a subject that small changes in everyday personal life are enough to help "save the climate."

Under no circumstances can the disruptive perspective of climate education reproduce the paradigmatic trap of such half-truths and omit from the curriculum the debate on the relationship between capitalism and climate, nor the power of the petro-dependence of the fossil economy in maintaining consumer society. To be disruptive, it is indispensable not to forgo curricular contents that direct climatic power/knowledge toward another correlation of forces capable of awakening a sense of subversion, indignation, organic collectivity, and awareness of the socio-environmental suffering deriving from solastalgia and eco-anxiety, to which the popular classes and youth, respectively, are exposed through the disaster of climate injustice.

It is the resolute refusal both of subservience to neoliberal logic and of the ideological concealment of the forms of knowledge that make it possible to understand the ecopolitical powers and interests at stake in this reconfiguration of socio-environmental class struggle. The capacities required to change the system and not the climate depend on overcoming the hegemonic idea of the virtuous and altruistic ecological subject, instrumentally docile and useful, in favor of the ecopolitical subject versed in the Freirean knowledge of indignation and rebellion and endowed with militant and combative collective power.

In this ultimate collapse warning, it is now the turn of political ecology to be seen as the decisive science, as the science of the struggle for survival. These are some of the keys to understanding the innovative meaning of the disruptive perspective, capable of guiding the formulation of the counter-hegemonic project of climate education, coherent with the dramatic existential context of the state of climate emergency, when it is no longer possible to go on doing more of the same that was done in "normal" times. It is within this existential framework that the indispensable work *Practices in Education for Climate Emergency*, calling for disruptive initiatives in climate education in order to establish an innovative reference point for the Anthropocene, capable of shaping pedagogical experience by bringing about a rebalancing of positive and negative effects, and of their potentials for climate action, within the identity of Environmental Education.

Brasília, Brazil Philippe Pomier Layrargues

References

Dumont, R. (1973). *L'utopie ou la mort!* Éditions du Seuil.
Ehrlich, P. (1971). *How to be a survivor*. Ballantine Books.
Falk, R. (1971). *This endangered planet*. Random House.
Foster, J. B. (2005). *Marx's ecology: Materialism and nature*. Monthly Review Press.
Latour, B., & Schultz, N. (2023). *Memorando sobre a nova classe ecológica*. Vozes.
Liebmann, H. (1976). *Terra: um planeta inabitável?* Editora da Universidade de São Paulo.

Marques, L. (2023). *O decênio decisivo e a ação climática*. Editora Elefante.

Mansholt, S. et al. (1973). *Ecologia: caso de vida ou de morte*. Moraes Editores.

McNeill, J. R., & Engelke, P. (2016). *The great acceleration: An environmental history of the Anthropocene since 1945*. Belknap Press.

Meadows, D. H., Meadows, D. L., Randers, J., & Behrens III, W. W. (1972). *The limits to growth*. Club of Rome.

Mészáros, I. (1996). *Produção destrutiva e estado capitalista*. Ensaio.

Pringle, L. (1971). *Ecology, science of survival*. Harper & Row.

Rockström, J., Steffen, W., Noone, K., Persson, Å., Chapin, F. S., Lambin, E., Lenton, T. M., Scheffer, M., Folke, C., Schellnhuber, H., Nykvist, B., De Wit, C. A., Hughes, T., van der Leeuw, S., Rodhe, H., Sörlin, S., Snyder, P. K., Costanza, R., Svedin, U., ... Foley, J.. (2009). Planetary boundaries: exploring the safe operating space for humanity. *Ecology and Society*, 14(2), 32. https://www.ecologyandsociety.org/vol14/iss2/art32/

Saito, K. (2021). *O ecossocialismo de Karl Marx: Capitalismo, natureza e a crítica inacabada à economia política*. Boitempo.

Snyder, E. E. (1978). *Please stop killing me!* New American Library.

Tanner, R. T. (1974). *Ecology, Environment and Education*. Professional Educators Publications.

Ward, B., & Dubos, R. (1972). *Only one Earth: Strategies for survival*. BBC and Collins.

Acknowledgments

This book originated within the framework of RESCLIM@TIEMPO project (PID2022-136933OB-C21), funded by MCIN/AEI/10.13039/501100011033 and by ESF Investing in your future. Júlio Conde acknowledges the State Research Agency (AEI) for his postdoctoral fellowship (Grant reference FJC2021-047345-I) funded by MCIN/AEI/10.13039/501100011033 and by the European Union NextGenerationEU/PRTR.

Contents

A Final Reflection

Contributors

Aitor Alonso-Méndez Universidade of Santiago de Compostela, Santiago de Compostela, Spain

Marla Arce Pimienta Instituto de Investigaciones en Ecosistemas y Sustentabilidad, Universidad Nacional Autónoma de México, Morelia, Michoacán, México

Mónica Arto Blanco Universidade da Coruña, A Coruña, Spain

José Manuel Gutiérrez Bastida ESenRED, Segovia, Spain

Juan Ramón Brambila-Topete Escuela Preparatoria Regional de Cihuatlán, Universidad de Guadalajara, Miguel Hidalgo, Jalisco, México

Fabiola Itzel Cabrera-García Instituto de Investigaciones en Educación, Universidad Veracruzana, Xalapa-Enríquez, México

Miriam Campos Leirós Teachers for Future Spain, Gondomar, Spain

José Antonio Caride Universidade of Santiago de Compostela, Santiago de Compostela, Spain

Irene Carniatto Universidade Estadual do Oeste do Paraná, Marechal Cândido Rondon, Brazil

Alicia Castillo Instituto de Investigaciones en Ecosistemas y Sustentabilidad, Universidad Nacional Autónoma de México, Morelia, Michoacán, México

Natasha Caverley Turtle Island Consulting Services Inc., North Saanich, Canada

Gisela Cebrián-Bernat Research Group EDIT – Educational Transformation, Leadership and Sustainability, Universitat Rovira i Virgili, Tarragona, Spain

Amy Cardinal Christianson Indigenous Leadership Initiative, Mashteuiatsh, Canada

M. Conceição Colaço University of Lisbon, School of Agriculture, CEABN-InBIO, Lisbon, Portugal

Júlio J. Conde Social Pedagogy and Environmental Education Research Group (SEPA-interea), Universidade de Santiago de Compostela, Santiago de Compostela, Galiza, Spain

Andrea Corres Faculty of Psychology and Education Sciences, Universitat Oberta de Catalunya, Barcelona, Spain

José Ignacio de Guzmán Alonso ESenRED, Segovia, Spain

Carolina Fernández-Castrillo Department of Communication and Media Studies, Universidad Carlos III de Madrid, Getafe, Spain

Antonio García-Vinuesa Social Pedagogy and Environmental Education Research Group (SEPA-interea), Universidade de Santiago de Compostela, Santiago de Compostela, Spain

Carmen Gloder Ramos Teachers for Future Spain, Gondomar, Spain

Hector Barros Gomes University of São Paulo, São Paulo, Brazil

Eugenia González del Castillo Instituto de Ciencias de la Atmósfera y Cambio Climático, Universidad Nacional Autónoma de México, Ciudad de México, México

Edgar J. González Gaudiano Instituto de Investigaciones en Educación, Universidad Veracruzana, Xalapa, México

Enrique González-Muñoz Departamento de Métodos de Investigación y Diagnóstico en Educación, Universidad de Granada, Granada, Spain

Aline Costa Gonzalez Universidade Estadual do Oeste do Paraná, Marechal Cândido Rondon, Brazil

Rita Gradaílle-Pernas Social Pedagogy and Environmental Education Research Group (SEPA-interea), Universidade de Santiago de Compostela, Santiago de Compostela, Spain

Alicia Guerrero-Fernández Departament of Didactics of Experimental and Social Sciences, University of Seville, Seville, Spain

María Puig Gutiérrez Departament of Didactics of Experimental and Social Sciences, University of Seville, Seville, Spain

Nallely Geraldine Hernández-Almanza Instituto de Investigaciones en Ecosistemas y Sustentabilidad, Universidad Nacional Autónoma de México, Morelia, Michoacán, México

Leigh Johnson Harkaway Primary School, Harkaway, Australia

Mercè Junyent-Pubill Department of Teaching of Mathematics and Experimental Sciences, Universitat Autònoma de Barcelona, Bellaterra, Spain

Lidia López-Lozano Departament of Didactics of Experimental and Social Sciences, University of Seville, Seville, Spain

María Elena Méndez-López CIHTI-Centro de Investigación en Ciencias de Información Geoespacial, Mérida, Yucatán, México

Pablo Ángel Meira Cartea Social Pedagogy and Environmental Education Research Group (SEPA-interea), Universidade de Santiago de Compostela, Santiago de Compostela, Galiza, Spain

Carmen Lúcia Melges Elias Gattás Institute of Biosciences, University of São Paulo—Communication and Education Center of NCE/ECA/USP – CECI Jaraguá, São Paulo, Brazil

Rosa Guadalupe Mendoza-Zuany Instituto de Investigaciones en Educación, Universidad Veracruzana, Xalapa-Enríquez, México

Inés Merayo-Fernández Social Pedagogy and Environmental Education Research Group (SEPA-interea), Universidade de Santiago de Compostela, Santiago de Compostela, Spain

Tania Merelas Iglesias Arabías S. Coop. Galega, Santiago de Compostela, Galicia, Spain

Jorge Conde Miguélez Universidade de Santiago de Compostela, Santiago de Compostela, Spain

Núria Monterde-Miralles Research Group EDIT – Educational Transformation, Leadership and Sustainability, Universitat Rovira i Virgili, Tarragona, Spain

Mariana Muñoz-Galván Environmental Culture Directorate, Ministry of the Environment of Mexico City, Mexico City, Mexico

Laura Muñoz Gloder Teachers for Future Spain, Gondomar, Spain

Miguel Pardellas Santiago Feitoría Verde S. Coop. Galega, Santiago de Compostela, Galicia, Spain

José Gutiérrez Pérez Departamento de Métodos de Investigación y Diagnóstico en Educación, Universidad de Granada, Granada, Spain

Vilmar Alves Pereira Universidade do Estado de Mato Grosso, Caceres, Brazil

Alejandra Atzin Ramírez Hernández Environmental Culture Directorate, Ministry of the Environment of Mexico City, Mexico City, Mexico

Xabier Riádigos-Couso Social Pedagogy and Environmental Education Research Group (SEPA-interea), Universidade de Santiago de Compostela, Santiago de Compostela, Spain

Fátima Rodríguez-Marín Departament of Didactics of Experimental and Social Sciences, University of Seville, Seville, Spain

Isabel Ruiz-Mallén Faculty of Psychology and Education Sciences, Universitat Oberta de Catalunya, Barcelona, Spain

Juan Carlos A. Sandoval-Rivera Instituto de Investigaciones en Educación, Universidad Veracruzana, Xalapa-Enríquez, México

Filka Sekulova Department of Economic History, Institutions, Politics and World Economy, Universitat de Barcelona, Barcelona, Spain

Rosana Louro Ferreira Silva Institute of Biosciences, USP Research group on Environmental Education and Teacher Training – GPEAFE, São Paulo, Brazil

Francisco Sóñora-Luna Universidade of Santiago de Compostela, Santiago de Compostela, Spain

Harumi Takano-Rojas Instituto de Investigaciones en Ecosistemas y Sustentabilidad, Universidad Nacional Autónoma de México, Morelia, Michoacán, México

Ferran Crespo i Torres Department of Language, Science and Mathematics Education, Universitat de Barcelona, Barcelona, Spain

Ximena J. Torres Instituto de Investigaciones en Ecosistemas y Sustentabilidad, Universidad Nacional Autónoma de México, Morelia, Michoacán, México

Marta Beatriz Esteban Tortajada Department of Pedagogy, Faculty of Educational Sciences and Psychology, Universitat Rovira i Virgili, Tarragona, Spain

Briony Towers Learning Ecologies for Action on DRR and Resilience (Leadrrr), Melbourne, Australia

Bethzabeé Velázquez-Martínez Antonio María Rivera High School, Xalapa, Mexico

Alondra Velasco-Morón Instituto de Investigaciones en Ecosistemas y Sustentabilidad, Universidad Nacional Autónoma de México, Morelia, Michoacán, México

Introduction

Climate Emergency and the Call for Disruptive Educational Approaches

Júlio J. Conde and Antonio García-Vinuesa

Abstract This introductory chapter lays out a foundation for rethinking climate change education within the context of climate emergency. Framing climate change as an emergency is not merely semantic but underscores its systemic roots in fossil-fuel dependency, growth-oriented economies and structural inequalities. The chapter critiques dominant deficit-based educational models, particularly education for sustainable development, for depoliticizing the crisis and diluting the transformative potential of educational initiatives. In contrast, it advocates for an education for climate emergency as an urgent expression of critical environmental education, one that positions learners as active agents capable of questioning hegemonic paradigms and envisioning alternative futures.

1 Introduction

As the scientific community has reached higher levels of confidence and consensus regarding the unprecedented disruption of the climate system and the threats it poses to societies and the environment, the ways in which the planet's climate situation is represented have evolved. Since the creation of the Intergovernmental Panel on Climate Change (IPCC) almost three decades ago, climate change has shifted from being conceived as one among many future risks within planetary boundaries, to being framed as a climate crisis, and, since 2019, to being consolidated as an urgent climate emergency (McHugh et al., 2021). This representational trajectory reflects the need to frame climate change in the most effective way possible to influence public perception and decision-making, given that "climate frames are thus an important advocacy tool for actors to exert influence over political agendas, affecting what policymakers and the public consider to be problems, or keeping items off the agenda" (p. 2).

J. J. Conde (✉) · A. García-Vinuesa
Social Pedagogy and Environmental Education Research Group (SEPA-interea), Universidade de Santiago de Compostela, Galiza, Spain
e-mail: julio.conde@usc.gal

J. J. Conde et al. (eds.), *Practices in Education for Climate Emergency*,
https://doi.org/10.1007/978-981-92-2304-6_1

The notion of climate emergency marks a turning point in how humanity perceives and responds to climate change. As of early 2025, over 2300 jurisdictions —together representing more than one billion people— have formally declared a climate emergency, reflecting a collective recognition that climate change is no longer merely a problem to be managed but a crisis demanding urgent and coordinated action (Segade-Vázquez et al., 2025). This wave of declarations signals a new era in which political institutions acknowledge the exceptional nature of this historical moment and increasingly frame their discourse around the rhetoric of emergency. This framing is reinforced by the concept of a critically diminishing "window of opportunity," repeatedly highlighted by the scientific community to describe the narrowing timeframe in which meaningful mitigation efforts must occur to prevent catastrophic climate impacts. In 2022, IPCC Working Group III co-chair Jim Skea warned that it will be impossible to limit global warming to 1.5 °C "without immediate and deep emissions reductions across all sectors" (IPCC, 2022a). According to the IPCC, limiting warming to approximately 1.5 °C requires global greenhouse gas emissions to peak before 2025 and decline by about 43% by 2030 to avoid severe consequences (IPCC, 2022b).

Without swift and substantial emission cuts, humanity risks triggering self-reinforcing warming pathways that could lead to irreversible transformations in Earth systems. Steffen et al. (2018) coined the term *Hothouse Earth* to describe a potential feedback process triggered if the global average temperature rises 2 °C above pre-industrial levels. Crossing this threshold could set off a cascade of changes —massive forest biomass loss, permafrost thaw, altered albedo, and reduced carbon absorption in soils and oceans— estimated to add approximately 0.4 °C of further warming. The urgency becomes even clearer when viewed through the lens of planetary boundaries, a concept developed to define the "safe operating space" for humanity. Beyond these boundaries, the Earth system may lose its capacity to self-regulate within the stable parameters that have prevailed throughout human history (Rockström et al., 2009). Anthropogenic climate change thus emerges not as an isolated crisis but as a core driver of transgressing multiple planetary boundaries, including those related to biosphere integrity, land-system change, and biogeochemical flows. Exceeding the climate boundary amplifies risks across these interconnected domains, reinforcing the urgency of systemic transformations to keep the Earth system within a habitable state.

However, this emergency is not merely a matter of emissions; it is rooted in the global production and consumption model anchored in fossil fuels and perpetual growth. From this perspective, the term climate emergency captures both the urgency to act and the deep systemic roots of the crisis. In this context, climate change education must be situated within the broader social and cultural systems that shape understanding, aiming to enhance citizens' environmental agency and socio-structural awareness. This contrasts with conventional approaches that focus on the transmission of factual knowledge or on media portrayals emphasizing catastrophic consequences. Such deficit-based models assume that better scientific information automatically leads to behavioral change; a premise repeatedly challenged in the literature, which highlights the roles of values, worldviews, emotions, and structural constraints in shaping everyday decisions (Kahan et al., 2012). Moreover, they tend to depoliticize climate change, framing it as a technical challenge rather than as a matter of justice and global inequality. As a result, learners are positioned as passive recipients of expert knowledge instead of as active agents capable of questioning dominant paradigms and envisioning alternative futures. The academic debate echoes this reflection, emphasizing that addressing climate change requires recognizing it as a socially and politically embedded issue, rather than a purely scientific problem (González-Gaudiano & Meira-Cartea, 2009).

Critical scrutiny is especially needed in relation to the hegemonic paradigm of education for sustainable development. According to Conde (2025), the current trajectory of education for sustainable development can be understood as a "controlled demolition" of the field of environmental education, consolidating a reformist orientation compatible with capitalist structures while neutralizing the systemic critiques that once gave environmental education its counter-hegemonic potential. The debate surrounding the relationship between environmental education and education for sustainable development reveals deep tensions that go beyond mere terminology, reflecting instead a political struggle over the framing of environmental issues. Scholars such as Berryman and Sauvé (2016) and Kopnina (2012) have warned that the predominance of education for sustainable development has progressively marginalized environmental education, eroding its critical orientation and subordinating it to a discourse that implicitly safeguards economic growth as a non-negotiable condition. Within this context, Conde (2025) calls for a climate change education that emerges as a subsidiary but urgent expression of environmental education, inheriting its transformative ethos while focusing specifically on the ethical, political and justice-oriented dimensions of climate emergency.

A socio-critical approach to climate change education —here framed as education for climate emergency— must position individuals and collectives as critical agents embedded in social systems, capable of questioning systemic inequities, exploring political dimensions, and envisioning just alternatives. As Jickling and Wals (2008) argue, education —including environmental education— should not be "just about social reproduction, but also, and perhaps foremost, about creating the ability to critique and transcend social norms, patterns of behavior, and lifestyles without authoritatively prescribing alternative norms, behaviors, and lifestyles," especially in a context where "education is seen more as a process of information transfer (mainly of scientific, technical and legislative types), which aims to ensure environmental 'conformity' to economically acceptable norms" (Sauvé, 1996, p. 21).

Education, as a transformative tool, should play a central role in shaping and regulating efforts to ensure adaptive responses to changing contexts, as well as participating in decision-making; contributing to the identification, assessment and prioritization of climate risks through inclusive and deliberative processes; and supporting the implementation and monitoring of socio-educational strategies across diverse scenarios. Aligned with these principles, this book brings together diverse climate change education initiatives that transcend traditional academic dissemination channels and aim for tangible impacts in their social contexts, promoting cultural shifts toward community development and fostering citizen participation in adaptation and mitigation strategies to embrace deep lifestyle changes.

Gathering such "disruptive initiatives" is crucial because mainstream climate change education often adheres —explicitly or implicitly— to the hegemonic sustainable development paradigm and to a scientific literacy model based on the information deficit theory. In contrast, these initiatives embrace a socio-critical perspective, recognizing that mitigating anthropogenic climate change requires profound transformations to the socioeconomic system and the global production-consumption model. Many innovative approaches —ranging from the development of teaching materials to community education projects, in-service teacher training, and even theatrical productions— remain underrepresented in global discussions, particularly those from Spanish- and Portuguese-speaking contexts. Highlighting these experiences not only broadens the epistemic diversity of climate education debates but also challenges the dominance of Anglo-centric frameworks that tend to universalize local perspectives as global solutions.

In a decade widely regarded as decisive for the future of humankind, these 19 contributions demonstrate that education can be far more than a vehicle for transmitting information. Taken together, the chapters show how educational practice may operate simultaneously as resistance and as proposition: resisting reductionist, technocratic, and depoliticized narratives of climate change, while proposing situated, justice-oriented, and culturally grounded responses to the climate emergency. By bringing into dialogue experiences developed across diverse settings, this volume highlights the capacity of education to nurture agency, strengthen public participation, and open pathways toward alternative ways of inhabiting the world under conditions of profound uncertainty.

2 Structure and Contents

This volume is intended for scholars, educators, policymakers, and practitioners interested in rethinking education as a key arena in the social and political struggle over climate futures, and in learning from initiatives that emerge from contexts that have too often remained peripheral in the international debates on climate change education. In this sense, the introduction section presents an additional chapter, written by Prof. Edgar González Gaudiano, discussing the politics of education for climate change[1] and reflecting on the importance of promoting initiatives that challenge dominant approaches.

[1] Scholars associated with the RESCLIMA research line (Conde et al., 2026) have proposed the term "education for climate change" to highlight a conceptual distinction between educating about climate and educating for change, understood as two related but analytically differentiated dimensions of climate education. For these authors, the significance of educating "for" change lies in its emphasis on a transformative educational framework that conceives of education as central to enabling critical and socially meaningful responses to the climate crisis (González-Gaudiano & Meira-Cartea, 2020).

The book is further structured into three main sections and a final reflection, offering a progressive narrative that moves from experiences and initiatives rooted in specific territories, through educational institutions, and culminating in organized civil action via networks and governance bodies. This structure underscores the multiple levels at which socio-educational actions can be developed to address and mitigate the consequences of the climate emergency: beginning in territories where climate impacts are experienced locally and often most acutely; advancing to school-based work supported by innovative pedagogies; and culminating in collective action and climate governance initiatives.

2.1 Voices from the Territories: Where Transformation Begins

This section gathers initiatives grounded in specific territories and localities —many rural or socially and environmentally vulnerable— where local actors, often women and Indigenous communities, lead socio-educational responses to climate emergencies. These initiatives draw on lived experiences, worldviews, local and ancestral knowledge, and socio-ecological struggles, giving rise to situated pedagogies rooted in climate justice and equity.

Chapter 3 – Juan Carlos Sandoval, Guadalupe Mendoza, and Itzel Cabrera (Universidad Veracruzana) present an educational initiative in rural indigenous communities across Veracruz, Yucatán, Mexico City, and Hidalgo. Using participatory and community-based approaches, the program integrates scientific and local knowledge to foster critical reflection and collective action. Its inclusive, ecofeminist, and decolonial framework addresses social inequalities and prioritizes the needs of the most vulnerable populations, linking climate emergency responses to broader struggles for justice.

> The project is rooted in the recognition that the climate emergency is inseparable from struggles for climate justice, requiring approaches that confront social inequalities and promote inclusive participation. (Chapter 3)

Chapter 4 – Fátima Rodríguez-Marín and her co-authors (Universidad de Sevilla) approach the climate emergency through the lens of degrowth, challenging the dominant sustainability paradigm. They argue that the pursuit of sustainable economic growth is incompatible with a just and equitable response to the current ecosocial crisis. Their proposals call for a transition toward agricultural models that are socially and environmentally sustainable, respectful of planetary biophysical limits, and capable of fostering resilient communities through adaptive and mitigation strategies.

The basic objective is to increase the resilience […] building horizontal, self-sufficient, self-managed, resilient, and mutually coordinated communities based on the common good, solidarity, and social justice. Resilience associated with the development of both adaptive strategies (adjusting to degrowth) and mitigation strategies that halt, in as much as possible, the depletion of resources, climate change, and the decline of biodiversity. (Chapter 4)

Chapter 5 – Aline Costa, Vilmar Alves, and Irene Carniatto (Universidade Estadual do Oeste do Paraná and Universidade Federal do Mato Grosso) examine the potential of popular environmental education to strengthen family farming in the climate emergency context. Based on documentary research and narrative interviews in western Paraná (Brazil) and Galicia (Spain), the study identifies needs, challenges, and pedagogical practices that merge scientific, local, and ancestral knowledge. Their proposals emphasize collective, participatory, and place-based practices that reinforce community belonging, reduce inequalities, and promote climate justice.

> In summary, the proposals include promoting the integration between the environment and society through collective, participatory, and contextualized practices with the realities of life, favoring the perception of belonging of the subjects and the place where they live. Ideas that seek to rescue values [...] because we are part of nature. Understanding that being all one, we are interdependent beings who inhabit the planet and urgently need everyone's contribution to prioritize quality of life, reduce social inequalities, and achieve climate justice. (Chapter 5)

Chapter 6 – Bethzabee Velázquez (Antonio María Rivera High School, Mexico) and Jorge Conde (Universidade de Santiago de Compostela) analyze the *Telebachillerato Comunitario* system in Veracruz, which combines distance learning with in-person tutoring to expand upper secondary education access for rural and marginalized youth. This model allows students to remain in their communities, strengthening local ties and reducing pressure for forced migration. The curriculum incorporates territory-linked content and projects that address local socio-environmental challenges, positioning the program as both an educational alternative and a platform for climate justice.

> In this context, the *Telebachillerato Comunitario* emerges not only as an alternative for rural youth to access upper secondary education, but also as a platform to address the climate emergency through locally relevant projects that integrate environmental education, community participation, and principles of climate justice and equity. (Chapter 6)

Chapter 7 – Hector Barros, Carmen Gattás, and Rosana Louro (Universidade de São Paulo) document the memories and resistance of the Guaraní Mbya Indigenous community in São Paulo, connecting cultural survival, environmental defense, and intergenerational education. Through participatory research and educommunication, Indigenous youth record and share community memories, knowledge, and narratives to affirm collective identity, denounce environmental racism, and confront the socio-ecological crisis. The chapter argues that addressing the climate emergency requires not only technological solutions but also the recognition of ancestral epistemologies, the protection of Indigenous territories, and the integration of climate and racial justice are presented as foundational pillars for sustaining life and envisioning alternative futures.

> The Guarani Mbya indigenous community shows that the climate emergency cannot be solved with technological innovations or merely through international agreements but requires a profound transformation in how relationships between living beings are established. Therefore, it cannot be separated from the ethical and political emergency, which demands a re-approach to indigenous values and worldviews. (Chapter 7)

2.2 *Educating for Action from Schools: Innovative Pedagogies, Art and Communication*

This section focuses on the school as a formalized setting that can and should represent a plural and community-oriented space in which to generate shared knowledge and proposals to respond to the challenges of climate emergency. These six chapters address initiatives that highlight pedagogical innovation, participatory methodologies, as well as the integration of arts and communication to foster the climate action capacity of students and teachers.

Chapter 8 – Núria Monterde, Gisela Cebrián, and Mercè Junyent (Universitat Rovira i Virgili and Universitat Autònoma de Barcelona) present school climate assemblies as an innovative educational strategy within the EDU4CLIM project, grounded in the socio-critical paradigm. The chapter outlines the methodological design highlighting their role in cultivating sustainability competencies such as critical thinking, systems thinking, and eco-citizenship. By engaging students in reflective dialogue, participatory deliberation, and democratic decision-making processes with peers and local stakeholders, the assemblies empower youth as active agents of social and ecological transformation, linking climate education with emancipatory and participatory pedagogies.

> Educational initiatives such as school climate assemblies are essential for creating structures and spaces for deliberative and participatory democracy in the school context, in which students can become agents of active change in response to environmental problems in their local areas. (Chapter 8)

Chapter 9 – Andrea Corres, Isabel Ruiz-Mallén, and Filka Sekulova (Universitat Oberta de Catalunya) examine the co-creation of nature-based schoolyards in Barcelona as a climate change education initiative. Using the GreenComp framework, the authors analyze how collaborative design and maintenance of green schoolyards foster sustainability competences among teachers, students, families, architects, and municipal technicians. By positioning schoolyard greening as both a pedagogical and socio-political process, the authors argue that co-creation can act as a disruptive intervention in climate change education, expanding opportunities for systemic reflection, experiential learning, and civic engagement.

> Our findings suggest that the process of co-creating nature-based and climate adapted schoolyards can enhance and promote relevant sustainability competences among the different stakeholders engaged in it. (Chapter 9)

Chapter 10 – José Manuel Gutiérrez Bastida and José Ignacio de Guzmán present the experience of ESenRED (Schools towards Sustainability Network), created in 2011 as a collaborative network of non-university educational institutions committed to making sustainability and environmental education the backbone of their pedagogical projects. The chapter highlights the role of cooperative networking, participatory action research, and student protagonism in addressing the climate emergency. By prioritizing real-world eco-social problems, student agency, and direct interventions in local contexts, the network demonstrates how collective educational practices can drive mitigation, adaptation, and communication in response to climate change.

> The urgent response to the climate emergency, the transition towards socially just and ecologically sustainable horizons, requires people who are trained to provide the necessary reply. (Chapter 10)

Chapter 11 – Conceição Colaço and colleagues present a shift in fire education and awareness communications from suppression-centric, top-down campaigns toward community-driven, culturally grounded approaches that recognize "good fire" and Indigenous fire stewardship. The chapter clarifies complementarities between risk education and environmental education and develops this argument through five case studies. Across cases, the authors distill principles for a fire-resilient society: know specific audiences; co-create with communities; ensure clarity and peer-review; and favor bottom-up or transversal methods.

> Fire is not the enemy. Understand and respect that wildfire is on a spectrum. It is a natural-force and one benefits from adopting a diplomatic approach to fire. (Chapter 11)

Chapter 12 – Carolina Fernández-Castrillo (Universidad Carlos III de Madrid) advances an educommunication framework for climate-crisis literacy grounded in (post) digital innovation through the "Transmedia Artivism i5" methodology. Centered on intermediality, immersion, interactivity, and intercreativity, the chapter positions students as "produsers" who co-create phygital (online/offline) interventions that blend experiential learning, cyberactivism, and media artivism. The work proposes that climate education must move beyond information transmission to include social-emotional and action-oriented learning that catalyzes civic participation and value transformation.

> Therefore, the learning paradigm must not depend exclusively on information transfer; it should also include social, emotional, and action-oriented learning. (Chapter 12)

Chapter 13 – Enrique González and José Gutiérrez (Universidad de Granada) present the co-design and implementation of two in-service teacher training programs on climate change education in Southern Spain, emphasizing game-based learning as a disruptive methodology. Developed through participatory action research with Teacher Training Centers, the programs sought to strengthen teachers' scientific understanding of climate change while equipping them with playful, emotionally engaging pedagogical tools such as escape rooms, board games, and role-playing games. The chapter details the iterative phases of needs assessment, theoretical grounding, and feedback-driven refinement, highlighting how collaborative design empowered teachers to adapt and evaluate the activities.

> Training teachers in these issues should be a priority, considering their capacity to influence numerous future generations. (Chapter 13)

Chapter 14 – Francisco Sóñora, Aitor Alonso, and José Antonio Caride (Universidade de Santiago de Compostela) present the Erasmus+ project "EduCinema Clima Tour Action" (2020–2024), which promoted film literacy, collective creativity, and community education for climate action and sustainable tourism. Using an interdisciplinary STE(A)M approach, the project produced three methodological e-books, a script, and the musical film *Climantopía*, engaging students from Spain, Italy, Poland, and Portugal in innovative, environmentally focused learning.

> Through an interdisciplinary social pedagogical approach, using STE(A)M methodology, the project outcomes are clearly novel in terms of both methodology (including the production of a film musical titled Climantopia) and the content dealt with. (Chapter 14)

2.3 The Power of Networks: Citizenship and Climate Governance

The fourth section broadens the focus on collective action, citizen participation, and governance initiatives that extend beyond formal institutions. The responsibility for climate action through education cannot be confined to schools or to isolated community projects. Civil society as a whole must build on intercommunity and intergenerational networks capable of scaling up impact and increasing the likelihood of success. The chapters in this section illustrate how networks, social movements, and participatory structures can shape climate policies and practices, thereby expanding the reach of socio-educational initiatives.

Chapter 15 – Mónica Arto (Universidade da Coruña), Tania Merelas (Cooperativa Arabías), and Miguel Pardellas (Cooperativa Feitoría Verde) present the CLIMACTIVA project, which identified and analyzed eleven citizen-led initiatives in Spain that address four climate risks: forest fires, floods, heat waves, and droughts. Their findings show how community action enhances adaptation and resilience in daily life, while identifying vulnerable groups facing inequalities linked to gender, age, and social status. The authors also examine interactions with public administrations to promote more inclusive climate governance, concluding that local, participatory, and gender-sensitive solutions complement public policies and help integrate adaptation into everyday life.

> The question is how to make adaptation part of our lifestyles, or, if we turn the question around, how to adapt our lifestyles to the climate emergency. (Chapter 15)

Chapter 16 – Ferran Crespo (Universitat de Barcelona) and Marta B. Esteban (Universitat Rovira i Virgili) contend that the climate emergency is a systemic and interdependent crisis —extending far beyond the biophysical dimension— that demands complexity-oriented approaches and transformative environmental education focused on action. From an ecofeminist perspective, they challenge the anthropocentrism and adult-centrism that marginalize both nature and children/youth, advocating for the recognition and strengthening of young people's agency as present-day ecological citizens. They develop the concept of co-agency, articulate a rights-based framework, and propose socio-educational support criteria and enabling contexts to cultivate competences, meaningful participation, and shared responsibility.

> Strengthening the moral and political agency of children and youth is essential for advancing a fair and inclusive eco-social transformation that addresses the climate emergency. (Chapter 16)

Chapter 17 – Miguel Pardellas (Cooperativa Feitoría Verde), Júlio Conde, and Pablo Meira (Universidade de Santiago de Compostela) present *Descarboniza! que non é pouco*, a project launched in Galicia in 2017 to bring climate change into everyday life and strengthen community resilience. Over time, it has evolved and diversified, incorporating intergenerational and gender perspectives, community risk management, and the collective creation of utopias through art. A consistent thread runs through all its activities: working from emotions, memory, and participation so that people not only understand the climate crisis but also feel capable of acting. The ultimate aim is to foster a "critical hope" that inspires just, sustainable, and shared futures.

> Harmonize the fourfold challenge of feeling, thinking, dreaming and acting; moving from present threats and risks toward the horizon of utopia and doing so through a form of collective educational action that is both intentional and organized. (Chapter 17)

Chapter 18 – Harumi Takano-Rojas (Universidad Nacional Autónoma de México), along with colleagues from the Universidad de Guadalajara and SECIHTI (Mexico), collaborated with a school to connect climate science with students and their community. Addressing a generational gap in local climate knowledge, the team developed exploratory workshops, a science fair, and a hands-on training program centered on a school weather station. The activities integrated scientific and local knowledge, aiming for students not only to understand climate change but also to relate it to their own surroundings and see themselves as agents of change. The project emphasized participatory methodologies, horizontal communication, and contextualized learning grounded in environmental justice.

> By positioning them not as recipients of information, but as co-participants in the construction of knowledge, the initiative fosters a sense of ownership and relevance that is often absent in traditional educational models. (Chapter 18)

Chapter 19 – Miriam Campos, Carmen Gloder, and Laura Muñoz describe Teachers for Future Spain, an independent movement of teachers that emerged between 2017 and 2019 to place eco-social education at the center of school life. Born from the need to build networks among teachers and to respond to the concerns of students who could not always join mobilizations, the movement created the program "28,000 for the Climate" as a framework for six lines of action —from "Zero Waste Recess" and "Eco-canteens" to eco-audits, safe school routes, and the rewilding of schoolyards. Their work combines open-access materials, horizontal collaboration, and **active** methodologies, aiming to embed sustainability into the daily routines of all Spanish schools. Their ultimate aspiration is to make themselves unnecessary once eco-social education is fully integrated into the 28,000 schools of the country.

> We are independent of any political party, religion or union, driven solely by our commitment to the planet, its inhabitants and the interdependent and eco-dependent relationships between them. (Chapter 19)

Chapter 20 – Mariana Muñoz and Alejandra Ramírez (Ministry of the Environment of Mexico City) present the city's four Environmental Culture Centers, which have become dynamic spaces for critical environmental education, community action, and socio-ecological regeneration. Located in protected natural areas, each center responds to its local context: from conserving the axolotl and chinampas in Xochimilco to defending biodiversity and promoting social cohesion in marginalized areas such as the Sierra de Santa Catarina. Through projects such as "My Garden, My Right" and "Gardens for Life", and through open community programming, the centers have strengthened intergenerational networks, revived traditional practices, and deepened the sense of belonging and socio-environmental justice.

> The Environmental Culture Centers are living spaces that foster critical environmental education, citizen participation and community empowerment, making them strategic nodes for building urban socio-environmental resilience. (Chapter 20)

Chapter 21 – Xabier Riádigos-Couso and colleagues analyze youth climate advisory councils in Galicia (Spain) as a model of participatory climate governance. They document the creation and functioning of the Youth Climate Advisory Councils within the European project People & Planet, designed to empower young people and strengthen local governments' capacity for climate action. Some municipalities adopted measures arising from the councils' work, demonstrating bridges for co-governance. The impacts include greater critical awareness, behavioral changes, increased youth agency, enhanced facilitation skills, and improved municipal appreciation of youth participation. Persistent challenges remain, including unappealing institutional communication channels, sustaining member engagement, logistical and language barriers, and the need for stable coordination and investment.

> Youth Advisory Climate Councils are presented as innovative and disruptive socio-educational structures that promote youth empowerment and facilitate their integration into local governance. (Chapter 21)

2.4 A Final Reflection

Chapter 22 – In the final chapter, Professor Pablo Meira (Universidade de Santiago de Compostela) critically reflects on the identity, trajectory, and current challenges of environmental education in the face of the climate emergency. The chapter argues that environmental education must move beyond conventional approaches that focus narrowly on the biophysical environment and instead transform the social, cultural, and economic relations underpinning ecological degradation. Highlighting the failure of hegemonic environmental education to address structural causes of unsustainability and the contradictions of sustainable development discourse, the author calls for a disruptive pedagogical praxis that embraces socio-ecological limits, justice, and degrowth as foundations for a viable future.

> The pool of humanity's problems is overflowing and continues to accumulate challenges for contemporary education. But anthropogenic climate change is not just another problem,

for it entails a structural fracture of the pool itself. How we respond to that fracture will determine whether it is possible to adequately address the other environmental and human issues floating within it. (Chapter 22)

References

Berryman, T., & Sauvé, L. (2016). Ruling relationships in sustainable development and education for sustainable development. *The Journal of Environmental Education, 47*(2), 104–117. https://doi.org/10.1080/00958964.2015.1092934

Conde, J. J. (2025). Education for sustainable development: the iron fist of capitalism in the era of climate emergency. *Australian Journal of Environmental Education, 41*(4), 985–1002. https://doi.org/10.1017/aee.2025.10075

Conde, J. J., García-Vinuesa, A., Bisquert, K. M., & Meira Cartea, P. Á. (2026). RESCLIMA: Two decades of research in educational and social responses to climate change. In W. Leal Filho, M. Sima, A. Lange Salvia, M. Kovaleva, & E. Manolas (Eds.), *University initiatives on climate change education and research* (pp. 2087–2118). Springer. https://doi.org/10.1007/978-3-031-59731-2_109

González-Gaudiano, E., & Meira-Cartea, P. Á. (2009). Educación, comunicación y cambio climático. Resistencias para la acción social responsable. *Trayectorias, 11*(29), 6–38.

González-Gaudiano, E., & Meira-Cartea, P. Á. (2020). Educación para el cambio climático: ¿Educar sobre el clima o para el cambio? *Perfiles Educativos, 42*(168), 157–174. https://doi.org/10.22201/iisue.24486167e.2020.168.59464

IPCC. (2022a, April 4). *The evidence is clear: the time for action is now. We can halve emissions by 2030.* https://www.ipcc.ch/2022/04/04/ipcc-ar6-wgiii-pressrelease/

IPCC. (2022b). *Climate change 2022: mitigation of climate change. Contribution of working group III to the sixth assessment report of the intergovernmental panel on climate change.* Cambridge University Press. https://doi.org/10.1017/9781009157926

Jickling, B., & Wals, A. E. J. (2008). Globalization and environmental education: looking beyond sustainable development. *Journal of Curriculum Studies, 40*(1), 1–21. https://doi.org/10.1080/00220270701684667

Kahan, D. M., Peters, E., Wittlin, M., Slovic, P., Ouellette, L. L., Braman, D., & Mandel, G. (2012). The polarizing impact of science literacy and numeracy on perceived climate change risks. *Nature Climate Change, 2*(10), 732–735. https://doi.org/10.1038/nclimate1547

Kopnina, H. (2012). Education for sustainable development (ESD): the turn away from 'environment' in environmental education? *Environmental Education Research, 18*(5), 699–717. https://doi.org/10.1080/13504622.2012.658028

McHugh, L. H., Lemos, M. C., & Morrison, T. H. (2021). Risk? Crisis? Emergency? Implications of the new climate emergency framing for governance and policy. *Wiley Interdisciplinary Reviews: Climate Change, 12*(6), e736. https://doi.org/10.1002/wcc.736

Rockström, J., Steffen, W., Noone, K., Persson, Å., Chapin, F. S., Lambin, E., et al. (2009). Planetary boundaries: exploring the safe operating space for humanity. *Ecology and Society, 14*(2), 32.

Sauvé, L. (1996). Environmental education and sustainable development: a further appraisal. *Canadian Journal of Environmental Education, 1*(1), 7–34.

Segade-Vázquez, M., García-Vinuesa, A., Rodríguez-Groba, A., & Conde, J. J. (2025). Characterization of educational research on climate change in the era of climate emergency (2017–2024). *Revista Española de Pedagogía, 83*, 179–198. https://doi.org/10.22550/2174-0909.4110

Steffen, W., Rockström, J., Richardson, K., Lenton, T. M., Folke, C., Liverman, D., et al. (2018). Trajectories of the earth system in the anthropocene. *Proceedings of the National Academy of Sciences, 115*(33), 8252–8259. https://doi.org/10.1073/pnas.1810141115

The Political and Politics in Education for Climate Change: New Wine in Old Bottles

Edgar J. González Gaudiano

Abstract The chapter centres on the analysis of *the political* and *politics* within proposals about education for climate change. It begins with the conception of *the political* as an ontological dimension constitutive of the social, and therefore of the educational; and of *politics* as the domain of what is instituted through the set of institutions, decisions, relationships, and political practices in society. Proposals from various multilateral organizations and scholars are discussed, as well as those from civil society organizations, in order to determine whether they are oriented toward the demands of the hegemonic system or are framed as disruptive alternatives. It concludes that most of the proposals circulating in specialized literature and reaching schools aim to hegemonize an institutionalized discursive configuration through a refined selection of meticulously chosen and polished elements that eliminate, exclude, or conceal others deemed unnecessary, irrelevant, or inconvenient. These exclusions correspond to the political, ethical, and social dimensions that bear upon the possibilities of constructing educational practices that disrupt the existing social order.

1 Introduction

Participating in a book that aims to present a set of initiatives understood as disruptive to the existing state of affairs in the field of Education for Climate Change (EfCC) compels us to ask: Why is it relevant to develop experiences that alter, fracture, or challenge the dominant pattern of a given situation? More specifically, what characteristics define the EfCC proposals circulating across different spheres of action and promoted by diverse social agents? In what directions do they lead? What conception of the subject and of subjectivity formation do they embody as an educational discourse? These and other questions should be raised in order to assess how disruptive the cases and reflections presented in this volume truly are.

E. J. González Gaudiano (✉)
Instituto de Investigaciones en Educación, Universidad Veracruzana, Xalapa, México
e-mail: egonzalezgaudiano@gmail.com

J. J. Conde et al. (eds.), *Practices in Education for Climate Emergency*,
https://doi.org/10.1007/978-981-92-2304-6_2

The social order, understood as a hegemonic construction articulated through political discourse, entails recognizing that such an order is neither homogeneous nor closed; on the contrary, it is contingent and permanently exposed to conflict among the forces that constitute it, based on operations of inclusion/exclusion that give rise to differences occupying dominant and subordinate positions. Therefore, the editors of this volume have deemed it important to provide a brief analysis of the centrality of the political dimension of education and of the policies proposed within EfCC, in order to account for the climate phenomenon when it is transposed into the educational domain.

Hence, for the purpose of developing this chapter on climate change policy and its concretization in educational processes, I have chosen to approach it on the basis of two analytical axes that, in my view, allow for the organization of the relevant ideas. These are:

1. Analysis of the political in its ontological character as instituting the social order and, therefore, also the educational domain.
2. Analysis of politics as the field of the instituted, following Claude Lefort (2004), through the set of institutions, decisions, relationships, and political practices in society.

Accordingly, drawing on this frame of reference in an extremely simplified version and from an integrated perspective, I will examine how the instituting has been translated into the instituted within the field of EfCC.

2 The Political

As a substantive component of the social, education is intimately intertwined with the political. Indeed, every educational proposal contains its own conception of the pedagogical subject within a framework of explicit or implicit political aims intended to institute and render intelligible a particular social order. This has been the case of environmental education promoted since the 1970s, as a means of giving expression to a set of socio-environmental concerns associated with the degradation and depletion of the environment, due to the forms and intensity of the use of natural resources.

Over the half-century during which environmental education has been established, *mutatis mutandis*, across different countries worldwide, significant shifts have taken place. Once the approach promoted by the UNEP-UNESCO International Environmental Education Programme (1975–1995) —which fostered ecological literacy through discrete, specific content in which a conception of the environment predominantly understood as nature prevailed— had been superseded, a brief period emerged in the 1990s during which environmental education received a strong boost. This occurred despite the fact that a process was already underway aimed at replacing it —particularly in its critical and disruptive expressions— with an approach more aligned with the prevailing capitalist system's patterns of production, distribution, and consumption, namely education for sustainable development.

Although this period lasted barely a decade, it laid the foundations in several countries for the construction of educational and organizational processes that have persisted, albeit with changes, to the present day, with particular relevance in some higher education institutions and civil society organizations. These processes have been built within the interstices of the social and institutional spheres, at the margins of the dominant current promoted by states and aligned organizations.

In this trajectory, the political —as the moment of institution of the social within the sphere of environmental education— has fostered diverse collective imaginaries (Castoriadis, 2007) of social change, each with its own centrality of signified and symbolic dimension. Within the range of positions that have emerged, we have, unfortunately, observed a limited number of experiences oriented toward the formation of environmental citizenship; toward articulating processes of decolonization and socio-environmental justice aimed at dismantling structures of power, knowledge, and being (epistemic and cultural); and toward recovering learnings from local experiences with high collective efficacy, as well as from situated strategies for territorial defence and resistance against community fragmentation, among others.

This type of disruptive approaches tends to foster the construction of signification, so that the population may select and interpret experiences through social and cultural mediation, recover invisibilized symbols and traditions, and construct their own narratives to organize their reality and position them within a network of meanings. Sauvé (2014) has, for decades, emphasized the need for an ecocitizenship grounded in the political dimension, a challenge that environmental education must assume on the basis of the imperative "of promoting appropriate public policies in the fields of education and environment, and identifying key elements for the development of a «political competence» in the population" (Sauvé, 2014, p. 12).

The formation of ecocitizenship consists of a social strategy aimed at empowering individuals in their relationship with the environment; it promotes comprehensive actions for environmental —and, in this case, climate— justice, so that, through effective organized participation, they can defend established rights and assume civic responsibilities. In this sense, it is essential for individuals to regain control over their lives, which requires acquiring competencies for action, local self-management, and critical reflection in order to be capable of making sustainable decisions within social and environmental spheres. It is evident that the development of ecocitizenship, as Sauvé asserts, entails ethical, critical, and political education.

3 The Politics

As we can infer, the institutional strategies developed to give meaning to the field of environmental education have been varied and span a broad spectrum of positions, ranging from *ecogestures* promoted to maintain the hegemonic social system with minor adjustments and inconsequential changes, to radical stances aimed at transforming the instituted social order, such as ecofeminism, to cite just one example.[1]

Education for Climate Change, as a constitutive part of environmental education —although with a more recent trajectory despite the thirty Conferences of the Parties (COPs) on Climate Change held since 1995— does not exhibit a different process. While some significant milestones are recognized over these three decades of negotiations —i.e., COP3, which approved the Kyoto Protocol; COP21, with the Paris Agreement; and COP26, with the Glasgow Climate Pact— appropriate response policies for the progressive elimination of fossil fuels, rather than merely their reduction, have not been supported, nor have clear commitments been made to build among different social groups a sense of urgency commensurate with the global emergency we are experiencing.

In that failed negotiation process, subordinated to the interests of the dominant system, we could not expect any significant educational policy. To verify this, it is sufficient to examine the terms in which Article 12 of the Paris Agreement is drafted: "Parties shall cooperate in taking measures, as appropriate, to enhance climate change education, training, public awareness, public participation and public access to information, recognizing the importance of these steps with respect to enhancing actions under this Agreement" (UNFCCC, 2015, p. 42). In this obscure declaration, there are no specific commitments or priorities, no binding deadlines, and it certainly does not recommend allocating financial resources to promote programs or projects. Not even the measures that would be appropriate to improve education, training, awareness, and public participation are defined. The policy of Education for Climate Change, as a field of the instituted, directly reflects the inadequacy of the political agreements reached to date.

While we acknowledge that social change to address climate change cannot occur without organized and effective social participation —which requires profound transformations in the attitudes and practices of the population that would need to be fostered through education— the instituted or emerging educational policy does not indicate that a trajectory has been adopted, or is about to be adopted, to construct the collective subjectivities and political subjects necessary to enable such change.

In a relatively recent study, González-Gaudiano et al. (2020) argued that, from our perspective, the best opportunity for everyday life spaces to rapidly transform their energy culture —both to transition toward a low-carbon society and to construct an emergency curriculum that facilitates the formation of ecocitizenship to address the complexity of the climate crisis— lies in a close integration of the scientific, political, and social spheres; an integration we have termed the *triple helix*. In this sense, we also stated that: "the majority of the population, although informed about the problem, is not activated to undertake the required changes with the necessary radicality, nor to exert sufficient pressure on governing groups to follow an alternative roadmap" (González-Gaudiano et al., 2020, p. 858).

---

[1] For a political mapping of different environmentalist currents, see Hopwood et al. (2005).

4 The Institutional Responses

In an effort to respond institutionally to these challenges, UNESCO periodically publishes proposals aimed at operationalizing educational change in a direction that passes through numerous filters to accommodate the criteria of countries and donor agencies. This is the case with *Greening Curriculum Guidance: Teaching and Learning for Climate Action* (UNESCO, 2025), which proposes integrating sustainability with climate action across all subjects, from preschool to higher education, with the intention of transforming educational processes toward the development of knowledge, skills, and values aimed at fostering resilience, climate justice, and sustainable lifestyles. The goal of this multilateral organization is to influence the educational systems of 90% of countries by 2030.

However, another document published by (UNESCO, 2026), *Climate Change and Sustainability in Science and Social Science Primary School Curricula*, raises questions about these ambitions. This report, produced by The Global Education Monitoring Report team (MECCE Project), emphasizes the need to accelerate efforts to "greening" the curriculum at this educational level. After reviewing more than 830 curricula from third and sixth grades in the fields of science and social sciences across 88 countries, the study found that 66% included at least one mention of the environment, sustainability, or climate change, with a low-to-moderate emphasis in 55% of cases and a low-to-absent emphasis in the remaining 44%. Regarding climate change specifically, although an increase was observed in curricula developed between 2013 and 2023, only 15% of those in the science and social science areas for third and sixth grades contained references to climate change. These results are cause for concern. They are for us as well, especially because we know that the premise of an information deficit as a means to transform behavioural patterns is not confirmed, even less so if it involves only scattered and superficial mentions.

Another relevant multilateral actor, UNEP (2022, p. 17), also puts forward proposals to support strategies with measures aimed at addressing the existing adaptation gap while simultaneously increasing mitigation efforts. The report's title alone (*Too Little, Too Slow*) allows one to infer its content, acknowledging that we are doing far too little and far too slowly to achieve climate-resilient development with net-zero emissions. The report recognizes that current adaptation practices fall short of what is necessary and proposes that "best practices in adaptation planning and implementation is needed to improve effectiveness."

With a focus on planning and financing, however, the report does not provide guidance on how to implement educational and organizational processes, develop leadership, or evaluate the outcomes of successful practices that could be adapted to similar contexts in areas the report highlights as priorities, such as vulnerability reduction, resilience strengthening, and the development of adaptive capacity. This is precisely what Reid et al. (2021) highlight regarding the fact that a growing number of scientists issue repeated warnings about the increasing severity of environmental problems, yet largely overlook educational processes, the humanities, the arts, and the social sciences as complementary, valuable, appropriate, and necessary contributions to support response policies to the climate crisis based on available scientific knowledge.

The World Bank also contributes to this discussion. In a document entitled *The Impact of Climate Change on Education and What to Do About It*, prepared by Sergio Venegas Marin, Lara Schwarz, and Shwetlena Sabarwal as the first part of the global report *Learning to Propel Climate Action*, the authors present how climate change is affecting education in multiple ways, including school closures, loss of instructional time due to extreme conditions, and erosion of educational processes. However, the report notes that:

> Despite these catastrophic consequences, education remains overlooked in the climate policy agenda. Education made up less than 1.3% of climate-related official development assistance in 2020 and mentioned in less than 1 in 3 Nationally Determined Contribution plans. This paper lays out four concrete ways in which governments can protect education systems from climate change so that their positive impacts on economic development, poverty alleviation, and social cohesion can be sustained and boosted. These are: (i) education management for resilience; (ii) school infrastructure for resilience; (iii) ensuring learning continuity in the face of climate shocks; and (iv) leveraging students and teachers as change agents. The paper presents an actionable agenda for each of these with operational examples in different contexts (Venegas Marin et al., 2024, p. 5).

Because the document is directed at countries highly vulnerable to climate change, a large number of its recommendations are based on increasing investments. However, like previous proposals, it calls for doing more and moving faster, recognizing that risks will continue to rise, and adds that the education sector must participate more actively in the climate debate to protect it from impacts and contribute to the large-scale "alleviation" of poverty.

Similarly, the European Commission (2020) has contributed to these reflections and proposals with *Climate Action for Empowerment and Its Transformative Potential in Latin America*, derived from the recommendations of COP26 in Glasgow. This study examines the situation in eight countries in the region, highlighting the elements suggested in Glasgow for climate empowerment, namely: policy coherence, coordinated action, tools and support, and finally, monitoring, evaluation, and reporting. It is noteworthy to observe the definition provided by the team responsible for empowerment through education, focusing on the construction of knowledge centred on action.

The concept of mainstreaming (*transversalization*) also appears central for curriculum designers: "Mainstreaming is not merely about introducing, naming, or incorporating climate change; it also involves problematizing content across diverse knowledge areas that are direct or indirect drivers of climate change (economics, industry, ethics, history, etc.)" (p. 81). Among the study's conclusions, it is particularly relevant to highlight that the authors acknowledge the imperative of not avoiding conflict in order to "address climate change in all its complexity and with responsibility… [and if this has not been the case until now] with the urgency and radicality recommended by the scientific community, it is largely because economic interests are at stake" (p. 90). I would add that political interests are involved as well.

From the academic perspective, some authors (Gillard et al., 2016) have also identified the need to strengthen the political component in educational responses to climate change. Ross Gillard and his collaborators recognize that socio-technical transitions and social-ecological resilience are two prominent normative agendas, each with clearly defined limits in the scope of their analyses, which have led to problematic governance prescriptions. The authors identify several blind spots in the majority of proposals, such as politics, power, agency, and ideas. They argue that governments and industries in most wealthy countries, as well as the international institutions at their disposal, have predominantly sought to reform rather than rethink the social and economic arrangements responsible for anthropogenic climate change —for example, through ecological modernization and green capitalism. Instead, they recommend applying social field theory and the political action approach to situate and explain the power of ideas and the role of agency above structure in times of uncertainty.

Although it does not directly address educational issues, a CLACSO (*Consejo Latinoamericano de Ciencias Sociales*) document seemed relevant to include in this chapter because of its emphasis on the social dimension and conflict, taking as its object of study climate change, public policies, and social movements in Latin America (Postigo, 2013). As the authors of this document point out, this triple relationship provides numerous insights for understanding the interactions and tensions between nature, society, and the state. They also note that resistance, denial, and lobbying against a phenomenon that can no longer be ignored are due to the fact that the findings of climate science directly challenge the energy matrix of the current global development model, constituting a new political arena that opens opportunities for social movements.

5 The Political in Education for Climate Change Policy: A Conclusion, as Always, Provisional

Something quite evident in most of the documents included in this chapter is the absence of references to the specific responsibilities that cause the climate crisis. This represents an attempt to discursively obscure a fundamental component of the prevailing issue regarding the climate emergency: political responsibility. Climate change is presented in these documents as a serious phenomenon that is currently growing, which will generate even more risks than those already occurring at present. There are, however, specific groups affected in several of the documents (the World Bank and UNEP), namely the "least developed" countries, which are therefore considered more vulnerable —or, more precisely, made vulnerable. From this perspective, we are all responsible for what is happening, but what is omitted is that not all of us are responsible in the same way or to the same extent. This implies an undifferentiated shared responsibility, which tends to recur in educational materials.

Anyone with some rudimentary knowledge of discourse analysis can recognize that omissions are also part of the discursive body and deserve to be questioned. Discourses are constituted not only by what is included but also by what is excluded. Here, we assume that the category of discourse is not restricted to oral, written, or sign language but, as Laclau and Mouffe (1987) propose, encompasses any action that carries signified, which is straightforwardly juxtaposed with the social, understood as a system of signification.

What we can observe in the dominant educational process for the climate emergency is the attempt to hegemonize an institutionalized discursive configuration through a carefully curated selection of meticulously chosen and refined elements that eliminate, exclude, or conceal others considered unnecessary, irrelevant, or inconvenient. In the case of Education for Climate Change, these exclusions correspond to the political, ethical, and social dimensions that revolve around the possibilities of constructing educational practices that are disruptive to the existing social order.

References

Castoriadis, C. (2007). *La institución imaginaria de la sociedad*. Tusquets.

European Commission. (2020). *La acción para el empoderamiento climático y su potencial transformador en América Latina*. Programa EUROCLIMA. https://doi.org/10.2841/457950

González-Gaudiano, E. J., Meira-Cartea, P. Á., & Gutiérrez-Pérez, J. (2020). ¿Cómo educar sobre la complejidad del cambio climático? Hacia un currículum de emergencia. *Revista Mexicana de Investigación Educativa., 25*(87), 843–872.

Gillard, R., Gouldson, A., Paavola, J., & Van Alstine, J. (2016). Transformational responses to climate change: beyond a systems perspective of social change in mitigation and adaptation. *WIREs Climate Change, 7*, 251–265. https://doi.org/10.1002/wcc.384

Hopwood, B., Mellor, M., & O'Brien, G. (2005). Sustainable development: mapping different approaches. *Sustainable Development, 13*, 38–52. https://doi.org/10.1002/sd.244

Laclau, E., & Mouffe, C. (1987). *Hegemonía y estrategia socialista. Hacia una radicalización de la democracia*. Siglo XXI.

Lefort, C. (2004). *La incertidumbre democrática. Ensayos sobre lo político*. Anthropos.

Postigo, J. C. (Ed.). (2013). *Cambio Climático, Movimientos Sociales y Políticas Pública. Una vinculación necesaria*. Consejo Latinoamericano de Ciencias Sociales. https://libreria.clacso.org/publicacion.php?p=759&c=5

Reid, A., Dillon, J., Ardoin, N., & Ferreira, J.-A. (2021). Scientists' warnings and the need to reimagine, recreate, and restore environmental education. *Environmental Education Research, 27*(6), 783–795. https://doi.org/10.1080/13504622.2021.1937577

Sauvé, L. (2014). Educación ambiental y ecociudadanía. Dimensiones claves de un proyecto político-pedagógico. *Revista Científica, 18*(1), 12–23. https://doi.org/10.14483/23448350.5558

UNEP. (2022). *Too little, too slow. Climate adaptation failure puts world in risk*. UNEP Copenhagen Climate Center. https://www.unep.org/resources/adaptation-gap-report-2022

UNESCO. (2025). *Greening curriculum guidance: teaching and learning for climate action*. https://doi.org/10.54675/CCBQ6474

UNESCO. (2026). *Climate change and sustainability in science and social science primary school curricula*. https://doi.org/10.54676/DENE4573

UNFCCC. (2015). *Paris agreement*. United Nations.

Venegas Marin, S., Schwarz, L., & Sabarwal, S. (2024). *The impact of climate change on education and what to do about it*. World Bank. https://doi.org/10.1596/41483

Voices From the Territories: Where Transformation Begins

Situated Learning to Address the Climate Emergency in Mexico: An Ecofeminist Educational Experience from the Global South

Juan Carlos A. Sandoval-Rivera, Rosa Guadalupe Mendoza-Zuany, and Fabiola Itzel Cabrera-García

Abstract The experiences of women from the Global South concerning living with climate change in their specific contexts, the situated responses they have developed to care for their physical and social environments, and the ways they have acted in the face of various socio-ecological challenges related to the climate emergency have inspired an alternative educational process for basic education in Mexico. This article reports on the results of its implementation through a collaborative research study conducted primarily with women from rural communities in Veracruz, Yucatán, Mexico City, and Hidalgo, Mexico, as well as with teachers from schools located in these contexts. From diverse ecofeminist perspectives, we recognise the value of the experiences of women who face the socio-ecological and climate crisis in their daily lives. Based on these experiences, we propose situated educational alternatives in schools to address the complexity surrounding the various social and ecological challenges and the climate emergency we face. As results, we present the methodological strategies through which families and teachers identify socio-ecological concerns, knowledge, and local practices to confront the impacts of climate change, as well as some concerns that arise from experiencing the climate crisis in their daily lives.

1 Introduction

The degradation of ecological systems that sustain human survival, which occurs differently across various geographical contexts around the planet, is becoming increasingly severe. This undoubtedly compromises the present and future of everyone, but particularly of populations that have historically faced significant socio-ecological challenges due to decades of unjust territorial expansion, dispossession, and colonization. These include Indigenous peoples, displaced communities, migrants, women in the Global South, communities affected by armed conflicts,

J. C. A. Sandoval-Rivera (✉) · R. G. Mendoza-Zuany · F. I. Cabrera-García
Instituto de Investigaciones en Educación, Universidad Veracruzana, Xalapa-Enríquez, México
e-mail: csandoval@uv.mx

J. J. Conde et al. (eds.), *Practices in Education for Climate Emergency*,
https://doi.org/10.1007/978-981-92-2304-6_3

and others. This issue has been widely discussed and documented within the field of environmental education (Brody et al., 2013; Caride & Meira, 2001; González-Gaudiano et al., 2020; Gutierrez-Bastida, 2018; Gutiérrez-Pérez et al., 2020), but little connected to questions emerging in the field of ecofeminism and gender studies as pointed out by Gough et al. (2024), Siegel (2024) or Puleo (2019). It is essential to highlight that, in most cases, challenges related to climate emergency can be directly linked to productivist logics, overexploitation, the expansion of extractive frontiers, and the trend toward monoculture production (González-Gaudiano, 2016; Sandoval & Mendoza, 2021), particularly in the Global South. Within this framework, we aim to revisit the experience of addressing the effects of climate change from the perspective of women in the Global South, the strategies they have developed to care for their natural and social environments, and their actions in response to socio-ecological concerns. Additionally, we examine the questions they are raising as they confront the various contemporary crises.

We recognise that, although climate change is a global phenomenon, its impacts, the experience of living through it, and the potential responses—whether to adapt to or mitigate its effects—are highly contextual. In the rural and Indigenous communities of Mexico where we have collaborated as part of this research, we have developed a collaborative process with primary school teachers to foster situated learning aimed at local sustainability and to better understand and prepare for the challenges posed by climate change in these contexts. In the communities where we have worked, socio-ecological challenges have been exacerbated by the climate change, deepened by socio-economic issues such as inequality, violence, lack of employment opportunities, migration, and more. Moreover, we have identified educational systems designed to sustain the status quo through an anthropocentric approach to education, which overlooks the socio-ecological crisis. These systems also reflect an androcentric bias, failing to acknowledge the experiences, knowledge, and practices women bring to the care of life (Puleo, 2010).

Thus, in response to a banking model of education (Freire, 1985) that has been reinforced in the wake of the pandemic, we ask how to develop processes of situated learning across ecological, social, cultural, and historical dimensions that transcend the anthropocentric, patriarchal, and neoliberal character of education. Such processes would aim to foster better conditions for adapting to the climate change and its effects. Based on this premise, we carried out collaborative research with teachers from rural and Indigenous regions in the states of Yucatán, Veracruz, Mexico City, and Hidalgo in Mexico. The research involved implementing a situated ecofeminist pedagogical approach focused on caring for the ecological and social environment in preschool, primary, and secondary schools. This was done in collaboration with teachers to develop capacities to better address the challenges posed by climate change.

In this chapter, we first revisit the theoretical foundations of ecofeminisms, which help us understand and highlight the concerns and responses of women to the impacts and challenges of climate change and the broader socio-ecological crisis. We also explore situated learning approach through a pedagogical strategy we named 'learning progressions', centred on learning to care for ourselves, others and surroundings from the school context, in collaboration with families. Subsequently, we present the collaborative methodology used with community members and teachers through a guided training process for the design and implementation of learning progressions in schools. As part of the results, we describe the methods employed by teachers to identify socio-ecological concerns, as well as the knowledge and practices specific to the contexts in which they work. This approach highlights the potential to facilitate the transformation of unfavourable and unjust conditions, which have been exacerbated by the climate change in rural communities across Mexico. Additionally, we outline the main socio-ecological concerns identified in the contexts where we have been collaborating: (1) factors related to nutrition and food production, and (2) factors associated with access to sufficient and quality water sources.

2 Theoretical Framework

Various ecofeminisms—as both theory and practice (Puleo, 2019)—have emphasised the imposed association between women, the natural world, and caregiving tasks, which has historically led to oppression and subordination (Warren, 1997). However, this patriarchally imposed relationship has also resulted in a wealth of valuable experience from which knowledge has been constructed and practices developed. On the one hand, these have served to care for immediate symbolic and material aspects, and on the other, to counteract oppression and subordination. Such knowledge and practices, along with the processes through which they have been created, undervalued and excluded from schools due to an andro-anthropocentric structure, where men (*andros*) and masculinity are the measure of all that is deemed valuable (Puleo, 2019). To address this, Puleo (2014) argues for the need to make women visible as protagonists of ecological alternatives and overcome the disdain for feminised attitudes related to compassion and care, which remain two neglected areas in most environmental education programmes. Conceptually, Warren (1997) warns against the oppressive conceptual framework underpinning patriarchy, which sustains hierarchical structures of domination and subordination among humans, other living beings, and Nature by arguing their inferiority. It is worth asking how these oppressive conceptual frameworks continue to influence the field of environmental educational research and shape the ways in which responses are sought to the climate emergency we face as humanity. Ecofeminisms help us recognise the persistence of anthropocentric and androcentric perspectives in environmental education and related fields and propose alternatives that view the care of both material and symbolic aspects as a shared responsibility of men and women (Mendoza et al., 2025). However, it is important to note that historically, the experience of caregiving has been accumulated and transmitted intergenerationally by women, as we have previously discussed. The diversity of experiences arising from living through climate change in daily life and proposing transformative strategies—many of which are fundamentally strategies of care—is as varied as the range of places women inhabit. For this reason, we consider it essential to situate teaching and learning processes aimed at addressing the climate emergency within their specific contexts.

Situated learning offers a pedagogical perspective well-suited to these objectives. It is grounded in constructivist theories, particularly the contributions of Vygotsky (1978), whose sociocultural theory provided the foundation for the development of a learning theory that emphasises its development within the framework of social interaction processes that emerge from everyday life. The most significant assumption of this perspective is that learning does not occur individually or independently, but rather through social interaction rooted in the problems, phenomena, and practices of daily life (Lave & Wenger, 1991). It is also proposed that learning is achieved within a specific sociocultural context, recognised by the learners, in which they develop skills and capacities to address concrete challenges. Pedagogically, situating care can enable the development of educational processes where individuals engage in critical analysis of their local context, using it as a starting point for environmental education aimed at transformation and preparation for the impacts of climate change in everyday life. This approach allows us to move away from the banking model of education that Freire (1985) criticised. Within the climate emergency framework, there persists a homogenisation of knowledge about climate change, which tends to promote uncritical and decontextualised adaptation while underestimating—or even invalidating—local knowledge and practices, particularly those created and transmitted by women (Cabrera, 2023; Sandoval & Mendoza, 2021). Lotz-Sisitka and Lupele (2017) refer to situated learning within specific sociocultural contexts as a form of "connecting learning" which serves as an interface between individuals' lived experiences in situated contexts—experiences that generate local knowledge and practices, especially among women—and the concepts traditionally addressed in curricula. Our proposal for learning progressions (Sandoval et al., 2021) provides a methodology for achieving situated and connected learning (Lotz-Sisitka & Lupele, 2017), with care for the natural and social environment at its core. This represents a pedagogical turn that broadens the socioculturally situated perspective to include the living elements of ecosystems and the physical environment that sustains them. It expands the sociocultural dimension to encompass the dialectical relationship between people and the physical place they inhabit, situating learning across social, cultural, historical, and ecological dimensions (Sandoval & Mendoza, 2021). Given the historical moment we are living in, where socio-ecological challenges are becoming increasingly acute, educational processes must also be ecologically situated. Situating learning in its historical dimension involves recognising how the past influences the present and how the present can shape the future. Through learning progressions, we propose a learning process that avoids the exclusion or subordination of local knowledge. Instead, it seeks to organically integrate knowledge tied to academic disciplines with that which emerges locally, while fostering active learning processes aimed at transforming oppressive and unjust contexts.

3 Methodology

This collaborative research is epistemologically grounded in critical, collaborative, and constructivist perspectives, and is oriented towards action and transformation (Denzin & Lincoln, 2008; Smith, 1999). We draw inspiration from Freire's (1985) proposals on the importance of integrating action and reflection as a pathway to transformation, as well as his critique of the epistemological error that diminishes and invalidates the wisdom of experience in formal education. Additionally, we take up hooks' (1994) call to recognise ourselves, from the Global South, as capable of theorising about our practices and experiences. Thus, we see the need to work towards the construction of our own theoretical frameworks (Denzin & Lincoln, 2008; Smith, 1999) to interpret our experiences and contexts through non-hegemonic theoretical categories that enable us to produce theory from us, for us. While engaging in dialogue with hegemonic eurocentric theory, this epistemological position is neither determined nor validated by it.

The research team involves the two academics coordinating the project, alongside doctoral and master's students in educational research conducting their thesis projects within the framework of the called CARE-México project. The project also involves teachers who voluntarily express interest in designing and implementing learning progressions, which we will explain further below. Teachers from various pre-primary, primary, and secondary schools in Mexico participate in both virtual and face to face guided training processes with our team. Throughout this process, collective and individual exercises are carried out to:

1. Identify: (a) Educational and socio-ecological challenges in the communities where teaching takes place, to determine the priority knowledge areas to be addressed in the learning progressions; (b) Knowledge and practices related to socio-ecological concerns and care for the physical and social environment.
2. Design and implement the learning progressions.
3. Systematise the experience to assess outcomes and make adjustments if necessary.

As part of the support provided by the thesis students and the project coordinators, we form a research team that continuously reflects on the process, learns from the participating teachers, and identifies relevant data. This data is drawn from session reports and observation logs, the designed learning progressions, observations of their implementation in schools, individual interviews with each teacher, and the systematisation exercises (Jara, 2001) conducted with groups of teachers. Our collaboration also extends to the families of the students, achieved through their participation in spaces we promote, such as meetings, interview-conversations, and activities conducted within the framework of the learning progressions. This collaboration primarily involves women, as we have observed that they have historically established strong connections with schools, often taking on the responsibility of supporting and accompanying their children, or, in their roles as grandmothers or aunts, serving as caretakers. It is essential for our project to recognise these women as interlocutors, knowledgeable contributors, creators and transmitters of knowledge, regardless of their level of formal education which, in rural communities, is significantly low. Through their involvement, teachers can better understand the community, enabling them to situate learning effectively. In this context, their participation is crucial both in the design of the learning progressions—where they serve as key sources of local knowledge and practices—and in identifying socio-ecological concerns. Furthermore, during the implementation phase, they collaborate with teachers in teaching and learning processes by actively participating in the so called "change challenges" posed within the learning progressions.

Regarding the pedagogical strategy we call "learning progressions" (Sandoval et al., 2021), these are structured into a sequence of four quadrants (Fig. 1). In Quadrant 1, the process begins with a narrative crafted by the teacher, which stems from the local concerns and knowledge they have identified. These narrative harnesses the power of storytelling to situate the central theme, evoking emotion, lived experiences, and an affective connection to what matters most. Additionally, this quadrant includes a section titled "Did you know…?", which horizontally integrates school-based and local knowledge, recognising their potential complementarities and/or contradictions, as well as the possibility of using local experiences as a foundation for understanding global issues. Quadrant 2 focuses on encouraging students to enquire their local context based on their own questions, delving deeper into the knowledge and practices of care at the local level. This serves as a basis for connecting these insights with knowledge and practices emerging from other contexts and ways of knowing, enabling an analysis of the effects of climate change in their own community. This process requires collaboration from families and local actors as sources of expertise, with particular emphasis on amplifying the voices of women.

In Quadrant 3, findings are shared, and the results of the enquiry are collectively analysed. This quadrant also deepens understanding through the "Did you know…?" section and plans actions to address the challenges identified in the earlier quadrants. Here, the groundwork is laid for implementing change challenges in the final quadrant. In Quadrant 4, these change challenges are planned with different timeframes and involve various stakeholders, always in collaboration with the families of the students. This is the quadrant where potential solutions are implemented to address the concerns or problems identified in the earlier quadrants. It is also where the concepts explored in the previous quadrants gain meaning through collective action, carried out via projects that can be scaled to the classroom, the school, or the wider community, and involve diverse participants.

Over the five years we have collaborated with teachers and families on this project, we have worked with approximately 100 preschool, primary, and secondary teachers who have designed and implemented 77 learning progressions. Additionally, six postgraduate students (four at the master's level and two at the doctoral level) have conducted their theses in schools, collaborating with teachers.

From the design and implementation of the teachers' learning progressions and the implementation carried out by students in collaboration with their families, the results we aim to share in this chapter have emerged: (a) The main methodological strategies for identifying socio-ecological concerns, knowledge, and practices that become

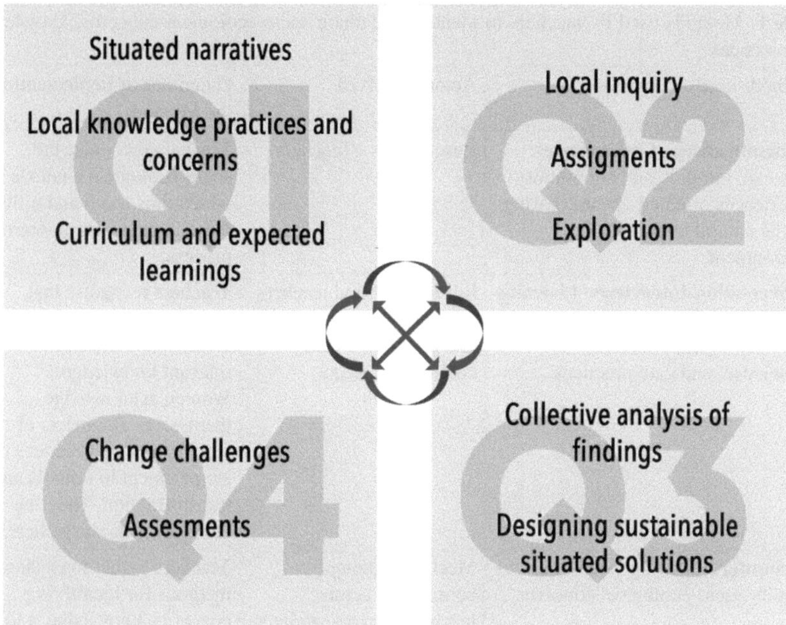

Fig. 1 Four quadrant model

the themes of learning progressions in schools. (b) The primary socio-ecological concerns identified in the contexts where we have been collaborating, and how they are addressed in school through learning progressions.

4 Methods Used with Teachers to Identify Socio-Ecological Concerns, Knowledge and Practices

In the accompanied training provided to teachers, a central challenge has been the identification of socio-ecological concerns, local knowledge, and practices to address the climate and socio-ecological crisis. The aim is to use these elements to design progressions of situated learning. This involves, on the one hand, ensuring that teachers recognise the importance of thoroughly understanding the socio-ecological contexts in which they teach, and on the other hand, transforming the relationship between schools, families, and communities. Traditionally, this relationship has been academically exclusionary, disregarding the fact that families and community members possess valuable knowledge, and that their understanding of the local context is crucial. In our accompanied training programme, we propose the use of qualitative methods—such as interviews and observations—to help teachers identify

Table 1 Methods used by teachers to identify and share socio-ecological concerns, knowledge, and practices

Methods used	Actors Involved	Outcomes of Implementing the Method
Remembrances of experiences: Personal, familial, and community experiences oriented towards caring for the natural and social environment.	Researchers - Teachers	Teachers recognise the socio-ecological contexts in which they teach and utilise these in the teaching-learning process.
Conversation-interviews: Listening to and sharing stories about socio-ecological concerns, knowledge, and care practices.	Researchers and teachers engaging in conversations and interviews with community actors	Teachers recognise that women in the community are creators and transmitters of relevant knowledge. Women acknowledge themselves as carriers of significant care practices that are pertinent to schools and transformation. They see themselves as storytellers.
Encounters: Sharing experiences to identify socio-ecological concerns.	Meetings among community actors, teachers, and researchers	Teachers gather to exchange methods for identifying concerns, knowledge, and care practices, as well as shared concerns. Women create a space for sharing experiences, which influences their daily lives and the learning process.

concerns, knowledge, and practices. Subsequently, the aim is to connect these findings with the impact of the climate change on these concerns. It is worth mentioning that throughout the training, which has involved approximately 100 teachers, we have observed innovations and adaptations in the methods applied in this project. The following table presents the main methods used, the stakeholders involved, and the implications and outcomes derived from their application (Table 1).

4.1 Remembrances of Personal, Domestic, and Community Experiences Oriented Towards Care

It has been primarily women teachers, trained in the design of learning progressions, who have drawn upon the remembrances of their experiences related to care of the natural and social environment. They have recorded these in personal journals and shared them in training spaces to reflect on the socio-ecological concerns they express, as well as the knowledge and practices connected to addressing the climate crisis in everyday life. Cabrera (2023) worked with women teachers from Veracruz who self-identify as indigenous peoples belonging to the Nahua ethnic group, the largest ethnic group in Mexico. These teachers relied on the remembrance of their own experiences, both of oppression and resistance, engaging them in exercises of dialogue and documentation of these memories. These involved “experiences and knowledge they possess not only as teachers but also as individuals from Indigenous and rural communities, speakers of Indigenous languages, with a particular legacy and position in the world” (p. 112). This process involves recalling forgotten knowledge, displaced ritual and daily practices, and once-heard stories to recover individual, familial, and community memories related to care. It is about reclaiming knowledge and practices of care that remain present in the community but have not been considered relevant in learning processes or in everyday life. Cabrera (2023) reflects on the significance of exploring socio-ecological remembrances:

> One of the most complex challenges for me during this phase was posing questions or trigger ideas that could go beyond a strictly anthropological perspective and help me uncover experiences related to the socio-ecological dimension. That is, to enable epistemic dialogue to encompass not only the cultural significance of biodiversity, for example, but also to understand the experience in biophysical terms […]. To expand these experiences, for instance, in relation to cultivation processes, I suggested they describe how these processes take place, how they learned them, how they experienced them, what they feel about them, what concerns they have, how they envision solutions, what changes they have observed in cultivation methods, and how they care for the process and themselves. Additionally, I asked them to describe the biological processes of grains from their own experience and knowledge. (p. 124)

Resorting to remembrances as a method allowed the teachers to recognize that the life experience, they have in their context, as well as the experiences of other women that appear in their memories, are a source for the identification of significant concerns and how care is or should be cared for.

4.2 *Conversation-Interviews: Listening to and Sharing Stories About Concerns, Knowledge, and Care Practices*

A key method for identifying the concerns, knowledge, and care practices of women in the communities where schools are located has been the conversation-interview. This method has been employed by the research team but primarily by teachers. Martínez-Bautista (2021) presents the conversation-interview as a method aimed at sharing stories that everyone, but especially women, can tell. In this context, these are stories that capture socio-ecological concerns, knowledge, and care practices, which emerge without being guided by predetermined themes. The process begins with prompts such as:

> Could you tell me a story about how you take care of yourself to feel well, how you care for your family, or how you care for your surroundings? You could also share a story about what problems or challenges you face daily and what care practices you use to address them. (Martínez-Bautista, 2021, p. 94)

This approach requires active listening and reciprocity: "It was an interaction where they said to me, 'I'll tell you my story, but I'd also like to know about you and your family'" (Martínez-Bautista, 2021, p. 99). This method is conducted mainly with women to counteract the marginalisation of their voices as sources of authorised knowledge. The conversation-interview weaves a tapestry of stories that incorporate four dimensions of situated learning: the articulation of tradition (historical dimension), innovation in response to socio-ecological challenges (social and ecological dimensions), and cultural practices rooted in the context, which provide meaning:

> [...] they shared stories about current socio-ecological problems and concerns, and how they apply local knowledge and practices to address challenges in their community. They also recalled how, during their childhood, socio-ecological problems were less prevalent. Above all, the community has always cared for the environment and themselves through local knowledge. Some of this knowledge was passed down by grandparents, parents, and others, while other local knowledge and practices have adapted and evolved over time. (Martínez-Bautista, 2021, p. 97)

Thus, the conversation-interview possesses an exploratory, pedagogical, and personal/emotional dimension that facilitates the creation of meaningful connections among participants. It also encourages more individuals to share, making it a method that strengthens the relationship between school, family, and community. This is also evident in the research by Luis-Cruz (2023), a preschool teacher and Nahua woman, who recounts her experience of being both a researcher and a teacher at the same educational centre:

> During the visits to the families, I was mostly welcomed by the mothers. I believe my presence in their homes meant a lot, as I was their child's teacher visiting their own house. I realised the importance of understanding the context, knowing where and how families live, to take this into account within the school environment and make better-informed decisions in my teaching practice. (Luis-Cruz, 2023, pp. 88–89)

The conversation-interview, therefore, is a method that provides valuable input for designing learning progressions. It enables teachers to recognise the women participating in these interviews as holders of significant care practices and knowledge, essential in addressing the socio-ecological crisis and climate emergency. Simultaneously, it fosters self-recognition among these women as knowledgeable experts in the care of their biophysical and social environments.

4.3 Encounters: Sharing Experiences to Identify Socio-Ecological Concerns

Encounters represent another method that facilitates the exchange of experiences to identify socio-ecological concerns, knowledge, and practices among community actors, teachers, and researchers. Within the CARE-México project, encounters have been organised to share and exchange socio-ecological concerns, knowledge, and practices for addressing them, as well as methods for identifying these concerns. This approach frames other collective and collaborative methods aimed at fostering interaction among participants, such as: sharing testimonies and experiences in plenary sessions and smaller groups; participatory mapping; deliberation; peer workshops on specific topics such as ecology; writing narratives for students that reflect identified concerns; and activities fostering emotional connection, such as dance, music and games.

This critical pedagogy method allows participants to gather, recognise one another, find common ground, deliberate pathways, generate knowledge, and plan actions. It benefits not only community actors but also teachers, who share their experiences of understanding the contexts in which they work—particularly when these differ from their place of origin—and strategies for deepening the identification of community concerns. Patraca-Rueda (2024) highlights the significance of an encounter among Afro-Veracruz women, collaborators in her doctoral research, as a pivotal moment in identifying socio-ecological concerns and defining their own educational agenda:

> The encounter among women, where we shared a sense of belonging as Afro-Veracruz women, was a turning point for this research. Not only did it allow us to meet face-to-face, connecting through our commonalities and our diversity, but its relevance also lay in the unconventional interaction dynamics. These dynamics facilitated the exchange of experiences, placing socio-ecological concerns at the centre beyond the constraints of a patriarchal, colonialist, and heteronormative system. (p. 191)

The convergence of women willing to share their feelings, fears, concerns, knowledge, and insights to contribute to situated educational processes creates a safe space. This space centres on elements excluded from androcentric education: affective ways of relating, care as a focal point in interactions, and dynamics and activities free from control or limitations imposed by organisers.

5 Addressing Socio-Ecological Concerns in Schools Through Learning Progressions

The teachers we collaborate with have strengthened their understanding and critical analysis of situated socio-ecological challenges within their contexts and have connected these with the phenomenon of climate change. Working with learning progressions has provided alternatives for addressing these challenges by placing care at the centre and, from this perspective, advancing the transformation of the biophysical and social environment to improve conditions for facing the effects of climate change. Below, we present how some of the most pressing issues raised by community actors have been addressed.

5.1 Factors Related to Nutrition and Food Production

A primary concern identified in all the context and exacerbated by climate change, is the association between nutrition and food production. This concern arises from the role of women as caretakers of their families' daily nutrition, encompassing the production, collection, purchase, and preparation of food. We found that the effects of climate change are creating significant challenges for families in accessing quality food due to limited economic resources. These challenges include rising food prices; the loss of key food species caused by the overuse of agrochemicals; reduced availability or production of food due to climate change impacts; the abandonment of local food production and subsistence farming; and the contamination of soil, water, and crops. These difficulties in accessing quality (albeit expensive) food have also led to increased consumption of low or non-nutritional products, such as ultra-processed foods and sugary drinks. This trend is compromising people's health, resulting in cases of malnutrition, obesity, cardiovascular diseases, and related health issues. Additionally, there is an increasing dependency on genetically modified seeds, which have displaced traditional seeds, reducing the diversity of locally produced, nutritious foods required for a healthy diet. Another factor linked to food is the impact on women's health caused by activities such as collecting heavy pieces of firewood, cooking on open stoves, and exposure to fertilisers. The learning progressions we have collaboratively designed to address these challenges in the classroom focus on recognising the concern and exploring how it is addressed locally. Teachers create narratives to foster both cognitive and emotional connections. These narratives incorporate "Did you know...?" prompts, which integrate school-based knowledge included in the curriculum (e.g., health risks associated with excessive consumption of processed foods, socio-economic and historical processes transforming agriculture, and calorie calculators); local knowledge documented by teachers and shared by students, such as the richness of food diversity in their communities and the need to preserve or recover this diversity by caring for seeds, soil, and other resources. Student participation is encouraged by drawing on their feelings, experiences, perspectives on healthy eating within their context, their relationship with climate change, and the questions they raise. These questions guide the inquiry process in the second quadrant of the learning progressions (Fig. 1). For example, students have documented traditional recipes and nutritious, forgotten ingredients with their grandmothers or explored current farming practices that are more resilient to climate change and do not rely on agrochemicals. Sharing each student's findings is a key step in the progression, fostering expansive learning (Engeström, 2001) not only for students and their families but also for teachers. The diversity of shared knowledge and practices, integrated with school-based knowledge across all subjects, prompts deliberation on ways to improve nutrition within the framework of local tensions and their global connections, such as the relationship with climate change.

To actively engage in the classroom, various change challenges have been implemented, aimed at transformation and requiring the integration of local and academic knowledge. These efforts are designed, depending on the context, to promote seed exchanges and the establishment of seed banks at the school, family, and community levels, which involve learning about the identification, collection, drying, and storage of seeds; to produce compost as a nutrient-rich substrate for fertilising crops and producing high-quality food even in areas with limited or no access to arable land; to reorganise, clean, and plan the use of household plots and backyards—spaces present in most students' homes—where families can produce their own food by combining traditional and agroecological techniques, practising beekeeping, and more; and to design diets using locally produced food and recipes that incorporate these ingredients. These change challenges represent actions that locally address the impacts of climate change on food systems, particularly for the most vulnerable groups. For example, they respond to the effects on seasonal crops caused by climate variability, the reduction of nutrient content in food crops and the decline of pollinators, associated with climate change and habitat loss, and the excessive use of chemical fertilisers to compensate for declining soil productivity, which poses risks to both human health and ecosystems.

5.2 Factors Associated with Access to Sufficient and Quality Water Sources

The second identified concern is the limited access to freshwater sources for domestic use. This issue is expressed in two key dimensions: quantity and quality. A recurring problem in the communities where we have worked is that, due to the effects of climate change, droughts are becoming increasingly prolonged, although extreme hydrometeorological events, such as floods, have also occurred. These changes in rainfall cycles compromise access to water, leading to various challenges, such as reduced agricultural production, skin and gastrointestinal health issues, and health risks associated with the transport and storage of water, which directly affect women and girls. Another concern is the scarcity linked to the overexploitation of aquifers by agro-industrial and industrial activities in nearby areas.

In Yucatán, it is notable that the main issue is not related to water availability, as sufficient water is accessible for most of the year, but rather to its quality. Pig farms in the region contaminate groundwater—the primary source of water for local communities—with animal waste. In all contexts, eutrophication caused by fertilisers and waste pollution is also a major concern. Although these problems are not directly caused by climate change, they stem from the same development model oriented towards production and exploitation that contributes to climate change and allows polluting industries to operate at the expense of community health and well-being. While such complex issues cannot be fully addressed or resolved within the school, it is important to critically engage with them to foster reflection and promote community action.

Through the use of learning progressions, we have addressed the topic of water from various perspectives. Firstly, we have examined the issue from the standpoint of the right to access water, which is often compromised by the prioritisation of agricultural and livestock use, the lack of regulation concerning waste disposal into aquifers, and the increasing water demands of agricultural production due to climate change. This approach aims to enable students to enquire and deliberate, both in the classroom and with their families, on how to look after water sources, manage water access points, and make water available to those experiencing water crises or struggling with poor-quality water. Another approach has involved understanding the water cycle and how it has been disrupted. Connections are made between the ecological dynamics of water, pollutants, and climate change. A gender perspective has been integral to all learning progressions addressing water conservation, recognising that women have traditionally been assigned the responsibility of fetching water for their households. This responsibility has been exacerbated by the climate crisis, as women and girls often have to walk increasingly long distances to access water for domestic use. Bringing these issues into the school setting, aims to spark debate about the implications of this culturally assigned role, not only at the local level but globally as well. To actively engage with this topic in the classroom, several change challenges have been implemented, requiring the integration of local and academic knowledge. These include creating artisanal water filters to provide water of sufficient quality for domestic use, though not for drinking; setting up rainwater harvesting systems; appointing monitors to detect and prevent water leaks and wastage, as well as guardians of water sources such as springs and rivers; and developing participatory proposals at the community level to maintain household water distribution systems and prevent leaks. The work with learning progressions has facilitated understanding and action in response to changing seasonal patterns that are making the planet warmer. These changes result in water stress, flooding, increased forest fires with subsequent biodiversity loss, and soil degradation, all of which undermine ecosystems' resilience. This, in turn, affects the ability to sustain productive cycles related to food production and the maintenance of the local water cycle.

6 Conclusions

Addressing these challenges has required teachers to understand and acknowledge the social, cultural, and ecological contexts in which they work. It has also required them to engage in dialogue with students' families to identify the symbolic and material aspects that have been neglected and have contributed to the emergence of complex problems in the local context. Additionally, teachers must identify the knowledge and care practices that should be brought into the school to design learning progressions that are locally relevant and provide a deeper understanding of issues that manifest locally but also occur on broader geographical scales. Students have learned to value the knowledge embedded in the experiences, practices, and histories of the women in their families. Discussions are encouraged about the responsibilities and roles that women hold in caring for families, animals, and plants, among other things, in various local and international contexts. Information and global data are also provided to highlight the gender injustice that is exacerbated by the global climate crisis, an issue that warrants critical deliberation. Simultaneously, families have actively participated in their children's educational processes, regardless of their own level of formal education, by sharing and teaching the knowledge they have inherited intergenerationally or developed through practical experimentation or everyday observation. This knowledge, which has ensured survival and offers valuable insights into tackling the contemporary climate emergency, has been integrated into the classroom. Addressing socio-ecological issues and the climate emergency in the classroom is breaking down disciplinary and thematic barriers, enabling the curriculum to be taught in connection with local concerns, knowledge, and everyday experiences. Learning progressions have proven to be an important pedagogical strategy for facilitating meaningful engagement with the content outlined in the official curriculum, integrating it with local knowledge, and fostering a critical understanding of the realities faced by students and their families in their specific contexts.

Perhaps the most important discovery from working with learning progressions is that teachers, students, and families are finding meaning in both scientific knowledge and the concepts addressed in the classroom. Through learning progressions, abstract content becomes relevant and meaningful, fostering a better understanding of phenomena such as climate change and informing future decision-making. We are well aware that many of the challenges identified and causing concern in communities extend beyond what schools can address. In terms of understanding the causes and effects of climate change, while it is a widely shared concern in the contexts where we collaborate—given that it is now impossible to operate outside the realities of the climate crisis—families and teachers we work with prioritise immediate survival issues. These include basic concerns related to health, access to food, and access to water. As families and communities are able to address these basic survival needs, they will be better positioned to engage with the complexities surrounding the climate emergency and to prepare for the challenges it poses now and in the future.

Thus, we agree with our collaborators that responding to the climate crisis is not a generic matter but a situated one. Moreover, while we share global challenges posed by the climate crisis with the rest of the planet's inhabitants, we face unique challenges shaped by geographical and political factors. As we attempt to demonstrate in this chapter, these challenges are also deeply tied to gender and structural injustices that have been culturally embedded and passed down through generations.

References

Brody, M. S., Dillon, J., & Wals, A. E. (2013). *International handbook of research on environmental education*. Routledge.

Cabrera, F. I. (2023). *Pertinencia cultural y socioecológica para el cuidado del entorno y de nosotras mismas: construyendo procesos de enseñanza-aprendizaje situados y ecofeministas con maestras de primarias indígenas de Veracruz, México [Cultural and socioecological relevance for the care of the environment and of ourselves: building situated and ecofeminist teaching-learning processes with indigenous primary school teachers in Veracruz, Mexico]* (Doctoral dissertation). Universidad Veracruzana Servicios CSC https://sistemas.colmex.mx/Reportes/Lingmex/22089.pdf

Caride, J. A., & Meira, P. Á. (2001). Educación ambiental y desarrollo humano [Environmental education and human development]..

Denzin, N., & Lincoln, Y. (2008). Introduction. In N. Denzin, Y. Lincoln, & L. Smith (Eds.), *Handbook of critical and indigenous methodologies* (pp. 1–20). Sage.

Engeström, Y. (2001). Expansive learning at work: toward an activity theoretical reconceptualization. *Journal of Education and Work, 14*(1), 133–156. https://doi.org/10.1080/13639080020028747

Freire, P. (1985). *Pedagogía del oprimido [Pedagogy of the oppressed]*. Siglo XXI.

González-Gaudiano, É. J. (2016). ESD: power, politics, and policy: tragic optimism from Latin America. *The Journal of Environmental Education, 47*(2), 118–127. https://doi.org/10.1080/00958964.2015.1072704

González-Gaudiano, É. J., Meira-Cartea, P. Á., & Gutiérrez-Pérez, J. (2020). ¿Cómo educar sobre la complejidad de la crisis climática? Hacia un currículum de emergencia [How can we teach the complexity of climate change? Towards an emergency curriculum]. *Revista Mexicana de Investigación Educativa, 25*(87), 843–872.

Gough, A., Ho, Y. C. J., Lloro, T., Russell, C., Walters, S., & Whitehouse, H. (2024). Ecofeminisms and education: repositioning gender and environment in education. *Gender and Education, 36*(4), 299–311. https://doi.org/10.1080/09540253.2024.2329289

Gutierrez-Bastida, J. M. (2018). *Educatio ambientalis: invitación a la educación ecosocial en el antropoceno [Educatio ambientalis: an invitation to ecosocial education in the anthropocene]*. Bubok Publishing.

Gutiérrez-Pérez, J., Meira-Cartea, P. Á., & González-Gaudiano, É. J. (2020). Educación y comunicación para el cambio climático [Education and communication for climate change]. *Revista Mexicana de Investigación Educativa, 25*(87), 819–842.

hooks, b. (1994). *Teaching to transgress: Education as the practice of freedom*. Routledge.

Jara, O. (2001). *Dilemas y desafíos de la sistematización de experiencias [Dilemmas and challenges in the systematization of experiences]*. ALFORJA.

Lave, J., & Wenger, W. (1991). *Situated learning: legitimate peripheral participation*. Cambridge University Press.

Lotz-Sisitka, H., & Lupele, J. (2017). Learning and quality education in Africa: learning today for tomorrow. In H. Lotz-Sisitka & J. Lupele (Eds.), *Schooling for sustainable development in Africa* (pp. 3–24). Springer.

Luis-Cruz, M. A. (2023). *Prácticas ecológicas y culturales en educación preescolar indígena [Ecological and cultural practices in indigenous preschool education]*. (Master's thesis, Universidad Veracruzana). Repositorio Universidad Verecruzana. https://www.uv.mx/mie/files/2023/08/Maria-de-los-Angeles-Luis-Cruz.pdf

Martínez-Bautista, P. (2021). *El uso de las narrativas indígenas sobre el cuidado de nuestro entorno en dos escuelas primarias de la Huasteca veracruzana [The use of indigenous narratives about caring for our environment in two primary schools in the Huasteca region of Veracruz]*. (Master's thesis, Universidad Veracruzana). Repositorio Universidad Verecruzana. https://cdigital.uv.mx/items/d3a08245-4a3f-4347-a137-65147a537c15

Mendoza, R. G., Cabrera, F. I., & Sandoval, J. C. (2025). Aprender a cuidar ante la crisis socioecológica: desestabilizando el androantropocentrismo [Learning to care in the face of the socio-ecological crisis: destabilizing androanthropocentrism]. *Revista Latinoamericana de Estudios Educativos, 55*(1), 115–140. https://doi.org/10.48102/rlee.2025.55.1.686

Patraca-Rueda, M. C. (2024). *Encontrar nuestras voces y nombrarnos a partir de la resistencia, el cuidado y la esperanza: aportes y experiencias socio-ecológicas de mujeres afroveracruzanas para la construcción de un proyecto educativo propio [Finding our voices and naming ourselves through resistance, care, and hope: socio-ecological contributions and experiences of afro-veracruzana women toward the construction of a self-defined educational project]* (Doctoral dissertation, Universidad Veracruzana). Repositorio Universidad Veracruzana. https://cdigital.uv.mx/items/d3a08245-4a3f-4347-a137-65147a537c15

Puleo, A. (2010). Ecofeminismo: la perspectiva de género en la conciencia ecologista [Ecofeminism: the gender perspective in ecological awareness]. In E. E. Acción (Ed.), *Claves del ecologismo social [Key Concepts of Social Ecology]* (pp. 169–173). Libros en Acción.

Puleo, A. (2014). *Ecofeminismo para otro mundo posible [Ecofeminism for another possible world]*. Ediciones Cátedra.

Puleo, A. (2019). *Claves ecofeministas: para rebeldes que aman a la Tierra y a los animales*. Plaza y Valdés.

Sandoval, J. C., & Mendoza, R. G. (2021). Aprender a cuidar y a cuidarnos. ¿Por qué es necesario hacerlo desde la escuela? [Learning to care and to care for ourselves: why is it necessary to do so from within the school?]. In R. G. Mendoza & J. C. Sandoval (Eds.), *Conocimientos y prácticas locales para el cuidado del entorno social y ecológico a través de procesos educativos situados [Local knowledge and practices for the care of the social and ecological environment through situated educational processes]* (pp. 1–43). Universidad Veracruzana.

Sandoval, J. C., Mendoza, R. G., Cabrera, F., Patraca, M. C., Martínez, P., & Pérez, M. (2021). *Aprendizaje situado para la sustentabilidad a partir de historias locales sobre preocupaciones, conocimientos y prácticas socio-ecológicas [Situated learning for sustainability through local stories of socio-ecological concerns, knowledge, and practices]*. Universidad Veracruzana.

Siegel, L. (2024). Ecofeminism ↔ Intraconnectivism: working beyond binaries in environmental education. *Gender and Education, 36*(4), 328–344. https://doi.org/10.1080/09540253.2024.2327429

Smith, L. T. (1999). *Decolonizing methodologies: research and indigenous peoples*. ZED Books LTD.

Vygotsky, L. (1978). *Thought and language*. MIT Press.

Warren, K. (1997). El poder y la promesa de un feminismo ecológico [The power and the promise of an ecological feminism]. In M. A. Romero (Ed.), *Ecología y feminismo [Ecology and feminism]* (pp. 117–146). Comares.

Environmental Education in and for Degrowth in the Face of Climate Change: Food and Its Socio-Environmental Impact

Fátima Rodríguez-Marín, Lidia López-Lozano, Alicia Guerrero-Fernández, and María Puig Gutiérrez

Abstract Climate change can be addressed from many perspectives and in turn focusing on different areas because it is a very complex socio-environmental issue. Many of the current educational proposals take sustainability as a theoretical reference, assuming that we still have time for an adjusted growth to the carrying capacity of the planet. In this paper we present the degrowth perspective based on the empirical fact that it is impossible to maintain economic growth and act in the face of climate change. In order to make this proposal concrete, we have chosen the topic of food and its socio-environmental impact, showing some results in the field of initial teacher training for Early Childhood and Primary Education teachers.

1 Introduction

The current eco-social crisis and the prevailing industrial agricultural model directly influence issues such as the reduced availability of natural resources, climate change, soil overexploitation, deforestation, water pollution, loss of biodiversity, changes in terrestrial systems, and the global phosphorus and nitrogen cycles (Moore, 2020; Willett et al., 2019). This agricultural model is a highly energy-dependent that alters the planet's biogeochemical cycles, thus showing that it cannot be sustained over time for one simple reason: the planet's resources are limited (Guerrero- Fernández, 2023; Rockström et al., 2023).

The food is a human activity that has a significant social and environmental impact, generating 26% of the greenhouse gases (Ritchie, 2019; Saynes Santillán et al., 2016) and accounting for approximately 70% of the global freshwater withdrawals (FAO and WHO, 2020). Furthermore, in 2021, nearly 2,300 million people experienced low food security (FAO, IFAD, WHO, WFP, and UNICEF, 2022), and 828 million

F. Rodríguez-Marín (✉) · L. López-Lozano · A. Guerrero-Fernández · M. Puig Gutiérrez
Departament of Didactics of Experimental and Social Sciences, University of Seville, Seville, Spain
e-mail: frodmar@us.es

J. J. Conde et al. (eds.), *Practices in Education for Climate Emergency*,
https://doi.org/10.1007/978-981-92-2304-6_4

suffered from hunger, and 931 million tons of food being wasted (ONU, 2021). This situation is increasing exponentially due to armed conflicts, inequalities, and extreme weather events. However, studies such as that by Willett et al. (2019) assert that it is possible to feed the entire world's population within limits if multisectoral and multidimensional systemic transformations are implemented (Dixon et al., 2023).

To ensure the food supply for the world population and, consequently, its survival, it is urgent to promote a change in the system that goes beyond partial proposals and reductionist, individual-based action strategies. The focus should be on diverse agricultural models and the promotion of those that are most efficient, thus ensuring a circular metabolism (Sempere, 2018) and better adapting to the biophysical limits, taking into account the entire food system, from the different agents involved to the different phases: production, processing, packaging, distribution, marketing, consumption, and waste management (Almazán et al., 2024; FAO and WHO, 2020; González et al., 2024; González & Almazán, 2023).

In our case, we believe the desirable model that should be opted for is an agricultural model based on permaculture (Mollison & Holmgren, 1978) because the practices, strategies, and principles on which it is based fit in with the model that has been advocated, as it is based on self-sufficiency, adapted to the natural cycles and the planet's limits, with low dependence on resources and energy, circular waste management, and the promotion of resilient communities (Smith & Stevenson, 2017). Permaculture has a variety of definitions, but we find it particularly useful: 'Conscious design of landscapes that mimic the patterns and relationships found in nature […]' (Holmgren, 2013, p.45). The term 'permaculture' (which refers to the idea of 'permanent agriculture') appeared in the mid-1970s to describe an innovative method of land design intended as a response to the planetary emergency caused by the socio-economic development model based on unlimited growth. Permaculture is more than an agricultural technology, being understood as a certain way of organising space and society, which aims to maximise its efficiency to create systems that are self-sustaining, have few needs and many products, without being exploitative or wasteful. Applying this vision to the garden, the following ecological concepts are used:

- The soil is not merely an inert substrate, but an "alive" ecosystem, in this sense it is a question of creating a "permanent" soil that is useful for agriculture (a process that lasts several years, in which the soil is neither ploughed nor disturbed except at the beginning of the process).
- Respect the natural nutrient cycle (organic matter not consumed must return to the soil).
- Use organic waste of all kinds for composting.
- Reproduce in the garden certain conditions existing in less humanised ecosystems (biomimicry): relevance of mulching to create a microclimate, retain humidity, prevent erosion, reduce the presence of weeds, use of ecological pest control, location of plants in the garden according to their degree of complementarity/antagonism (associated plants) with the creation of 'food forests', etc.

2 The Perspective of Environmental Education on Degrowth: Critical Environmental Literacy

This didactic proposal, which aims to promote critical Environmental Literacy (EL), is developed from the perspective of education in and for degrowth (Rodríguez-Marín, et al., 2024). We consider that posing this perspective when studying about food is a disruptive initiative because this notion is a minority for treating socio-environmental problems such as climate change in the field of education (Díez-Gutiérrez et al., 2025; Rodríguez-Marín, et al., 2024).

Although a priori it seems that there is more diversity in "educations" (Scalabrino et al., 2022), this one is apparent, because most of the proposals are located in sustainability as a theoretical framework (Gil Pérez & Vilches, 2023, 2019).

When there is a hegemonic paradigm in a particular scientific community, people in that community assume a particular theoretical frame of reference as the 'norm', resisting evidence that points to a paradigm shift. This process of evolution of science, characterised by Kuhn (1971), is applicable, at the present time, to the case of the research field of environmental education and/or education for sustainability.

In this case, assuming this model assumes that there is still time for growth adjusted to the carrying capacity of the biosphere, based on the protagonism of conventional economic agents and technological innovation. Within this position we could include most of the institutional proposals (Agenda 2030 and SDGs, European Green Pact, Green New Deal).

However, this positioning obviates a central debate in political ecology, understood in a wide context, where the notion of sustainable development is criticised from the perspective of the emerging degrowth paradigm (Arias, 2020; Daly et al., 2019; González & Almazán, 2023; Gudynas, 2023; Latouche, 2009; Moore, 2020; Santiago, 2024; Sempere, 2018; Taibo, 2016; Tejero & Santiago, 2019; Trainer, 2021, 2024).

The degrowth perspective is based on an empirical fact: the inability of institutional policies, based on maintaining economic growth, to halt the ecosocial crisis. In this respect, the data provided by the environmental sciences are conclusive: we continue to deplete material and energy resources, climate change is accelerating and the brutal decline in biodiversity is accelerating (Bordera et al., 2024; Bordera & Puig, 2022; Caesar et al., 2024; Hickel, 2020; Hickel et al., 2022; IPCC, 2023; Kallis et al., 2024; Kongshøj, 2023; Meadows, 2022; Ripple et al., 2024; Turiel, 2020; Valero & Valero Delgado, 2021; Valladares, 2023; WWF, 2024). Hence, degrowth is seen as inevitable (Fernández & González, 2014; González & Almazán, 2023). From this perspective, compatibility is sought between degrowth as a "desire" and degrowth as a "necessity", with two complementary strategies: to curb the causes of the crisis and, above all, to promote adaptation, in the fairest and most resilient way possible, to a reality of decreasing resources (García et al., 2019).

In response to this, it is an emerging line of research in the field of Education. In this sense, authors such as García et al. (2019) have reviewed the terms associated with degrowth, with "the concepts of biophysical limits, degrowth, and collapse" and “the need to express these major principles in the form of educational content” (p. 1101.2). Díez-Gutiérrez and Jarquín-Ramírez (2025) believe that “to advance towards an authentic society with a culture of degrowth, what is needed is a real educational and social revolution and an ambitious program of ethical education and counter-hegemonic politics so as to transform us from being neoliberal subjects and societies into post-capitalist ecological subjects and societies” (p. 199) and Massip Sabater and Espinet (2024) emphasize that “degrowth is understood as decolonial as it is an anti-capitalist movement that builds non-capitalist alternatives” (p. 6). These authors agree that it is a matter for the common good, the only way to guarantee human survival in a just manner.

We are facing a reality of inevitable degrowth that could turn out to be chaotic (in the manner of civilisational collapse) (Riechmann, 2019). This is due to the difficulty for this degrowth to be planned and fair, taking into account barriers such as the resistance to change of the ruling classes, as they limit themselves to reformist proposals and short-term strategies that do not work on community resilience and do not question the dominant socio-political framework. In short, we subscribe to this idea as we consider degrowth to be an inevitable fact centered on the current dominant sociopolitical system being the causal agent of the crisis (Valladares, 2023).

Likewise, the incorporation of the degrowth perspective in the framework of environmental literacy means that we move from understanding EL as the understanding of the functioning of natural systems and the implementation of strategies that seek to respect, maintain and improve them, taking into account the reality from a systemic perspective and the interrelationships that are established between humans and the biosphere (Álvarez-García et al., 2018; Mello O’Brien, 2007; Roth, 1992; Stables & Bishop, 2001; Uncer et al., 2014). It addresses aspects linked to knowledge, know-how and action, however, it does not address the causes of the ecosocial crisis, nor does it make explicit the need to make the population environmentally literate to guarantee their adaptation, something we consider essential.

Furthermore, little mention is made of the role of emotions in this framework (Guerrero Fernández, 2023). To propose a critical EL for resilience which highlights the need to adequately understand that the crisis is associated with the logic of the functioning of the capitalist system itself. In this regard, it is proposed that education enables the transition from a causality centred on individual responsibility (institutionally encouraged) to a causality centred on the current dominant socio-political system as the agent of change. It is a tool that allows structuring content around three dimensions: knowledge and skills, attitudes and emotions, and behaviours (Guerrero Fernández et al., 2022).

These dimensions, which acquire meaning and functionality in an interrelated way, should make it possible to design teaching proposals from a more complex perspective. The aim is to address these dimensions in order to focus on the causes and roots of the ecosocial crisis, considering them indispensable not only to mitigate the situation but also to train a critical and resilient citizenry capable of adapting to a limited scenario in terms of resources and energy.

3 Sustainable Food from the Perspective of Education on and for Degrowth

Achieving critical environmental literacy from the perspective of degrowth entails rethinking two of the major areas of a didactic proposal. On the one hand, the selection and organization of content based on its intended purpose (Solís Ramirez & Rivero García, 2024) and, on the other, the method (Hamed Al-Lal & Pineda Alfonso, 2024).

Regarding content, it is necessary to discuss and question which knowledge is a priority and essential to ensure efficient adaptation, taking into account the limitations of knowledge such as everyday and mythical knowledge.

We consider it necessary to consider various types of knowledge in an integrated manner as well as establishing gradations (Rodríguez-Marín et al., 2014) in order to generate complex thinking that is linked to fostering creativity and a critical spirit. Therefore, sustainable food is presented through a problem-based question: "*Is our food assured?*" and "*How can we reinvent our food to make it ecological, fair, and safe?*"

When working on the impact of what we eat, it is necessary to address different types of content from simple to more complex levels, selecting key content such as: social metabolism, circularity, the biophysical limits of the planet, the depletion of natural resources, the ecological footprint, the rate of energy return, resilience, self-sufficiency and food security and sovereignty (Guerrero Fernández, 2023).

Furthermore, it is essential to start from the integration of different perspectives and fields such as ecology, taking into account the macrocosm (the circulation of matter and energy flow in the biosphere or the dynamics of agro-ecosystems), the mesocosm (perception of the food pathway or social customs related to food) and the microcosm (metabolic aspects of food in our organism or food production). It is necessary to go beyond the mesocosm by paying attention to the micro and macro, based on the use of scientific knowledge, in order to ensure a complex approach to the subject.

It is developed through three subproblems, whose strategic content is the circular social metabolism of the biosphere regarding the nutrient and water cycle, the carbon footprint and its connection to climate change, and food sovereignty.

Regarding attitudes and behaviour, also formulated as hypotheses, the objective is to foster a commitment to solve the problems that stem from a deep awareness and pro-environmental behaviour as a collective and at the community level.

In relation to the methodological proposal, an approach focused on pupil problem-solving (Tsivitanidou et al., 2018) is considered essential, as it prioritizes action-based learning and reflection and "know-how". This approach, in turn, develops research competencies (the ability to recognize, confront, and solve problems creatively) given the progressive increase in new and uncertain situations.

The basic objective is to increase the resilience (understood here as the capacity to reorganize and remain stable in the face of adverse circumstances, not only as individuals but as a collective) of the population, enhancing collaborative work skills and building horizontal, self-sufficient, self-managed, resilient, and mutually coordinated communities based on the common good, solidarity, and social justice. Resilience associated with the development of both adaptive strategies (adjusting to degrowth) and mitigation strategies that halt, in as much as possible, the depletion of resources, climate change, and the decline of biodiversity (López-Lozano et al., 2024).

Embracing these precepts, in this research we present experiences in the context of initial teacher training for pre-primary and primary education within the framework of subjects that correspond to the areas of Didactics of Experimental and Social Science.

4 Experiences in Initial Training for Pre-primary Education

Under the challenge of reinventing our food so as to reduce its socio-environmental impact, and focusing on the question: *Is our food assured*?, two specific subproblems were developed in parallel. The first, called "*How can we reinvent our food to make it more ecological?*", focuses on the subproblems 1 and 2 in Fig. 1, and delves deeper into the environmental impact. The second subproblem, "*How can we make our food more fair and secure?",* addresses the role of inequality, highlighting the growing gap between countries and social classes, and food insecurity, with 29.3% of the population having experienced this situation in 2021 (FAO, FIDA, OMS, PMA, & UNICEF, 2022).

The comprehensive training proposal has as objective to address this problem in an integrated manner, as well as any possible alternatives from a degrowth perspective, leading to a resilient production model adjusted to the planet's biophysical limits. It was developed with 56 students organized into 12 groups over the four weeks they studied the subjects "Knowledge of the Social Environment in Pre-primary Education" and "Teaching about the Natural Environment at the stage of 0 to 6 years", which are obligatory subjects in the first quarter of the third year of Degree in Pre-primary Education at the University of Sevilla. A methodological model based on school inquiry was followed (Puig & Rodríguez-Marín, 2021) indicating in each of the sequences implemented the phases of said model, depending on whether it is an activity of motivation, contrasting information, synthesis or review of what has been learnt, or an action that implies going a step further and showing a commitment to the learning acquired.

Subproblem 1 addresses three aspects of food: the water footprint, the carbon footprint, and organic waste. Through the analysis of these elements, the problem and its alternatives are addressed, taking into account the principles of biomimicry and circular social metabolism. It consists of eight activities. The presentation/motivation activity uses chickpea and bean seeds to represent resources, exploring their availability between generations from a synchronous and asynchronous perspective. To explain the initial ideas, a group and individual reflection activity is proposed. At the

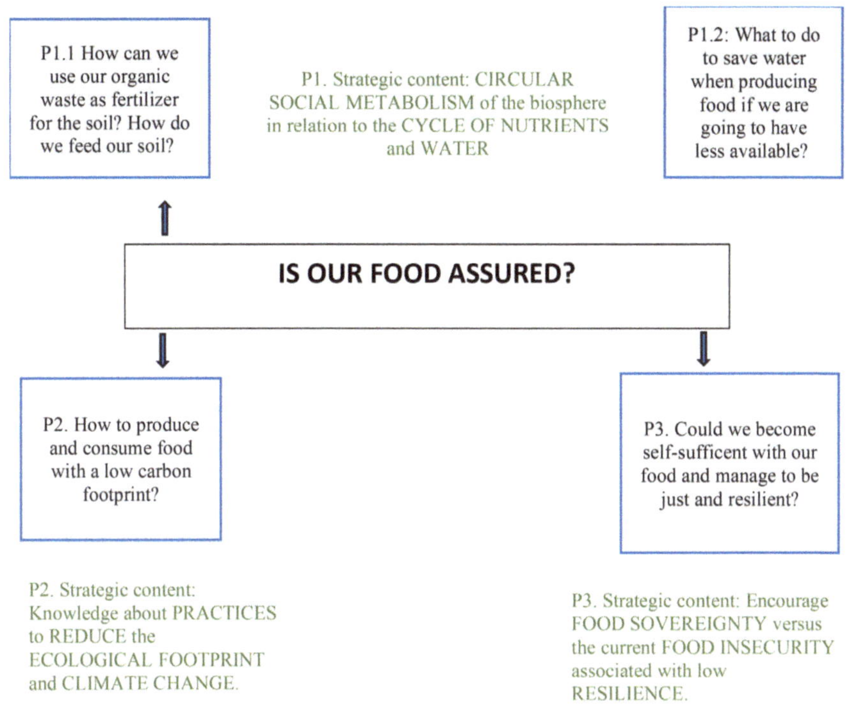

Fig. 1 Map of problems and strategic content of the training proposal Based on the work of Solís Ramirez and Rivero García (2024)

heart of the proposal are four contrasting activities: a game to address the water footprint, a workshop to analyze the kilometers food travel from production to consumption, an activity about waste, and a visit to the faculty and collaborating organization's organic garden. This is followed by an activity based on the students' ideas at this final stage, similar to the one conducted at the beginning. Finally, the students are challenged to design a device that will help them become more self-sufficient regarding the issues addressed, as a presentation at a science divulgation event such as Science Week.

The results indicate that 83% of the students achieve a high level of knowledge in the three variables analyzed. The carbon footprint is the variable where the greatest depth of knowledge can be observed, both in its diagnosis and in the proposed alternatives. However, some difficulties were detected, such as the weak relationship perceived between food and its environmental impact. Few teams mentioned the ecological benefits of a plant-based diet. Regarding the reduction of water consumption, the initiatives were usually linked to the use of more efficient irrigation systems, while knowledge about the plant requirements or adaptations was less widely used. Whereas regarding waste, most teams recognized the importance of organic waste to be transformed into compost, thus achieving waste reduction.

Regarding subproblem 2, it consists of six activities primarily aimed at fostering critical thinking by highlighting the inequalities that occur in relation to food and offering tools and action strategies that will promote more resilient behaviour. The subproblem begins with a motivational activity involving the viewing of a video about hunger, which then leads to a discussion. Subsequently, the *restaurants of the world* activity (Puig et al., 2023) were performed, which included creating a photo album of their own food waste. This activity helped raise awareness about the current consumption pattern. The sequence was followed by a visit to the classroom of a representative of the NGO "Justicia Alimentaria" to address issues related to fair trade and workers' rights. Finally, there was a visit to the city's gardens with different agricultural models, which gives the students the opportunity to learn about alternatives to more traditional models. This set of activities allows them to explore the social aspect of a topic that, at first glance, might be considered more natural in nature, and focusing on the fact that natural phenomena have a social component and vice versa.

The results of the teamwork show that the most prominent concepts were fair trade, food sovereignty, and food security, with 58.3% relevance. Together with these main concepts, food waste (16.7%) and food justice (16.7%) were also highlighted. The first of these, although a widely known term, has been highlighted for its scope and precision, as some groups reported being unfamiliar with its meaning and impact. In the case of the concept of food justice, it appears linked to that of food sovereignty.

One of the most striking results is how the group work demonstrates the interrelationship between the highlighted concepts, focusing the importance of all of them in achieving fairer and safer food. Furthermore, 83.3% of the groups mentioned how the class work made them aware of the reality of food worldwide and the impact their actions can have on it. Finally, it is relevant to note that 75% of the reports proposed actions that contribute to reinventing fairer and safer food options, such as conscious and responsible consumption (58.3%), reducing food waste (50%), and purchasing seasonal and fair-trade products (33.33%).

In light of these results, it can be considered that the proposed sequence has been positive and has allowed the established objectives to be met, thus promoting simple everyday actions that seek a more ecological, fair, and safer food option.

5 Experiences in Initial Training in Primary Education

Sharing the challenge and teaching principles proposed in the previous case, we present for Primary Education an experience using the school garden as a training resource to address the socio-environmental impact of food. It was developed with 33 students enrolled in the compulsory subject "Didactics of Experimental Sciences" in the second year of the Degree in Primary Education at the University of Sevilla.

After carrying out the activities of the first three phases listed in Table 1, as the final activity in the contrast phase, the students were asked: "What can we work on with Primary Education pupils in the garden?" This was done to initiate a process of school inquiry and to understand the hypotheses they propose. This data was collected in the teacher's journal, and a summary of the predominant ideas was extracted, which served as a starting point to explore the garden as a resource to address the impact of food.

After posing this question and exploring their initial hypotheses, they dedicated the next two months (2 hours per week) to designing and implementing a permaculture garden, focusing on strategies and practices that are key to ensuring an agricultural model that is more aligned with the planet's biophysical limits. At the end, they were asked to reflect upon the role of the garden as a teaching resource. This served to analyze and contrast it with their initial hypotheses, in order to determine the extent to which they were able to link the use of the school garden to addressing issues such as the impact of food.

To carry out the study, four areas were considered as emerging categories to be analyzed—health, biology, issues, and education—and three levels of argumentation based on the depth and complexity of the written statement under information units (IU) (see Table 1).

At the beginning (47 IU in total), the importance of the garden to work on general aspects related to education (53%, 25 IU) stands out, such as pupil cooperation and motivation. The next predominant category is linked to health-related aspects (23%, 11 IU), such as when they indicate that the garden serves to promote healthy eating and to consume less processed and sugary products. This is followed by the importance of this resource to address topics in the field of biology (19%, 9 IU), such as understanding the parts and the growth of a plant or the role of photosynthesis. Finally, only 4% (2 IU) relate the garden to socio-environmental issues, referring to ideas such as the importance of consuming local, nearby, and seasonal foods. This summary shows a low level of depth (L1), which was useful as a starting point.

After developing the teaching proposal about the garden, we found the following categorization from the analysis of their final reflections: 27 IU related to health (L1 56%, 15 IU), 19 IU related to biology (L2 47%, 9 IU), 68 IU related to education (L3 47%, 9 IU), and 99 IU related to environmental issues, with statements from L2

Table 1 Categories of analysis and levels used. Based on the work of López-Lozano et al. (2024)

<table>
<tr><th colspan="2">Categories of Analysis</th><th>Levels</th></tr>
<tr><td>Health</td><td>The garden as a resource that allows to work on healthy eating, nutrition, and/or disease prevention.</td><td rowspan="4">N1. No in-depth coverage
N2. Shallow and generic aspects are mentioned
N3. Greater depth and complexity</td></tr>
<tr><td>Biology</td><td>The garden as a resource that allows to work on biology content linked, for example, to the plant growth process or photosynthesis.</td></tr>
<tr><td>Issues</td><td>The garden as a resource that allows to work on socio-environmental issues, specifically the impact of human nutrition.</td></tr>
<tr><td>Education</td><td>The garden as a resource that allows to work on educational aspects such as teamwork, cooperation, and the interdisciplinary nature of different areas of knowledge.</td></tr>
</table>

predominating in the latter (40%, 40 IU), followed by L1 and L3. Examples of this are the following quotes:

> Some of the techniques I have learned are, for example, mulching and biological pest control. With mulching, you can use the branches that fall from the plant itself to cover the soil to retain moisture. It is also important to note that irrigation is drip or manual. [...] spaces for composting organic waste, which helps to reduce the amount of waste ending up in landfills and promotes the circular economy. (sic) (L2).

> Not only does it teach the pupils about sustainable food production, but it also provides them with practical skills and knowledge to adopt a more environmentally conscious and responsible lifestyle. (sic) (L1).

> Ecological models reduce the dependence on external inputs and maximize the use of local resources, utilizing techniques such as crop rotation and composting. Permaculture is also directly related to this latter model. It fosters species diversity, attempting to mimic the structure of a forest with plant associations and the presence of animals. This model is characterized by the absence of fuels and by being a closed, circular system in the nutrient cycle. It is an alternative to industrial agricultural models. (sic) (L3).

As evidenced by the results, initially, the students were unable to understand the garden as a resource to address environmental issues but rather saw its potential to address aspects linked to biology (Marín Gil, 2015), health (Maldonado Aragón & Valadez Ramírez, 2021), or a combination of different aspects such as cooperative work or motivation (Conde Núñez et al., 2018). This could be due to the limited prior experience many of the prospective teachers have with a garden, their academic background, or even the influence of the information they find in various media outlets (social media, news, etc.). However, following the training proposal implemented, where this resource took on a central role in addressing environmental issues and as a tool to work on content linked to the different dimensions of Environmental literacy (knowledge and skills, attitudes, emotions, and behaviour) (Guerrero Fernández et al., 2022), the students acquired a greater awareness of the garden's role in addressing and responding to problems.

Furthermore, it is noteworthy that the highest-level categories (L2) are those of problems and biology, as both were experienced and explicitly addressed in the different sessions of the proposal related to the use of the garden (especially the first one).

We conclude that conducting experiential practices in a permaculture garden, where prospective teachers develop activities to design and implement strategies adapted to the eco-social crisis and the biophysical limits, has a positive influence on understanding how the agricultural model implemented generates a decisive impact on current socio-environmental problems, something that can influence the mobilization of behaviour at both the individual and collective levels.

6 Conclusions

The experiences presented allow us to reach some conclusions that should be subject of reflection and can guide the development of new educational proposals within the framework of education in and for degrowth. Understanding the garden as a resource is key to addressing complex issues such as the impact of food, as well as to address fundamental concepts such as degrowth and circular social metabolism. It is an accessible resource that offers students the opportunity for direct experimentation and ongoing reflection as they implement action strategies, which are essential to achieve EL from a critical and comprehensive perspective. To achieve this, it is essential that these strategies transcend the individual and emphasize collectiveness as the best way to generate profound, systemic change.

In turn, the results of this study highlight the urgency to involve pupils in real-life contexts, such as the school garden, using a didactic approach based on school inquiry, as they significantly influence the EL processes. This is because this didactic model offers a teaching-learning context that makes it possible to understand world problems from a comprehensive perspective, where the social and environmental aspects are interconnected, something that is essential to avoid a fragmented and biased view of reality.

Ultimately, this type of applied proposal for initial teacher training allows to foster awareness of their role as agents of change, both personally and professionally, preparing them for their future roles as pre-primary and primary education teachers.

Acknowledgements Project "Enviromental literacy. A challenge for the formation of the teachers in the 21st Cnetury" (indentifier PID2020-114171GB-I00) from the 2020 convocatory of the State Programs to Generate Scientific and Technological Knowledge and Strength of the R&D + i System.

References

Almazán, A., Garrido Martínez, J., Alonso López, P., Álvarez González, V., & González Reyes, L. (2024). *Técnicas humildes para el decrecimiento*. Ecologistas en Acción.

Álvarez-García, O., Sureda Negre, J., & Comas Forgas, R. (2018). Assessing environmental competencies of primary education pre-service teachers in Spain: a comparative study between two universities. *International Journal of Sustainability in Higher Education, 19*(1), 15–31. https://doi.org/10.1108/IJSHE-12-2016-0227

Arias, A. (2020). *La batalla por las ideas tras la pandemia: crítica del liberalismo verde*. Catarata.

Bordera, J., & Puig, F. (2022). *El crecimiento se va a detener, por una razón o por otra (entrevista a dennis meadows)*. Fundación Contexto y Acción. https://ctxt.es/es/20220701/politica/40230/dennis-meadows-crecimiento-limites-colapso-crisis-ecologicadecrecimiento.htm

Bordera, J., Turiel, A., & Valladares, F. (2024). *El final de las estaciones: razones para la rebelión de la ciencia y el decrecimiento*. Revista Contexto (Escritos Contextatarios).

Caesar, L., Sakschewski, B., Andersen, L. S., Beringer, T., Braun, J., Dennis, D., et al. (2024). *Planetary health check: a scientific assessment of the state of the planet*. Potsdam Institute for Climate Impact Research (PIK).

Conde Núñez, M. C., Mariscal Díaz, P., & Sánchez Cepeda, J. M. (2018). La metodología en el trabajo de huerto escolar y coherencia con la ambientalización curricular. Análisis de una práctica docente. *Didáctica de las ciencias experimentales y sociales, 35*, 113–126. https://doi.org/10.7203/DCES.35.12799

Daly, H., Vettese, T., Pollin, R., Burton, M., & Somerville, P. (2019). *Decrecimiento vs green new deal*. Traficantes de Sueños.

Díez-Gutiérrez, E. J., & Jarquín-Ramírez, M. R. (2025). Un nuevo currículo y una nueva enseñanza en y para el decrecimiento. *Revista Latinoamericana de Estudios Educativos, 55*(1), 181–200. https://doi.org/10.48102/rlee.2025.55.1.683

Díez-Gutiérrez, E.-J., Trujillo Vargas, J. J., Perlado-Lamo de Espinosa, I., Palomo-Cermeño, E., Mateos-Toro, L.-M., et al. (2025). Is degrowth education an alternative in the minds of educators in the face of the serious eco-social crisis and global warming? *Sustainability, 17*(4), 1668. https://doi.org/10.3390/su17041668

Dixon, K. A., Michelsen, M. K., & Carpenter, C. L. (2023). Modern diets and the health of our planet: an investigation into the environmental impacts of food choices. *Nutrients, 15*, 692. https://doi.org/10.3390/nu15030692

FAO & OMS. (2020). *Dietas saludables sostenibles: principios rectores*. Food and Agriculture Organization of the United Nations. https://doi.org/10.4060/ca6640es

FAO, FIDA, OMS, PMA & UNICEF. (2022). *El estado de la seguridad alimentaria y la nutrición en el mundo 2022: adaptación de las políticas alimentarias y agrícolas para hacer las dietas saludables más asequibles*. https://doi.org/10.4060/cc0639es

Fernández, R. y. González, L. (2014). *En la espiral de la energía*. Libros en Acción. Baladre.

García, J. E., Fernández, J., Rodríguez, F., & Puig, M. (2019). Más allá de la sostenibilidad: por una educación ambiental que incremente la resiliencia de la población ante el decrecimiento/colapso. *Revista de Educación Ambiental y Sostenibilidad, 1*(1101), 1–15. https://doi.org/10.25267/Rev_educ_ambient_sostenibilidad.2019.v1.i1.1101

Gil Pérez, D., & Vilches, A. (2023). Cómo avanzar en la necesaria transición a la sostenibilidad. *Ciência & Educação (Bauru), 40*(2), 221–230. https://doi.org/10.1590/1516-731320230027

Gil-Pérez, D., & Vilches, A. (2019). La comprensión e impulso de la Sostenibilidad: un requisito imprescindible para una acción educativa y ciudadana eficaz. *Revista de Educación Ambiental y Sostenibilidad, 1*(2), 2101.

González, L., & Almazán, A. (2023). *Decrecimiento: del qué al cómo. Propuestas para el estado español*. Icaria.

González, L., González, E., & Almazán, A. (2024). *Transición ecosocial en Cataluña. Una propuesta decrecentista*. CGT. https://cgt.org.es/presentacion-del-informe-transicion-ecosocial-en-cataluna-una-propuesta-decrecentista/

Gudynas, E. (2023). *Desarrollos alternativos. Alternativas al desarrollo: una guía ante las opciones de cambio*. Ediciones desde abajo.

Guerrero Fernández, A. (2023). *Grado de alfabetización Ambiental de los futuros maestros y maestras en la formación en ciencias: una perspectiva decrecentista*. (Tesis doctoral). Universidad de Sevilla. https://hdl.handle.net/11441/147936

Guerrero Fernández, A., Rodríguez Marín, F., Solís Ramírez, E., & Rivero García, A. (2022). Alfabetización ambiental del profesorado de educación infantil y primaria en formación inicial: conocimientos, actitudes y comportamientos mediante el cuestionario de dimensiones ambientales. *Revista Interuniversitaria de Formación del Profesorado, 97*(36.1), 92434. https://doi.org/10.47553/rifop.v97i36.1.92434

Hamed Al-Lal, S., & Pineda Alfonso, J. A. (2024). Una propuesta metodológica para favorecer la alfabetización ambiental del profesorado en formación. In F. Rodríguez-Marín, E. García-González, & A. Rivero (Eds.), *Alimentación ecológica y justa ante el decrecimiento: una propuesta para la formación de docentes*. Octaedro.

Hickel, J. (2020). *Less is more: how degrowth will save the world*. William Heinemann.

Hickel, J., Kallis, G., Jackson, T., O'Neill, D. W., Schor, J. B., Steinberger, J. K., et al. (2022). Degrowth can work — here's how science can help. *Nature, 612*, 400–403.

Holmgren, D. (2013). *Permacultura: principios y senderos más allá de la sostenibilidad*. Kaicron.

IPCC (2023). Sixth assessment report.. https://www.ipcc.ch/assessment-report/ar6/

Kallis, G., Mastini, R., & Zografos, C. (2024). Perceptions of degrowth in the European parliament. *Nature Sustainability, 7*, 64–72. https://doi.org/10.1038/s41893-023-01246-x

Kongshøj, K. (2023). Social policy in a future of degrowth? Challenges for decommodification, commoning and public support. *Humanities and Social Science Communications, 10*, 850. https://doi.org/10.1057/s41599-023-02255-z

Kuhn, T. S. (1971). *La estructura de las revoluciones científicas*. Fondo de Cultura Económica.

Latouche, S. (2009). *Pequeño tratado del decrecimiento sereno*. Icaria Editorial.

López-Lozano, L., García Díaz, J. E., Cubero Pérez, R., Rodríguez Marín, F., Guerrero Fernández, A., & Puig Gutiérrez, M. (2024). Educar en tiempos de decrecimiento: alfabetizar ambientalmente a los maestros y maestras del siglo XXI. In M. C. P. Fuentes (Ed.), *Exploración de variables psicológicas y educativas: avances en la investigación escolar* (pp. 523–534).

Maldonado Aragón, M. S., & Valadez Ramírez, A. A. (2021). Efectos de un huerto escolar y talleres en la ingesta de comida saludable en preescolares. *Revista Electrónica de Psicología Iztacala, 24*(1), 325–346.

Marín Gil, J. (2015). El huerto escolar: mágica fuente de conocimiento y sabiduría. *Aula, 239*, 44–47.

Massip Sabater, M., & Espinet, M. (2024). Delearning capitalism: is degrowth a decolonial perspective in social science education? *JSSE-Journal of Social Science Education, 23*(4), 7127. https://doi.org/10.11576/jsse-7127

Meadows, D. (2022). Are there limits to economic growth? It's time to call time on a 50-year argument. *Nature, 603*, 361.

Mello O'Brien, S. R. (2007). *Indications of environmental literacy: using a new students knowledge, and attitudes ofuniversity-aged survey instrument to measure awareness*. (Doctoral thesis). Iowa State University. https://lib.dr.iastate.edu/cgi/viewcontent.cgi?article=16054&context=rtd

Mollison, B., & Holmgren, D. (1978). *Permaculture one: a perennial agricultural system for human settlements*. Corgi.

Moore, J. W. (2020). *El capitalismo en la trama de la vida*. Traficantes de Sueños.

ONU. (2021). Informe anual del PNUMA (Programa de Naciones Unidas para el Medio Ambiente). https://wedocs.unep.org/bitstream/handle/20.500.11822/37946/UNEP_AR2021_SP.pdf

Puig, M., Guerrero, A., López-Lozano, L., & Ramos, M. N. (2023). Un viaje gastronómico por el mundo: investigando sobre alimentación sostenible en la formación inicial de docentes de educación infantil. In M. E. En, A. R. Cambil, Fernández, & N. de Alba (Eds.), *La didáctica de las ciencias sociales ante el reto de los objetivos de desarrollo sostenible* (pp. 997–1005). Narcea.

Puig, M., & Rodríguez-Marín, F. (2021). La enseñanza del entorno desde la coordinación docente: ciclo de mejora de aula integrado. In R. Porlán, E. Navarro-Medina, & A. F. Villarejo (Eds.), *Ciclos de mejora en el aula, año 2021: experiencias de innovación docente de la universidad de Sevilla* (pp. 1127–1147). Editorial Universidad de Sevilla. Instituto de Ciencias de la Educación (ICE).

Riechmann, J. (2019). Capitalismo verde. Nuestra bandera. *Revista de debate político, 244*, 189–199.

Ripple, W. J., Wolf, C., Gregg, J. W., Rockström, J., Mann, M. E., Oreskes, N., et al. (2024). The 2024 state of the climate report: perilous times on planet earth. *Bioscience, 0*, 1–13. https://doi.org/10.1093/biosci/biae087

Ritchie, H. (2019). *Food production is responsible for one-quarter of the world's greenhouse gas emissions*. Published online at OurWorldinData.org. Retrieved from: https://ourworldindata.org/food-ghg-emissions

Rockström, J., Gupta, J., Qin, D., Lade, S. J., Abrams, J. F., Andersen, L. S., et al. (2023). Safe and just earth system boundaries. *Nature, 619*, 102–111. https://doi.org/10.1038/s41586-023-06083-8

Rodríguez-Marín, F., Fernández-Arroyo, J., & García-Díaz, J. E. (2014). Las hipótesis de transición como herramienta para la educación ambiental. *Enseñanza de las Ciencias, 32*(3), 303–318.

Rodríguez-Marín, F., García-González, E., & Rivero, A. (2024). *Alimentación ecológica y justa ante el decrecimiento: una propuesta para la formación de docentes*. Octaedro.

Rodríguez-Marín, F., Rodríguez-Rojas, C., Mora, W. M., & García, J. E. (2024). ¿Está presente el debate sobre el decrecimiento en el ámbito de la educación ambiental y la didáctica de las ciencias experimentales? In I. G. Dufranc, E. M. G. Terceño, R. B. Toma, & J. O. Revilla (Eds.), *31 encuentros internacionales de didáctica de las ciencias experimentales: hacia una educación científica alineada con la agenda 2030* (pp. 175–180). Universidad de Burgos.

Roth, C. E. (1992). *Enviromental literacy: its roots, evolution and directions in the 1990s*. ERIC Clearinghouse for Science, Mathematics, and Environmental Education.

Santiago, E. (2024). *Contra el mito del colapso ecológico*. Arpa.

Saynes Santillán, V., Etchevers Barra, J., Paz Pellat, F., & Alvarado Cárdenas, L. (2016). Emisiones de gases de efecto invernadero en sistemas agrícolas de México. *Terra Latinoamericana, 34*(1), 83–96.

Scalabrino, C., Navarrete, A., & Oliva, J. M. (2022). A theoretical framework to address education for sustainability for an earlier transition to a just, low carbon and circular economy. *Environmental Education Research, 28*(5), 735–766. https://doi.org/10.1080/13504622.2022.2031899

Sempere, J. (2018). *Las cenizas de prometeo*. Pasado y presente.

Smith, G. A., & Stevenson, R. B. (2017). Sustaining education for sustainability in turbulent times. *The Journal of Environmental Education, 48*(2), 79–95. https://doi.org/10.1080/00958964.2016.1264920

Solís Ramirez, E., & Rivero García, A. (2024). Finalidad, contenidos e hipótesis de transición sobre la alimentación humana y su impacto socioambiental. In F. Rodríguez-Marín, E. García-González, & A. Rivero (Eds.), *Alimentación ecológica y justa ante el decrecimiento: una propuesta para la formación de docentes*. Octaedro.

Stables, A., & Bishop, K. (2001). Weak and strong conceptions of environmental literacy: implications for environmental education. *Environmental Education Research, 7*(1), 89–97. https://doi.org/10.1080/13504620125643

Taibo, C. (2016). *Colapso: capitalismo terminal, transición ecosocial, ecofascismo*. Los Libros de la Catarata.

Tejero, H., & Santiago, E. (2019). *¿Qué hacer en caso de incendio? Manifiesto por el green new deal*. Capitan Swing.

Entrenador, T. (2021). Decrecimiento: ¿cuánto se necesita? *Biofísica Econ. Sust., 6*, 5. https://doi.org/10.1007/s41247-021-00087-6

Trainer, T. (2024). Una crítica (amistosa) al movimiento por el decrecimiento. *Rebelión*. https://rebelion.org/una-critica-amistosa-al-movimiento-por-el-decrecimiento/

Tsivitanidou, O. E., Gray, P., Rybska, E., Louca, L., & Constantinou, C. P. (Eds.). (2018). *Professional development for inquiry-based science teaching and learning*. Springer.

Turiel, A. (2020). *Petrocalipsis: crisis energética y cómo (no) la vamos a solucionar*. Alfabeto.

Uncer, G., Boone, J. W., Tuzun, O. Y., & Oztekin, C. (2014). An evaluation of the environmental literacy of preservice teachers in Turkey through Rasch analysis. *Environmental Education Research, 20*(2), 202–227. https://doi.org/10.1080/13504622.2013.768604

Valero, A., & Valero Delgado, A. (2021). *Thanatia: Los límites minerales del planeta*. Icaria.

Valladares, F. (2023). *La recivilización*. Destino.

Willett, W., Rockström, J., Loken, B., Springmann, M., Lang, T., Vermeulen, S., et al. (2019). Food in the anthropocene: the EAT-lancet commission on healthy diets from sustainable food systems. *The Lancet, 393*, 447–492.

WWF. (2024). *Informe planeta vivo 2024: un sistema en peligro.* WWF. https://wwfes.awsassets.panda.org/downloads/informe_planeta_vivo_2024-completo.pdf

Pedagogical Practices of Popular Environmental Education in Family Farming

Aline Costa Gonzalez, Vilmar Alves Pereira, and Irene Carniatto

Abstract Climate Change can have negative consequences on food production. In this perspective, the goal of this study is to propose pedagogical practices in which Popular Environmental Education is inserted, directed to the agents of the land to face the climate crisis. Through documentary research via the Internet and narrative research by listening to family farmers in the Western Region of Paraná (Brazil) and Galicia (Spain), it was possible to relate demands and challenges pointed out by them and by the literature, and to present proposals that should be considered to improve the quality of life in the productive units. The proposals include the promotion of integration between the environment and society, through collective, participatory and contextualized practices with the lived realities, favouring the perception of belonging of the subjects and the place where they live. Ideas that seek to rescue values for the understanding that we are part of Nature, and as interdependent beings inhabiting the planet, we urgently need the contribution of everyone, so that the quality of life is privileged, social inequalities are reduced, and climate justice is achieved.

1 Introduction

"Agriculture is the art of cultivating the Sun." This Chinese proverb expresses how dependent this activity is on natural phenomena and how fascinating it is to transform solar energy into organic matter and energy to feed us. Farmers developed the ability to observe changes in the weather and organize their activities based on the seasons to plant and harvest. However, it is increasingly difficult to understand how and to what extent climate phenomena are occurring to plan a good harvest.

It is already known that the growing risks imposed by the socio-environmental crisis that we are experiencing come from the misuse of land, the unlimited use of

A. Costa Gonzalez (✉) · I. Carniatto
Universidade Estadual do Oeste do Paraná, Marechal Cândido Rondon, Brazil

V. Alves Pereira
Universidade do Estado de Mato Grosso, Caceres, Brazil

J. J. Conde et al. (eds.), *Practices in Education for Climate Emergency*,
https://doi.org/10.1007/978-981-92-2304-6_5

natural resources, the emission of greenhouse gases and the degradation of ecosystems (UNDRR, 2021a), and that extreme events have been recorded in different parts of the world with greater intensity and frequency in the last decade (Jacobi & Giatti, 2017; Patriota, 2010; Sevillano, 2021; United Nations, 2022).

One of the sectors that has been heavily impacted by climate change in terms of socioeconomic and environmental issues is agriculture. Recent field research has shown that family farming in the western region of Paraná (Brazil) and Galicia (Spain), which are regions with great potential in this area, has been suffering from the impacts of climate instability, long periods of drought, and rising temperatures. Losses caused by the increased occurrence of strong winds, storms, and hail in the western region of Paraná and an increase in fires and pests in crops in Galicia were also reported (Gonzalez, Pereira, & Oliveira, 2024).

Furthermore, in Brazil, several public policies have been dismantled or neglected in recent years, and this has exacerbated the country's social and economic inequalities, thus revealing the urgency of finding alternatives so that human life and dignity are at the center of political and educational decisions, and one of these points is directly linked to food security. Caring for and improving the lives of people who work to ensure food for everyone means that this care extends to other people in society, improving the ecology of the environment as a whole.

According to Milanez and Fonseca (2011), the consequences of Climate Change, unfortunately, do not have great popular appeal, and there is no association among the population between Climate Change and socio-environmental disasters. Nor is there an understanding that tragedies increase the vulnerability of those who are least able to deal with their consequences. This hinders the engagement necessary for social mobilization, which, together with the lack of interaction between scientists and decision-makers, makes it difficult to develop public policies and mitigation strategies (Lampis et al., 2020). Thus, with the increase in the occurrence of extreme events, the socio-environmental and economic costs will be greater, especially in the most vulnerable areas, such as family farming.

Although all these events make it clear to governments, decision-makers and large economic groups that the only solution to end the environmental impacts that threaten all living species on the planet will only occur if we reduce greenhouse gas emissions, through major changes in production and consumption patterns, and the energy base coming from fossil fuels, this harms the interests of such representatives, making them ignore the proposals for changes, as the idea directly affects the capitalist model, whose profit for these few is above all (Boff, 2017; Chomsky et al., 2020; Marques, 2018). This neoliberal logic assumes individual freedom above any values, disregarding policies that attempt to promote distributive social justice, which benefits the collective (Sparemberger & Hartwig, 2023).

There have already been several meetings to discuss the environment and the climate crisis, such as the various Conferences of the United Nations Environment Programme; Earth Summit in Rio de Janeiro in 1992; World Summit on Sustainable Development in Johannesburg in 2002; Rio + 20 in 2012; Climate Action Summit in New York in 2019, among others (UNEP, 2022), however, between some failures and many agreements signed by most countries, the threat of collapse remains alive, and shows that we urgently need practical actions, both global and local, to mitigate and prevent disasters and their impacts.

Research data show the need for changes in the understanding of disaster risks and their consequences, as well as improvements in the actions of different sectors of society to reduce damage (UNDRR, 2021b). In other words, the crisis is urgent and, in addition to the lack of public policies aimed at family farmers to face the Climate Emergency, there is also a lack of practical and systematic actions of an educational and pedagogical nature, legal devices and more in-depth and decentralized debates on this crisis in the context of family farming (Ferreti & Nepomoceno, 2023; Tanure, 2020).

Therefore, this study seeks to propose pedagogical practices that include Environmental Education, as well as to contribute to Public Policies aimed at field agents facing the climate crisis. Through documentary research (Fonseca, 2002) and narrative investigation (Carniatto, 2002; Clandinin & Connelly, 2011), whose findings were based on official documents researched on the internet on intergovernmental websites, selected, reviewed and organized according to Fonseca (2002) and conversations with Galician and Brazilian farmers, whose listening was attentive and recorded for later analysis of what was narrated, as part of the doctoral research and academic exchange in Galicia, in the year 2022.

That said, the study presents a descriptive and qualitative approach, focusing on a universe of multiple meanings, beliefs, values and aspirations (Minayo, 1994), to contribute to the visibility of these less favored social actors and, above all, to the right to food security and a dignified life for the entire population, given the imminent low food production for the coming years.

1.1 Public Policies and their Aims

Public Policies, in general, aim to promote the well-being of society and solve collective problems through government actions. They seek to meet the needs of the population, guarantee rights, and promote socioeconomic development, improving people's quality of life (Vázquez & Delaplace, 2011).

It is worth noting that Public Policies may vary according to the context and characteristics of each country. In the case of Brazil and Spain, the places studied for this research, there are significant differences in terms of political structure, government authorities, and challenges faced.

In Brazil, the main objective is related to reducing social inequalities, as it is a country that faces a greater degree of poverty, lack of access to basic services, and violence in various forms, seeking to alleviate these problems through social programs and social inclusion. In Spain, the focus is more on economic development and improving the quality of life. The country seeks to solve problems such as immigration and environmental sustainability, stimulating economic growth, education, research, and the preservation of biodiversity (Van Der Ploeg, 2011).

Although these two countries have their approaches and priorities for Public Policies aimed at family farming and climate change, both recognize this agricultural sector as strategic for food security and income generation, have programs to encourage organic and agroecological production and the generation of renewable energy, in addition to being signatory countries of the Paris Agreement, whose commitment is to reduce greenhouse gas emissions in the coming years (Brazil, 2023; European Council, n.d.).

One of the documents that encompasses agriculture and agriculture and the climate issue in Spain is the Common Agricultural Policy (CAP), created in 1962 to establish a partnership between the agricultural sector, society, and farmers in the European Union. This Policy has undergone several changes over the years, and the latest version, from 2023, points out in your text more relevance to small farmers and agroecological practices. In Brazil, the related document is the National Policy on Agroecology and Organic Production, which had been terminated by the previous government and was resumed in 2023.

Even so, the implementation of the proposed actions does not always occur as written in the documents, and it also varies according to the priorities of each government administration (Vázquez & Delaplace, 2011).

1.2 Environmental and Popular Education

The existence of Public Policies is undoubtedly extremely important. However, education in the context of the climate emergency is essential for behavioral changes to occur in favor of improving the quality of life of society and the planet, and that is why we will advocate it in this work.

Leff (2001) reports that to solve complex environmental problems and reverse their causes, a paradigm shift is necessary in the system of knowledge, rational behaviors, and values, based on economic aspects such as social growth. Therefore, education, with its power to dialogue with various instances of human development and its transformative potential, comes to assist in this process.

From this perspective, Environmental Education as an area of knowledge has great potential for these paradigm shifts to occur, acting as a mediator between the State and society, promoting problematizing activities on the socio-environmental crisis, and restructuring pedagogical practices and social action (Tamaio, 2010). According to Sorrentinoet et al. (2005, p. 287) "deals with a paradigm shift that implies both a scientific and political revolution" and "by educating for citizenship, it can build the possibility of political action, in the sense of contributing to forming a community that is responsible for the world it inhabits".

If aimed at family farmers, we have the recent Popular Environmental Education that can be more effective, since in addition to the democratic and participatory approach, the concept of interdependence between the natural, socioeconomic and cultural environment, the pluralism of ideas and recognition of individual and cultural diversity that Environmental Education advocates (Brazil, 1999), Popular Environmental Education recognizes the importance of popular and ancestral knowledge rooted in different communities and their political perspectives, enabling their actions in the world, since it is based on popular education (Pereira et al., 2022).

Popular education as a point of view, addressed in Popular Environmental Education, brings a proposal of criticality to neoliberal, authoritarian and technocratic philosophy, as it deals with a liberating conception, of reflective autonomy, of solidarity, emancipatory and humanizing social transformation, acting as an alternative to the oppressive model, in the fight against social inequalities (Freire, 1987). The contribution of this knowledge differs from traditional education, giving space to a more problematizing, sensitive, creative, dialogical, egalitarian, and transformative education (Claro & Pereira, 2019; Pereira et al., 2022), playing a fundamental role in confronting the Climate Emergency. This is because the horizon of Popular Environmental Education is directly associated with the defense and achievement of a societal project. Hence, from the educational perspective of confronting a society whose climate context presents numerous anomalies.

Furthermore, these educational approaches use social pedagogy practices, which value local knowledge, traditional practices and the exchange of experiences as a scientific educational process that goes beyond school walls, being successful in different social spaces, in everyday life and in guaranteeing active citizenship (Caride, 2005; Freire, 1992) and can contribute to strengthening the identity and autonomy of farmers.

2 Demands of Family Farming and Challenges to be Faced

Given several demands raised in recent research carried out in Western Paraná (Brazil) and Galicia (Spain) in the year 2022 (Gonzalez et al., 2024), we were able to compile here those that make life most difficult for family farmers: 1) Lack of appreciation and interest in family farmers, compared to agribusiness farmers; 2) Difficult access to government managers and Public Policies aimed at agriculture; 3) Lack of information and better communication about climate change, sustainable technologies and food awareness, both for farmers and for society.

Given this, although we recognize that Environmental Education can contribute to these issues, the various challenges that contribute to the climate crisis becoming increasingly worse are visible. One of them is the fact that human beings are disconnected and uprooted from Mother Earth. According to Boff (2015), the capacity to feel a sense of belonging to the organism Gaia has been lost. From the moment that human beings dissociate themselves from nature, see it and treat it as an object and a right of dominion, forgetting the finiteness present in natural resources and the interdependence of all living beings that inhabit this environment, they disconnect from the meaning of life as a common existence.

Another major challenge is the ideas promoted by neoliberalism about sustainable development, which presents a "green discourse" that is unfeasible given the critical point we are at. The idea of sustainability is not bad, but the way to promote it needs to be reviewed and reorganized, as it goes beyond the consumerist, individualist and extractivist discourse, in which money is more important than life (Layrargues, 1998). People influenced by these ideas talk about change, but they do not want to give up profit and will only move towards a cause if they have direct benefits for themselves, especially if they are financial gains.

We can also point out the fact that content is taught in schools in a linear and dismembered way. When we focus on the parts, we tend to separate them from the whole, making it difficult to understand the human-nature relationship. It is a great challenge to understand the complexity of life, to understand the way we are producing life in society, with its political, economic and social processes that determine the socio-environmental crisis (Morin, 2000).

Another problem is related to Public Policies. Although there are several socio-environmental policies on paper in both countries studied, we noticed that there is often a lack of action and investment for existing programs and projects to happen. The fragility of policies in the face of discontinuation between one government administration and another is evident, as well as the lack of intersectoral thematic articulation and integration between policies, hindering the progress of several areas of society (Lotta & Favareto, 2016).

To propose effective actions, it is necessary to diagnose the region in which such activities are to be implemented so that the dynamics occur by the demands, challenges, and characteristics of that place. However, when researching two places like Brazil and Spain, which are quite different in terms of culture, history, economy, and climate, we were able to see that some proposals can work in different environments where family farming is found. Based on these challenges and demands found in the research, we thought of some pedagogical practices. Disruptive for Environmental Education interventions.

3 Pedagogical Practices of Popular Environmental Education in Family Farming

Popular Environmental Education in the context of the climate emergency and family farming is based on the premise of "ecological, historical, psychological, political and ethical awareness of a climate reality that is often external to human concerns" (Pena-Vega, 2023) and the active engagement of farmers, which can help develop skills and improve people's ability to face socio-environmental challenges. It seeks to promote knowledge through a critical-dialectical epistemology, adding to it a scientific and existential historical-cultural dimension for the apprehension of knowledge and thus culminating in the mental and social transformation and reconstruction of a complex reality (Freire, 1983; Pena-Vega, 2023).

The idea is to dedicate activities aimed at the active participation of the rural community, encourage dialogue, knowledge sharing, inclusion and social mobilization, so that they understand the impacts of climate change on their daily activities and are able to adopt measures to mitigate, prevent and adapt to these effects, improving the quality of life of rural families, maintaining household income and food production.

On the horizon of Popular Environmental Education, as a result of research PhD, in which data on occurrences of natural disasters were analyzed (Gonzalez et al., 2024), the perception of family farmers about climate change (Gonzalez et al., 2024) and environmental and educational legislation in force in the regions studied, we believe that some practices can be taken into consideration, as:

3.1 Family Farmers in Schools

Bring farmers into classrooms and promote environmental campaigns, cultural interactions, and discussion groups between educators, students, and farmers, seeking a collective and interactive construction of knowledge between the countryside and the city.

It is possible to take advantage of these moments to contextualize the land structure and history of agriculture in the region, land disputes, forms of soil management and strategies used for production, as well as the importance of ensuring food and nutritional security.

The dialogue between the different social figures establishes connections, bringing the reality experienced by farmers closer to the school proposal and its teaching method, obtaining types of teaching-learning in line with traditional values and practices of the countryside. In this sense, Arroyo (2007) states that:

> A frequently raised hypothesis is that our school system is urban, only thought of in the urban paradigm. The formulation of educational and public policies, in general, thinks of the city and urban citizens as the prototype of subjects with rights. There is an idealization of the city as the civilizing space par excellence, of coexistence, sociability and socialization, of the expression of political, cultural and educational dynamics. This idealization of the city corresponds to a negative view of the countryside as a place of backwardness, of cultural traditionalism. These images that complement each other inspire public, educational and school policies and inspire most legal texts. The urban paradigm is the inspiration for the right to education. (p. 158)

These social relations between the countryside and the city bring visibility to the rural agent, who often feels distant and even frowned upon by society, and can be used to improve curricula and Political Pedagogical Projects so that they recognize the transversality of Environmental Education and are truly included in socio-environmental educational actions throughout the students' academic journey (Gonzalez et al., 2024).

3.2 Meetings Between Family Farmers

Promote meetings between farmers in the region, which could be seminars, field days and the use of commemorative dates in the municipality or town, so that they can share their experiences and activities carried out in their production units, so that they can visualize possibilities for reinterpreting their daily tasks.

In one of the conversations with a Brazilian farmer, he had treated his corn plantation with neem oil, an organic product that replaces pesticides. In the speech, he said: *"I am treating it with natural oil, but the neighboring sites still use a lot of poison. It's very annoying because it can affect my cultivation"*.

Then the agricultural technician who accompanied the farmer on his visit, who was also present during the conversation, suggested that the man make a video of the well-tended crop and how he had used the oil to combat pests so that she could show it to the other farmers during their visits to their properties. She said that her own words don't always work, but if they are talking to them, the idea is easier to accept.

The suggestion given by the agricultural technique may be a good proposal, as it is based on the technique of rapport, a French term used in psychology to gain another person's trust and consequently convince them of something through mirroring and empathy (Tompkins & Lawley, 1994).

Considering the idea of collectivity present among small producers, it is possible to stimulate the creation of cooperation networks to share knowledge and good practices related to sustainable agriculture and tackling the climate emergency since many have not yet adopted these practices and have difficulties concerning mitigation strategies in the face of disaster risks.

In addition, cooperative and union organizations work to preserve the ecology and economy of rural life and defend these individuals against large producers. Pereira et al. (2023), in their work, reinforce the relevance of cooperativism/associativism for the development of family farming, including in facing the challenges arising from the logic of agribusiness.

3.3 Production Units as an Educational Space

The goal is to promote sustainable and agroecological rural tourism on properties, using ecological trails, agroforestry systems, organic gardens, and ecological animal husbandry with visits open to schools and tourists.

In two conversations with a Brazilian fish farmer and a Galician cattle rancher, they reported their interest in reopening their properties to visitors, especially for children, so that they could learn about their sustainable production and promote the importance of preserving the environment. According to the fish farmer: *"I wish there was a partnership between us and someone from the city hall, or some organ, to open another visit to my property, because now I have a project to cultivate fish together with hydroponics. This is interesting for people to know, that with the circulation of water I can produce organic fish and lettuce, and I would also work on environmental education there.*"

In line with this statement, the Galician cattle rancher said, *"There is a lack of environmental awareness among young people today. Here, I produce quality meat in an ecological way. Many children have already come here to learn about it and creater this consciousness, but these visits did not last long. It is important what administration, the Galician Board, for example, promote these approximations".*

These two producers reported who put this idea into practice a few years ago and enjoyed receiving visitors and seeing people's joy as they got involved in the natural and cultural environment they provided. However, the ecotourism activity did not last long due to a lack of support.

Understanding and using rural properties as educational spaces is enriching, since they offer several learning opportunities, as being in this environment allows: 1) Connection with nature, allowing people to learn about biodiversity, natural cycles and environmental preservation; 2) Learning practical activities such as planting, harvesting and caring for animals; 3) Appreciation of local culture using typical cuisine, traditional agricultural techniques and craft techniques to acquire knowledge and preserve cultural identity; and 4) Integration with the community, being able to promote educational and interactive events such as fairs, courses and workshops, where hosts can share their knowledge with guests (Gadotti, 2000; Matarezi, 2005; Pereira et al., 2023).

3.4 Agroecology as an Instrument of Environmental Education

Encouraging the transition to sustainable production is essential and extremely beneficial for the environment, in addition to generating more income for families. The ecological practices of this agricultural model work with soil management, use of natural resources and biological growth of animals in a responsible manner, according to the local characteristics of the environment, respecting the web of life (Primavesi, 2008), having great potential in rescuing interactions between farmers and the agrosystem, in a context social, political, temporal and economic, enabling balance in small-scale rural development.

Using this dynamic composed of daily and interconnected activities reinforces the degree of belonging to nature and the family's autonomy. Therefore, it is necessary to provide financial incentives and easy-to-understand information so that they can make the transition and maintain agroecological production, considering that several properties visited during the research period did not work with this dynamic.

Furthermore, raising awareness among the population/consumer about healthy eating, basing the construction of the socio-educational dimension of agroecology on critical Environmental Education, focusing on citizen mobilization, strengthening relationships and articulating alternatives to replace the hegemonic agro-food system, through collective and participatory processes in existing agroecology units, democratizing information and knowledge in order to increase the consumption of these products, as well as pressuring the implementation of public policies aimed at this segment (Bisquert, 2021).

Although agroecology has been presented for some time as a solution for mitigation and adaptation against climate change (Intergovernmental Panel on Climate Change, 2023) and of great importance in the prevention of forest fires and pests, by integrating woody vegetation, crops and extensive livestock farming (Damianidis et al., 2021), the relevance given to social programs and public policies does not always materialize. However, with a more incisive presence in the new Common Agricultural Policy (PAC) and with the resumption of the National Policy on Agroecology and Organic Production, both starting in 2023, we hope that the search for initiatives with socio-educational perspectives will be more viable, as well as the increase in financial incentives so that farmers can make the transition and maintain their properties.

3.5 Extension Visits and Institutional Partnerships

There is no sustainable rural development without thinking about social justice. Everyone needs to participate, and therefore, there is a need to acquire a critical sense on the subject in order to truly collaborate. Through partnerships between universities, farmers, government, public and private companies, communities, and social movements, it is possible to promote collective actions such as networks and exchanges between these segments, which aim to understand the root of the problem and sharpen critical thinking to mitigate environmental impacts.

Extension projects and partnerships strengthen the resilience of rural communities, access to updated information on climate change, public policies and incentive programs, encourage cooperation, solidarity and the construction of joint solutions to existing challenges, considering institutions as mediators so that there is dialogue on initiatives and public policies, favoring practical and efficient actions that transcend the social perception that the Climate Emergency is a complex and abstract issue.

Pereira et al. (2023) state that extension practice, when related to family farming, agroecology and organic food production, can:

> Assuming very important meanings from the perspective of an extension that recognizes itself in the close relationship with popular knowledge. This place of popular horizon extension redefines postures, and associates extension with cooperative society projects in the humanizing and transformative direction of reality. (p. 3)

This interaction provides debates and experiences that achieve a new way of conceiving the extensions carried out, as well as understanding the possibilities they offer as a space for commitment, stimulation of protagonism, identity strength and the fighting movements of these subjects, towards the necessary transformations (Pereira et al., 2023).

3.6 Public Policies and Integrated Actions

Articulate the integration and adaptation of existing public policies in the economic, agricultural, environmental, climate, educational and food security fields, to unify and execute in the best possible way the actions already proposed, which are often restricted to their portfolios, without much dissemination or understanding on the part of the agents of the institutions involved, on how to proceed.

Tighten public policies to monitor the use of pesticides and carbon emissions, given the numerous scientific data on the damage caused to health and the environment, in addition to contributing to climate injustice, in which those most affected are those who use the poisons least and emit greenhouse gases.

One of the most common things farmers said was the extremely different way in which small and large producers were treated by the government. According to a Galician farmer: *"Family farmers have a small piece of land and work there. Now, the big farmers, they really harm the land, the environment, and the organs* (public) *close your eyes the for this. The norm is for the minority"*. You talk like that show the dissatisfaction with Public Policies, to the realize that for agribusiness there is all the necessary assistance to meet market interests, while the needs of family farming remain below expectations.

Furthermore, if they cannot count on public policies to maintain basic needs such as food, housing, health and education, these individuals' concerns regarding the climate emergency obviously take a back seat, as their concerns about eating and paying their bills are immediate, while the effects of the climate crisis are seen as medium and long term., as we can see in the words of a Brazilian farmer: *"I lack a little concern in informing myself about these things because we are always involved in work. It's work, work, it's not possible to the to think too much about tomorrow. Live as you can, I'm just in this job role, work and I'm not stopping to the to look at other things"*.

In this context, it is necessary to include Environmental Education and the population in the debates for implementing actions, because according to Tamaio (2010), the policies proposed in most countries are generally for economic and technological incentives, often disregarding the participation of society and references to Environmental Education. However, the educational process plays a fundamental role in presenting the causes and consequences of the climate emergency, as well as in building viable alternatives for changing the capitalist economic model, expanding responsibility beyond the government, involving the entire population.

Environmental Education interventions must be supported by studies that delve deeper into how the crisis affects daily activities and the real need for a change in values, improving the understanding of concepts, approaching the population emotionally, so that there is involvement as participatory citizens who belong to nature, to face possible future critical scenarios (Guimarães, 2004).

Furthermore, three of the farmers interviewed, one Brazilian and two Galician, said that public authorities should do what we were doing, as the following speech shows: *"They should listen to the farmers, just like you are doing. Whoever makes the laws needs to get out of the administrative and come to know everyday life of farmers, to from there then take decisions that fit the reality of small farmers"*. In this sense, it is necessary to promote the opening of public spaces for the voices of these farmers, in order to value, recognize and organize active participation in decision-making, strengthening the negotiating power of this social group, allowing them to defend their interests and needs.

Finally, it is worth highlighting that these are some possibilities envisioned from the immersion carried out in the production units and in the work context of the actors in the field and climate change, in which it was possible to resignify thoughts and learnings regarding the cooperative, productive and resistance environment present in the lives and history of these Brazilian and Galician subjects.

Being in contact with the territories and promoting attentive listening to the farmers' speeches provided the opportunity to share academic and popular knowledge, culminating in fruitful reflections and conversations about the climate, the fundamental role and daily struggle of family farmers, the hegemonic logic of capitalist societies, the challenges to be overcome in the face of the power of agribusiness and the various human practices that are harmful to the environment.

4 Final Considerations

Given the undeniable need for behavioral changes in human beings in the face of the socio-environmental crisis, the work sought to aggregate ideas and reflections in light of the sharing of knowledge, expansion and systematization of knowledge on the subject. Although there are several demands and challenges to be faced, we believe that education, with its strength and social function, is the most appropriate path at this time. The pedagogical dimension and Environmental Education reach society at all levels and, therefore, must be contemplated in daily activities for transformation to occur.

In short, the proposals encompass the promotion of integration between the environment and society through collective, participatory practices that are contextualized with the realities of life, favoring the perception of belonging of the subjects and the place where they live. Ideas that seek to rescue values, through critical and incisive thinking, but also philosophical and caring for life, because we are part of nature. Understanding that being all one, we are interdependent beings inhabiting the planet and we urgently need everyone's contribution to prioritize quality of life, reduce social inequalities and achieve climate justice.

We understand the power that exists in educational processes so that individualism can be broken and human beings can begin to see themselves as a community and act as a collective, whose interests in life take precedence over financial interests, so that international agreements are truly fulfilled and public policies are implemented in favor of those most in need.

Finally, it was possible to perceive that although the right to basic education, an ecologically balanced environment and healthy food are part of the Public Policies of both countries analyzed, the reality does not always match what is established by the local government. Often there is a lack of detail about what is intended, who it is aimed at and how to successfully implement such policies, showing that we do not necessarily need to create Public Policies, but at least guarantee access to existing policies for populations such as family farmers.

References

Arroyo, M. G. (2007). Policies for the training of rural educators. *Cedes Notebooks, 27*, 157–176. https://doi.org/10.1590/S0101-32622007000200004

Bisquert, K.M. (2021). *The socio-educational dimension of the agro-ecological movement in Galicia: Diet and the agri-food system as axes of sustainability* (Doctoral dissertation). University of Santiago de Compostela.

Boff, L. (2015). *Rights of the heart: How to green the desert*. Paulus.

Boff, L. (2017). *Sustainability: What it is – what it is not*. Voices Limited Publishing House.

Brazil. (1999). Law 9,795, of April 27. Which provides for environmental education, establishes the national policy for environmental education and contains other provisions. Available at: http://www.planalto.gov.br/ccivil_03/Leis/L9795.htm. Accessed on: July 13, 2023.

Brazil. (2023). Ministry of finance. Agricultural policy and environment. Available at: https://www.gov.br/fazenda/pt-br/assuntos/politica-agricola-e-meio-ambiente/atuacao-spe/mudancas-climaticas. Accessed on: July 2, 2023.

Caride, J. A. (2005). *The frontiers of social pedagogy: Scientific and historical perspectives*. Gedisa.

Carniatto, I. (2002). *The formation of the subject teacher: Narrative research in science/biology*. Edunioeste.

Chomsky, N., Pollin, R., & Polychroniou, C. J. (2020). *Climate crisis and the global green new deal: The political economy of saving the planet*. New Farm.

Clandinin, D. J., & Connelly, F. M. (2011). Narrative research: Experience and history in qualitative research. *EdUFU*.

Claro, L. C., & Pereira, R. A. (2019). Popular environmental education as a formative concept: Interweaving of extension, teaching and research. *Electronics Rev. Master's Degree in Environmental Education, 36*(3), 8–23. https://doi.org/10.14295/remea.v36i3.9676

Damianidis, C., Santiago-Freijanes, J. J., Denherder, M., Burgess, P., Mosquera-Losada, M. R., Graves, A., Papadopoulos, A., Pisanelli, A., Camilli, F., Rois-Díaz, M., Kay, S., Palma, J. H. N., & Pantera, A. (2021). Agroforestry as a sustainable land use option to reduce wildfires risk in European Mediterranean areas. *Agroforestry Systems, 95*, 919–929. https://doi.org/10.1007/s10457-020-00482-w

European Council (n.d.). *Common agricultural policy*. https://www.consilium.europa.eu/pt/policies/the-common-agricultural-policy-explained/

Ferretti, K. D., & Nepomocene, T. A. R. (2023). Mapping of public policies related to climate change in the context of family farming in Brazil. *Grifos Magazine, 33*(61), 01–18. https://doi.org/10.22295/grifos.v33i61.7117

Fonseca, J. J. S. (2002). *Scientific research methodology*. UEC.

Freire, P. (1983). *Education and change*. Peace and Land.

Freire, P. (1987). *Pedagogy of the oppressed*. Peace and Land.

Freire, P. (1992). *Pedagogy of hope*. Peace and Land.

Gadotti, M. (2000). *Pedagogy of the earth*. Peiropolis.

Gonzalez, A. C., Carniatto, I., & Pereira, V. A. (2024). Impacts of socio-environmental disasters in Western Paraná between 2010 and 2020. *Environment & Society, 26*, e0085. https://doi.org/10.1590/1809-4422asoc20220085r2vu2023L4AO

Gonzalez, A. C., Pereira, V. A., & Oliveira, I. C. (2024). Perceptions and demands of family farmers (Brazil and Spain) in relation to climate change. *Ibero-American Journal of Studies in Education, 19*, e024060. https://doi.org/10.21723/riaee.v19iesp.1.18225

Guimarães, M. (2004) Critical environmental education. In P. Pomier (Coord.) *Identities of Brazilian environmental education* (pp. 25–34). Ministry of Environment.

Intergovernmental Panel on Climate Change [IPCC]. (2023). *Summary for Policymakers. Climate Change and Land: An IPCC special report on climate change, desertification, land degradation, sustainable land management, food security, and greenhouse gas fluxes in terrestrial ecosystems*. IPCC. https://www.ipcc.ch/srccl/

Jacobi, P. R., & Giatti, L. (2017). Extreme events, emergencies and climate change. *Environment & Society, 20*, 1–6. https://doi.org/10.1590/1809-4422asoceditorialv2032017

Lampis, A., Towers, P. H. C., Jacobi, P. R., & Leone, A. L. (2020). The production of risks and disasters in Latin America in the context of a climate emergency. *The Social in Question, 23*(48), 75–96. https://www.redalyc.org/journal/5522/552264320003/html/

Layrargues, P. P. (1998). *The smoke screen: Green corporate discourse and the ideology of economic rationality*. São Paulo, Annablume.

Leff, E. (2001). *Environmental epistemology*. Cortez Editora.

Lotta, G., & Favareto, A. (2016). Challenges of integration in new institutional arrangements of public policies in Brazil. *Journal of Sociology and Politics, 24*, 49–65. https://doi.org/10.1590/1678-987316245704

Marques, L. (2018). *Capitalism and environmental collapse*. Unicamp Publishing House.

Matarezi, J. (2005). Educational structures and spaces: When structures and spaces become educators. In L. A. Ferraro (Ed.), *Meetings and paths: Training of environmental educators and educational collectives*. Directorate of Environmental Education.

Milanez, B., & Fonseca, I. F. (2011). Climate justice and extreme weather events: An analysis of social perception in Brazil. *Third Included Magazine, 1*(2), 82–100. https://doi.org/10.5216/teri.v1i2.17842

Minayo, M. C. S. (1994). *Social research: Theory, methods and creativity*. Voices Limited Publishing House.

Morin, A. N. D. (2000). *The seven skills needed for the education of the future*. UNESCO.

Patriota, A. (2010). Haiti: Challenges and opportunities in the post-earthquake period. *Bulletin of international economics and politics*. IPEA. 2010: https://repositorio.ipea.gov.br/bitstream/11058/4727/1/BEPI_n2_haiti.pdf

Pena-Vega, A. (2023). *The seven skills needed for climate change education*. Cortez Publisher.

Pereira, V. A., Brandalize, M. d. S., Grosko, S., de Oliveira, I. C., Zitkoski, J. J., Corbari, F., & Zonin, W. J. (2023). Extension, popular education and agroecology in family farming: Experience report of postgraduate students and teachers. *Journal of Social and Environmental Management, 17*(7), e03479–e03479. https://doi.org/10.24857/rgsa.v17n7-008

Pereira, V. A., da Silva, R. F., & Ramirez-Sánchez, M. Y. (2022). Popular environmental education in Latin America and the Caribbean and education for sustainable development: Inconsistencies and challenges. *Scientific Journal of the Faculty of Education and Environment, 13*(1), 92–113. https://doi.org/10.31072/rcf.v13i1.1050

Primavesi, A. M. (2008). Agroecology and soil management. *Agriculturas Magazine, 5*(3), 7–10. https://aspta.org.br/article/agroecologia-e-manejo-do-solo/

Sevillano, E.G. (2021). Storms cause flooding in Germany, leaving dozens dead and missing. *The Country*.. https://brasil.elpais.com/internacional/2021-07-15/inundacoes-na-alemanha-deixam-dezenas-de-mortos-e-desaparecidos.html.

Sparemberger, R., & Hartwig, E. (2023). The tension between neoliberal development and the right to an ecologically balanced environment: The counter-hegemonic theory of human rights as an alternative. *Paths of Law, 20*, e202441–e202441. https://doi.org/10.18623/rvd.v20.2441

Sorrentino, M., Trajber, R., Mendonça, P., & Ferraro Junior, L. A. (2005). Environmental education as public policy. *Education and Research, 31*(2), 285–299.

Tamaio, I. (2010). *A public policy proposal: Parameters and guidelines for environmental education in the context of climate change caused by human action.* Ministry of the Environment.

Tanure, T. M. do P. (2020). Climate change and agriculture in Brazil: Regional economic impacts and impacts on family and employer cultivation (Doctoral dissertation). Federal University of Minas Gerais.

Tompkins, P. & Lawley, J. (1994). Report: The magic ingredient. Personal Success.

United Nations [UN] (2022). UN: Challenges remain in Philippines after Typhoon Haiyan. https://news.un.org/pt/audio/2013/12/1084951

United Nations Environment Programme [UNEP] (2022). *Environmental milestones: A timeline of the UN's 75th anniversary.* https://www.unep.org/pt-br/news-and-stories/story/environmental-moments-un75-timeline

United Nations Office for Disaster Risk Reduction [UNDRR]. (2021a). *Words into action: Nature-based solutions for disaster risk reduction.* UNDRR. https://www.undrr.org/publication/words-action-nature-based-solutions-disaster-risk-reduction

United Nations Office for Disaster Risk Reduction [UNDRR]. (2021b). *Regional assessment report on disaster risk in Latin America and the Caribbean.* UNDRR. https://www.undrr.org/publication/undrr-roamc-regional-assessment-report-disaster-risk-latin-america-and-caribbean-rar

Van der Ploeg, J. D. (2011). Trajectories of rural development: International comparative research. *SociologieS, 13*, 114–140. https://doi.org/10.1590/S1517-45222011000200006

Vazquez, D., & Delaplace, D. (2011). Public policies from a human rights perspective: A field under construction. *Rev International Human Rights Institute, 8*(14), 34–65.

Teaching at the Edge: How Rural Schools Face the Climate Crisis – The Case of TEBAEV

Bethzabeé Velázquez-Martínez and Jorge Conde Miguélez

Abstract This chapter analyzes how the Telebachillerato of the State of Veracruz (TEBAEV)—an educational model designed for rural and peri-urban areas in Veracruz, Mexico—addresses the climate emergency through a disruptive educational approach. TEBAEV operates in communities marked by high marginalization and environmental vulnerability, making it a key space for implementing Climate Change Education (CCE) that is situated, critical, and transformative. Based on a mixed-methods study involving 3698 teachers, the findings reveal that although there is a clear willingness to integrate environmental content, current practices are mostly limited to awareness-raising, lacking scientific depth and continuity in community-based projects. Drawing on the theoretical frameworks of CCE, Critical Pedagogy, and Socioconstructivism, this study proposes an educational approach that not only transmits information but also fosters critical awareness, collective action, and the co-construction of local solutions. The chapter identifies structural barriers—such as limited infrastructure and insufficient teacher training—and proposes strategies to strengthen CCE: ongoing professional development, contextualized instructional materials, and inter-institutional partnerships. It concludes that TEBAEV could become a model of resilience and socio-environmental justice if it is consolidated as a space for situated learning that empowers young people and their communities to act in the face of climate change.

1 When Education Confronts the Climate: A Response from Veracruz

In Mexico, the climate emergency has assumed particular urgency in recent years due to the country's exposure to extreme hydrometeorological events, biodiversity loss, and the growing socioeconomic vulnerability of large segments of the

B. Velázquez-Martínez (✉)
Antonio María Rivera High School, Xalapa, Mexico

J. Conde Miguélez
Universidade de Santiago de Compostela, Santiago de Compostela, Spain

J. J. Conde et al. (eds.), *Practices in Education for Climate Emergency*,
https://doi.org/10.1007/978-981-92-2304-6_6

population (INECC, 2019; IPCC, 2022). Although national educational policies have incorporated references to climate change within the curriculum, significant gaps persist in how these topics are integrated into classrooms, especially in rural and peripheral regions. In this context, the Telebachillerato del Estado de Veracruz (TEBAEV) system becomes a key arena for addressing this challenge, as it serves rural communities characterized by high levels of social marginalization and vulnerability to environmental threats such as flooding and deforestation (INEGI, 2020; Velázquez-Martínez, 2024).

This chapter treats Climate Change Education (CCE) as a comprehensive approach that not only seeks to provide scientific knowledge about climate change but also promotes critical thinking, informed decision-making, and active participation in local initiatives aimed at mitigation and adaptation (Anderson, 2012; Monroe et al., 2019).

The study focuses on TEBAEV, a subsystem of upper secondary education in Mexico designed to expand educational coverage in rural and peri-urban communities marked by high socioeconomic marginalization and environmental vulnerability. This educational model employs video-based lessons and teaching guides specifically developed for these contexts, enabling a flexible approach adapted to local conditions that impede conventional in-person instruction (SEV, 1999; Velázquez-Martínez, 2024).

Recent literature acknowledges that, while important advances in CCE have been achieved at the international level (Anderson, 2012; Monroe et al., 2019), research gaps remain regarding how teachers effectively translate this approach into their daily practices—particularly in communities with limited economic and pedagogical resources (Bello et al., 2017; Busch et al., 2019). According to Velázquez-Martínez (2024), curricular materials and teacher training within TEBAEV present substantial opportunities to incorporate a climate-responsive perspective; however, these efforts are hindered by resource scarcity, a lack of interdisciplinary training, and the dispersal of climate-related content throughout the curriculum.

The disruptive nature of the TEBAEV experience lies precisely in situating climate change instruction within a context of vulnerability and socioeconomic inequality, where young people must balance their studies with labor activities and confront the absence of basic infrastructure (Espinoza, 2014; Estrada & Alejo, 2018). Such conditions demand a critical pedagogical vision (Freire, 1970; Giroux, 2011) and a socioconstructivist approach (Vygotsky, 1978) that links civic engagement with an understanding of the climate emergency. Through these lenses, students are empowered to address the impacts that climate change will have on their communities. The integration of these frameworks not only necessitates reinterpreting teaching dynamics but also fosters both critical reflection on socio-environmental realities and their transformation through practical resilience projects (Velázquez-Martínez et al., 2023).

Here, "disruption" is understood as an intentional break from traditional educational approaches that reproduce the status quo, paving the way for a transformative paradigm that confronts the climate emergency and socioeconomic inequalities in a contextualized and critical manner. From a Critical Pedagogy standpoint

(Freire, 1970; Giroux, 2011), such disruption involves questioning unjust structures and forms of domination, thereby fostering awareness and collective action to address the root causes of the environmental crisis. Complementing this is the CCE perspective (Anderson, 2012; Monroe et al., 2019), which seeks to drive innovative processes—such as community resilience projects—moving beyond the mere inclusion of scientific-environmental content to effect immediate local change. Finally, from a Socioconstructivist and transformative learning perspective (Mezirow, 1997; Vygotsky, 1978), disruption also entails reconceptualizing power relations in the classroom and promoting active student participation, generating epistemic breaks that enable learners to critically reinterpret their socio-environmental context and collaborate on actions that challenge the inertia of conventional school systems.

This chapter presents an analysis of how TEBAEV teachers integrate the pedagogical and didactic elements of CCE into their everyday practices, taking into account the influence of three major theoretical frameworks. The deliberate selection of these approaches—CCE, Critical Pedagogy, and Socioconstructivism—responds to the need for a deeply transformative educational response to the climate emergency. CCE allows us to move beyond traditional models focused solely on transmitting scientific knowledge, toward pedagogies oriented to practical action and local engagement with climate change (Anderson, 2012; Monroe et al., 2019). Critical Pedagogy (Freire, 1970; Giroux, 2011), in turn, introduces a rupture with conventional methods by openly challenging the socioeconomic and political structures that generate and sustain environmental and social vulnerability, thus encouraging active, conscious, and committed student participation. Lastly, the Socioconstructivist approach (Vygotsky, 1978) reinforces disruption by relocating learning outside the traditional classroom into real, collaborative contexts where students and teachers jointly construct practical solutions to concrete socio-environmental problems.

By combining these approaches, the study both offers an educational framework suitable for addressing the climate emergency and identifies the tensions, strengths, and opportunities present in implementing localized CCE. This implementation must be intersectional in responding to the challenges faced by marginalized communities—communities that rely heavily on natural resources, bear little responsibility for the climate crisis, yet will suffer its consequences most acutely, and exhibit high dropout rates, especially in the vulnerable rural areas served by TEBAEV (Velázquez-Martínez, 2024). Velázquez-Martínez's (2024) research revealed that, although the TEBAEV model is flexible and has integrated environmental content into the official curriculum, scientific literacy regarding climate change—when not aligned with the objectives of a climate-emergency-oriented CCE—remains insufficient and fails to sustain transformative changes in attitudes and cultural practices.

From the perspective of "disruption" in CCE, this work reaffirms the importance of:

- Generating awareness and collective action that transcend isolated sensitization activities.
- Strengthening teacher training in scientific-environmental and critical pedagogical domains.

- Integrating school and community naturally so that learning is genuinely contextualized, relevant, and emancipatory.

We begin from the premise that Mexico's rural school environments—often rendered invisible by public policies—can be transformed into sites of resistance and agency, where teachers and students co-create concrete solutions to the socio-environmental challenges that affect them.

The following pages outline the methodological design used to gather information from 3698 TEBAEV teachers, the analytical categories that emerged from their qualitative responses, and their association with the aforementioned theoretical frameworks. Subsequent sections discuss the key findings, highlighting structural barriers and pedagogical opportunities for CCE in disadvantaged contexts. We then reflect on the practical and policy implications required to strengthen disruptive initiatives in the face of the current climate emergency in regions such as Veracruz. Finally, we aim to provide empirical evidence and concrete proposals that will help TEBAEV consolidate a transformative educational model and serve as a reference for other upper secondary (equivalent to grades 10–12 in the U.S. K–12 system) experiences in Mexico.

2 Veracruz on Alert: How Climate is Redefining Rural Education

The climate emergency, in Mexico, is reflected in multiple indicators of vulnerability: extreme hydrometeorological events, biodiversity loss, high greenhouse gas emissions, and the persistent socioeconomic inequality between urban and rural regions (INECC, 2019; IPCC, 2022). This situation impacts most severely those Mexican states with great geographical diversity and high levels of marginalization—such as Veracruz—where human activities, including deforestation, agricultural expansion, and unsustainable land use, have aggravated environmental degradation and increased the frequency and intensity of floods; these impacts are further intensified by extreme precipitation and other climate-driven events (INECC, 2016). Starting from the premise that education is a fundamental pillar for societies confronting the climate challenge, the formal incorporation of environmental content into various educational plans (SEP, 2017) and dedicated research in the field of CCE remains scarce (García-Vinuesa & Meira, 2019), particularly at the upper secondary level and in rural contexts, where structural deficiencies hinder the adoption of comprehensive educational approaches (Estrada & Alejo, 2018).

In Veracruz, ecological diversity and rural communities' dependency on natural resources increase their vulnerability to climate change. In this context, forest overexploitation acts as an aggravating factor that intensifies the problem, while droughts, water scarcity, and reduced food security emerge as direct consequences of climate disturbances (Velázquez-Martínez, 2024). These conditions demand educational policies and practices oriented not only toward scientific literacy but also toward fostering an active citizenry capable of understanding, debating, and acting upon climate risks (Monroe et al., 2019). However, various studies demonstrate that environmental content in Mexican upper secondary education is often scattered and superficially articulated, lacking a critical perspective on the structural causes and consequences of the problem (Bello et al., 2017; González & Maldonado, 2013).

TEBAEV constitutes a subsystem designed to provide educational coverage for young people in rural or peri-urban areas commonly excluded from conventional upper secondary education (SEV, 1999). These communities exhibit high levels of marginalization and marked socioeconomic lag (Espinoza, 2014), translating into obstacles to continuing their studies—even at basic levels—and, consequently, limiting their active participation in climate change mitigation and adaptation. Nevertheless, TEBAEV features a flexible model that includes teaching guides and video-based lessons, which have been identified as opportunities to integrate CCE more deeply and contextually (Velázquez-Martínez, 2024). Within this scenario, TEBAEV positions itself as a transformative and essential educational proposal for serving rural youth in Veracruz, ensuring their access to equitable and high-quality education.

Nevertheless, the structural conditions of these schools are limited—lack of equipment, limited Internet access, absence of scientific laboratories—which makes adapting the curriculum to Veracruz's climatic context a challenge (Blanco, 2009; Estrada & Alejo, 2018). Moreover, research has underscored the limited specialized training of teachers in climate change topics, which impedes the planning of learning experiences grounded in local realities (García-Vinuesa & Meira, 2019; González & Meira, 2020; Robertson & Barbosa, 2015). Consequently, students often perceive environmental content as irrelevant or disconnected from their daily lives, weakening their motivation for action and school retention (Velázquez-Martínez, 2024).

This challenge is framed by the tension between socio-environmental urgency and educational inequity in the Mexican context: on one hand, CCE requires a multidimensional approach that articulates the scientific, social, cultural, and emotional aspects of climate change; on the other, persistent gaps in infrastructure, funding, and teacher training prevent the effective implementation of curricular initiatives, particularly in rural areas (Blanco, 2021). In Veracruz, these issues are exacerbated by the dispersion of environmental competencies and content throughout the study plans, resulting in weak appropriation of the phenomenon and educational practices that do not transcend initial awareness-raising (González & Maldonado, 2013).

Yet, TEBAEV continues to serve as a valuable pedagogical laboratory for exploring disruptive educational strategies that promote climate resilience. Its presence in over a thousand communities (see Fig. 1) (INEGI, 2021) and its vocation to serve historically excluded populations make it a key system for studying how knowledge and actions regarding the climate emergency are—or are not—constructed. In this sense, the appropriation of CCE in these settings requires, first, a deeper understanding of Veracruz's socio-environmental reality and, second, a critical, contextualized pedagogy that encourages the participation of both teachers and students in transforming their own communities.

The intricate climate-educational situation in Veracruz demonstrates that the urgency of seeking educational responses to the climate emergency converges with

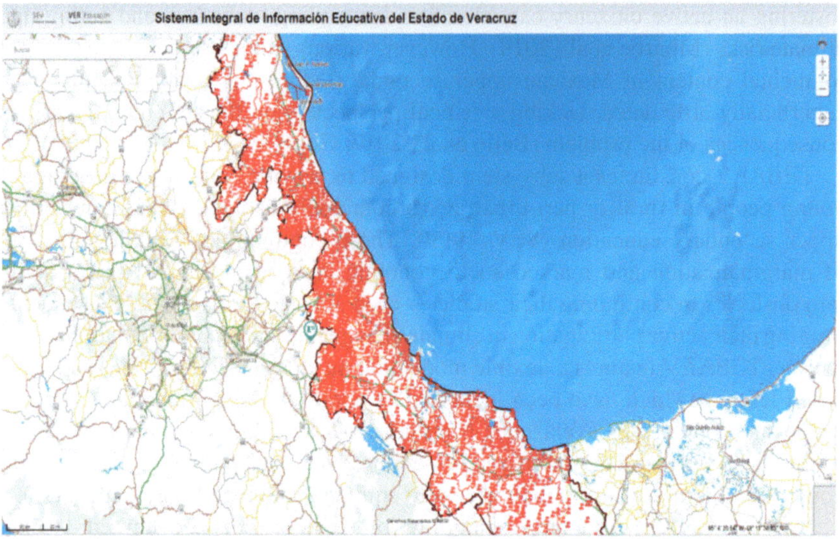

Fig. 1 Distribution of TEBAEV across the state of Veracruz. *Source* Comprehensive Educational Information System of the State of Veracruz, Mexico. INEGI, 2021

profound socioeconomic inequalities, resulting in insufficient educational coverage. Despite institutional efforts to incorporate an environmental perspective into the curriculum, TEBAEV faces constraints that demand specific teacher training strategies and appropriate instructional resources. Assuming that a situated CCE can drive community resilience and empowerment, the following sections aim to describe and understand how these tensions are reflected in the daily practices of TEBAEV schools and how they could be reoriented to foster both adaptation and socio-environmental transformation in Veracruz.

3 TEBAEV: A School Reaching Where Others Can't

The TEBAEV model is conceived as an educational alternative aimed at expanding upper secondary education coverage in rural and semi-rural contexts, where opportunities for schooling at this level are limited. Drawing on the *Telesecundarias* (or Junior High School) experience, TEBAEV combines video-based lessons and specially designed teaching guides to offer a curriculum that addresses the geographic and sociocultural diversity of Veracruz's communities.

Although the model initially sought to address the lack of infrastructure and academic resources for in-person teaching, it has gradually demonstrated that self-directed learning and flexibility appeal to young people living in highly marginalized areas, where the right to education is often constrained. However, this potential is frequently challenged by structural obstacles—such as insufficient technological equipment and the geographic dispersion of campuses in hard-to-reach regions—which complicate teaching efforts and disrupt students' continuity of study. At the same time, professional development for instructors is hampered by the demands of travel and a paucity of targeted support for training in emerging topics like the climate emergency.

Nevertheless, TEBAEV plays a crucial role in reducing educational lag. In most cases, school governance relies on parents' committees (*Patronatos de Padres de Familia*), which manage resources and address immediate community needs. When aligned with pedagogical initiatives that foster collective action, this governance model can strengthen local belonging and civic participation. Yet, budgetary instability and heterogeneous school conditions generate disparities in teaching quality across the network.

From a curricular standpoint, TEBAEV's constructivist approach promotes collaborative, situated learning with activities designed to contextualize theory. In practice, however, alignment between official content and community dynamics varies widely. Some campuses leverage flexibility to develop productive projects or involve students in solving socio-environmental problems; others face training limitations that confine teaching to mere content reproduction.

Pedagogically, the model relies on printed and audiovisual materials, expecting teachers to serve as facilitators of the learning process. This expectation clashes with heavy workloads, a lack of continuous support, and students' need to balance studies with temporary work or household responsibilities. Given Veracruz's high exposure to the climate crisis, targeted training in teaching methods that encourage critical reflection and collective environmental action becomes especially pertinent.

TEBAEV represents a strategic subsystem for upper secondary education in Veracruz by delivering schooling to historically excluded regions—hence its disruptive and emancipatory character. Its transformative potential lies in contextualizing learning experiences and linking them to each community's realities, thereby strengthening students' capacity to respond to climate-related risks. Realizing this potential, however, requires focused investment in teacher preparation, the provision of contextualized materials, and the effective integration of CCE for the climate emergency into pedagogical processes. In doing so, TEBAEV could establish itself as a benchmark of resilience and educational equity in environments marked by environmental and social vulnerability.

4 Small Schools, Big Possibilities: Educating to Weather Climate Change

One of TEBAEV's strengths is its flexible model: teachers can adapt video lessons and teaching guides to local realities, creating classroom projects that leverage region-specific environmental data. For example, the Ecology course can be linked to identifying nearby pollution sources or to developing practical solutions for reforesting vulnerable areas. In this way, students can connect theory with firsthand experience of environmental impacts in their own community, reinforcing the relevance of their learning.

Nevertheless, these opportunities are frequently constrained by insufficient teacher training in climate-specific topics and by a lack of resources for field activities. Although isolated cases exist in which teachers have launched sustainable agriculture initiatives or community water monitoring projects, the approach does not always move beyond awareness-raising, due to the absence of robust epistemological and methodological support. This highlights the need to strengthen collaboration networks among educational institutions, environmental authorities, and local citizens so that contextualized scientific information is promoted—and students are engaged in mitigation and adaptation actions against climate change.

Another relevant aspect lies in harnessing communities' traditional knowledge (Mendoza & Sandoval, 2021). Integrating local practices around seed management, rational land use, or gathering traditions enhances students' empathy with their natural surroundings and fosters a critical view of socio-environmental transformations. Likewise, the productive projects driven by the *Telebachillerato* model could become true spaces of reciprocal learning, intertwining scientific literacy with cultural practices and creating stronger school–community linkages.

The primary challenge, then, is to consolidate an approach that weaves technical climate-change knowledge together with social and cultural dimensions while involving students in the search for collective solutions. Although TEBAEV possesses unique conditions to materialize these proposals—thanks to its presence in biodiverse localities and its close ties with families through the parents' committees—its impact depends on systematically coordinating pedagogical efforts and providing ongoing professional development for teachers to adapt content to their local contexts.

5 Three Strategies for Teaching in Times of Climate Crisis in Veracruz

In response to the urgency of the climate crisis in marginalized rural territories, this section builds on three disruptive streams—Climate Change Education, Critical Pedagogy, and Socioconstructivism—that, by combining scientific rigor, socio-environmental action, and the critique of unequal power structures, depart from traditional methods and guide participatory, empowering, and deeply contextualized educational proposals designed to address TEBAEV's unique challenges.

5.1 Climate Change Education (CCE)

CCE is conceived as an emerging field that aims to develop students' scientific, critical, and practical competencies to understand and act upon the climate crisis, without neglecting experiential dimensions and active participation. Anderson (2012) identifies three core components of this approach:

- *Scientific understanding:* Equipping students with key concepts—such as the greenhouse effect, carbon footprint, and environmental vulnerability—to grasp the magnitude and complexity of the phenomenon.
- *Critical thinking:* Cultivating the ability to analyze local and global social and environmental challenges, evaluating possible sustainable solutions.
- *Change-oriented action:* Encouraging engagement in mitigation and adaptation initiatives, including reforestation campaigns, water-saving efforts, or resilient agriculture projects.

In the TEBAEV context, CCE becomes essential for linking each community's socio-environmental realities with curricular components, fostering knowledge appropriation that goes beyond rote memorization and leads to tangible commitments.

5.2 Critical Pedagogy

Critical Pedagogy (Freire, 1970) emphasizes reflection on socio-environmental injustice and student agency. Under this framework, education does not merely transmit information; it seeks to unmask power structures and inequalities that perpetuate vulnerability to climate risks. Three dimensions are highlighted:

- *Critical consciousness:* Students must identify the social, economic, and political drivers that worsen climate impacts in their communities, assess the associated risks, and act from a critical-reflective stance.
- *Dialogue and participation:* Teaching and learning become dialogic acts in which teachers and students co-construct knowledge and plan collective actions that produce visible community impacts.
- *Transformation of reality:* Upon recognizing the structural causes of the crisis, students are motivated to lead initiatives that improve local conditions, driving change from the ground up.

Through TEBAEV, a critical lens is fostered to challenge socioeconomic inequalities surrounding schools and to propose collective solutions that enhance resilience in the face of climate emergencies.

5.3 Socioconstructivism

Socioconstructivism (Vygotsky, 1978) foregrounds collaboration and contextualized learning. Within this paradigm, knowledge is co-constructed by students, teachers, and the wider school community through interaction with cultural and environmental contexts. Key elements include:

- *Social construction of knowledge:* Learning thrives when students work together to solve authentic problems, sharing experiences and ideas.
- *Zone of Proximal Development:* Teachers or more knowledgeable peers provide guidance that helps students progressively overcome cognitive challenges.
- *Contextualization:* Learning activities are designed around local realities, integrating community practices and knowledge to facilitate comprehension and ownership of content.

At TEBAEV, this approach offers pathways for meaningful integration of climate-change discussions into projects that link school and territory—such as community monitoring activities and environmental protection campaigns.

5.4 Integrating the Three Approaches at TEBAEV

Far from operating in isolation, CCE serves as an integrative paradigm that draws upon Critical Pedagogy and Socioconstructivism to meet TEBAEV's challenges. Reid (2019) argues that CCE not only conveys scientific concepts but also promotes civic agency and empowerment in the face of climate injustices (critical dimension), while supporting collaborative, contextualized learning (constructivist dimension), thereby generating concrete mitigation and adaptation actions within communities (see Fig. 2).

This theoretical articulation provides a robust framework for reorienting teaching practice at TEBAEV so that education becomes a relevant and emancipatory process—one that empowers students and strengthens their communities' resilience in the face of the climate emergency.

6 First-Hand: How TEBAEV's Teachers Teach (and Live) the Climate Emergency

To understand how TEBAEV's teaching staff integrate climate-change content into their pedagogical practice, a mixed-methods study with an emphasis on quantitative techniques was designed. First, an online survey was administered to teachers across all municipalities in Veracruz, covering rural, peri-urban, and highly socio-environmentally vulnerable regions. The instrument comprised 23

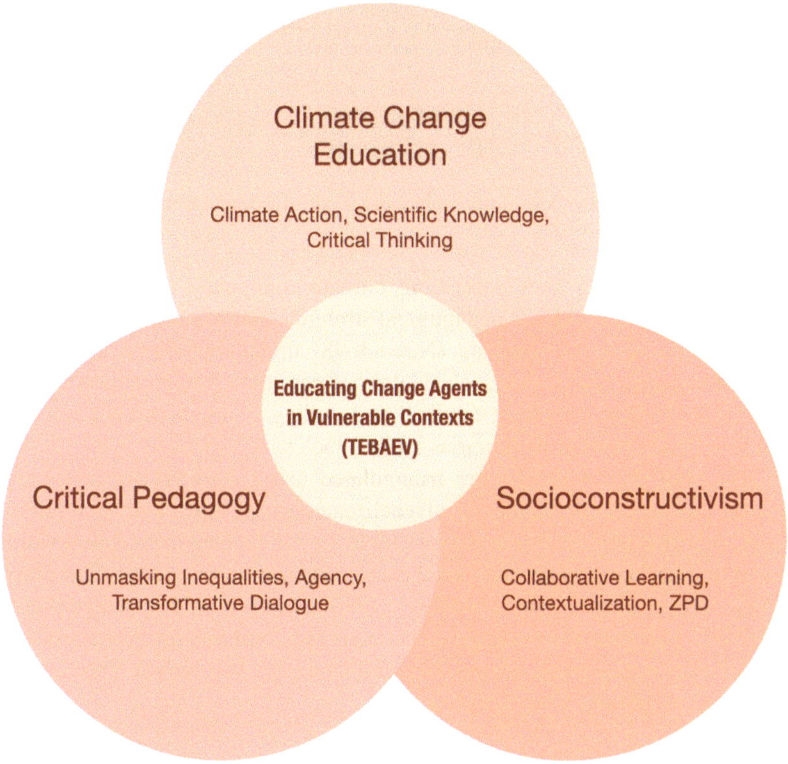

Fig. 2 Diagram showing the synergy of three theoretical–pedagogical frameworks in TEBAEV: Climate Change Education (CCE), Critical Pedagogy, and Socioconstructivism

items—including multiple-choice, closed-ended, and open-ended questions—that probed both sociodemographic aspects (academic background, years of experience) and pedagogical dimensions (teaching strategies, use of curricular materials, and perceptions of CCE). The survey was conducted during TEBAEV's mandatory virtual training week—a continuous professional development event—thereby facilitating broad teacher participation. In total, 3698 valid responses were collected, representing approximately 77% of the teaching population.

Concurrently, semi-structured interviews were held with a focus group of 25 teachers who had extensive experience designing and implementing environmental projects. Their open-ended format enabled deep exploration of the criteria and motivations underpinning the instructional strategies they employed.

In accordance with the exploratory sequential mixed-methods design of Creswell and Creswell (2017), qualitative and quantitative data were analyzed concurrently and in an integrated manner (Fig. 3). Interview transcripts were processed following Bardín's (1996) methodology: initial codes were generated and then iteratively refined into broader categories that aligned with CCE. From this coding

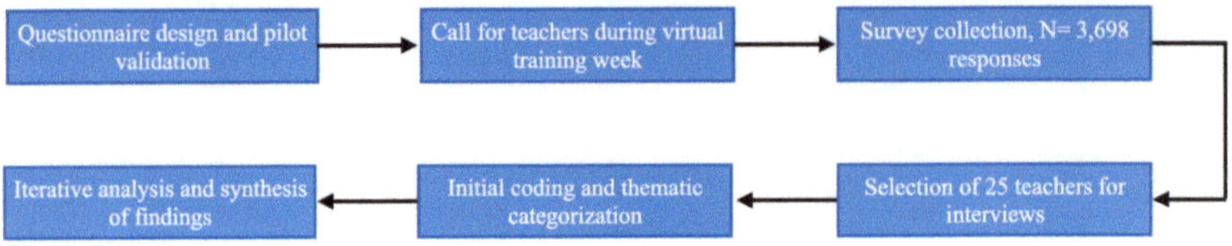

Fig. 3 Diagram showing the flow of the exploratory sequential design methodology

emerged several action axes describing climate-emergency teaching praxis and clearly identifying TEBAEV's pedagogical strengths and weaknesses.

At the same time, survey data (N = 3698) underwent descriptive analysis, including frequency and percentage calculations per item, disaggregation by contextual variables (region, teaching tenure), and coding of open-ended responses to capture teachers' reported motivations and obstacles.

Finally, both data sources were triangulated using convergence matrices that compared the numerical incidence of each category in curricular materials with teachers' reported perceptions and uses (Table 1). This robust intersection validated the identified action axes and refined the characterization of pedagogical practices by integrating qualitative and quantitative findings into a unified analysis of CCE at TEBAEV. Together, the findings reveal a strong emphasis on awareness-raising and community engagement, alongside significant gaps in scientific literacy and the development of sustained impact projects.

7 From Talk to Action: How TEBAEV's Teachers are Tackling Climate Change

The analysis of teachers' responses reveals the paradox confronting TEBAEV: although there is clear willingness to bring CCE into classrooms as a response to the climate emergency, practices fall short in scientific depth and in fostering the cultural transformation required for climate resilience. In regions where the climate crisis is not a distant threat but everyday reality—flooding, drought, or soil degradation—CCE should serve as a cross-cutting axis of the curriculum, promoting relevant and emancipatory learning (Anderson, 2012; Freire, 1970).

Scientific literacy emerges as an essential component for students to grasp the complexity of climate change and not merely describe isolated events—such as a flood or a drought—as anecdotal examples. Instead, they learn to frame these occurrences within a broader analysis of physical processes (greenhouse effect, water cycles) and structural dynamics (global emissions, land-use change). Consequently, situated narratives gain depth and become springboards for connecting the local to the planetary scale of climate change. In instances where teachers incorporate school-based research strategies or simple experiments, increased motivation and understanding of the processes driving the climate emergency have been observed.

Table 1 Main Qualitative and Quantitative Findings of the CCE Categories in TEBAEV

Category	Quantitative finding	Qualitative finding	Verbatim excerpt
Awareness raising and critical consciousness	High frequency of informative mentions (> 75% of materials)	Reflection on human impact without transformative actions or structural questioning	"It is limited to presenting data; we do not question the structures that generate environmental inequality."
Scientific literacy	Only 20% of documents include basic scientific concepts (greenhouse effect, carbon footprint)	Theoretical knowledge without research practices or environmental data management	"We talk about the greenhouse effect, but we don't know how to investigate it with students."
Community participation	40% of materials suggest community activities	Reforestation and cleanup campaigns with families, but lacking resources for continuity	"We organize cleanup campaigns, but we lack resources to sustain them."
Collaborative projects	15% of teaching guides propose school gardens or monitoring activities	Valuable situated learning, although with limited implementation and lack of long-term evaluation	"The school garden is excellent, but we have no way to evaluate learning over time."
Socio-environmental action	Less than 10% of documents include concrete mitigation/adaptation actions	Targeted initiatives for nurseries and waste management without impact indicators	"Students created nurseries, but there is no indicator to measure their effect in the community."

This finding underscores the need to build competencies in data analysis and problem solving, a core objective of CCE (Bybee, 1997).

Regarding community participation and collaborative projects, these results confirm the potential of Socioconstructivism at TEBAEV by fostering situated, dynamic learning in which the school serves as a hub linking it with the local community. Nevertheless, their transformative impact is limited by lack of institutional follow-through and scant resource support. In an emergency climate context, such projects could scale up to bolster community resilience and self-sufficiency in the face of increasingly extreme events.

The findings indicate that mere information or awareness-raising is insufficient; it is necessary to weave together environmental consciousness, scientific understanding, and socio-environmental action more tightly so that CCE at TEBAEV meets the demands of the climate emergency. This entails:

A. *Designing integrated projects* in which students, after identifying a local problem (awareness-raising), apply basic scientific methods—measuring variables, conducting simple experiments, analyzing meteorological data—and then plan and implement community interventions (e.g., managing school gardens, water-saving campaigns, or ecosystem restoration workshops).

B. *Fostering Freirean critical reflection* spaces where students question the structural causes (political, economic, and social) underlying their environmental reality, promoting genuine dialogue among peers and with local stakeholders.

C. *Supporting Socioconstructivist collaborative learning* by organizing mixed teams of students, teachers, and community members so that knowledge is not merely transmitted but co-constructed through real-world experiences, generating situated, sustainable solutions.

In this way, the school would shift from "teaching about" climate change to becoming a living laboratory of transformation, where each student develops scientific competencies, critical spirit, and organizational capacities to act as an agent of change in their own Veracruz community.

8 Breaking the Mold: Practical Ideas to Transform Climate Education from Within

Based on the preceding discussion, three key points emerge for the initial and feasible integration of CCE into TEBAEV:

1. Emphasize environmental awareness and community actions, limiting scientific literacy to the most essential concepts while prioritizing the cultural embedding of sustainable practices.
2. Strengthen teacher preparation and provide contextualized instructional resources so that climate concepts can be appropriated in a practical and meaningful way.
3. Harness the disruptive potential of an education that—beyond merely delivering content—questions socio-environmental injustice, fosters local participation, and transitions to concrete collective actions.

From this perspective, the disruptive elements refer to the intentional break with traditional approaches—those that emphasize rote memorization and lecture-style teaching—in order to promote critical, situated learning oriented toward transforming the environment. Given its flexibility and its proximity to highly vulnerable populations, TEBAEV could become a space of resistance in the face of the climate emergency, provided that improvement strategies focused on co-constructing knowledge and long-term community action are implemented.

Following, the specific improvement proposals and recommendations for TEBAEV drawn from this study's findings are synthetized:

1. Strengthen Teacher Training

- Design ongoing professional-development programs in CCE that include both scientific literacy and project-based methodologies, use of environmental data, and evaluation of community-level climate actions.
- Encourage collaboration among teachers from different disciplines at campuses with multiple instructors to enrich critical perspectives and develop interdisciplinary proposals.

2. Revise Instructional Materials

- Incorporate greater emphasis on addressing local challenges (water management, agroecology, recycling, etc.) into guides and video lessons, integrating practical case studies and fieldwork to promote socio-environmental reflection, systematic documentation of experiences, and cultural appropriation of sustainable actions.

3. Promote Collaborative Projects and Support Networks

- Foster partnerships with higher-education institutions, environmental NGOs, and local authorities to support projects that integrate research and action around the climate crisis.

4. Establish Evaluation and Monitoring of Climate Practices

- Define clear indicators to measure project impacts (e.g., number of trees reforested, volume of water saved, or behavioral changes among students).
- Provide feedback to teachers and students based on results, fostering continuous improvement and dissemination of successful experiences.

5. Secure Institutional Support and Funding

- Mobilize resources to provide basic infrastructure (laboratories, Internet access, field equipment) and enable ongoing professional development at every campus.
- Raise awareness among educational authorities about the fundamental importance of CCE in the curriculum, so that it is recognized as a priority in regions with high exposure to climate change.

By integrating these initiatives, TEBAEV's disruptive potential becomes a suite of truly holistic educational experiences. In this way, TEBAEV can lead a process of pedagogical innovation grounded in collaboration, critical reflection, and transformative action—addressing the climate emergency in the territories it serves and generating meaningful learning for Veracruz's communities.

9 From Classroom to Action: Toward Empowering Climate Education in Veracruz

The imperative to strengthen CCE at TEBAEV becomes urgent in light of the advancing climate emergency and the persistent socio-environmental inequities affecting numerous Veracruz communities. In a context marked by social and environmental vulnerability, limited specialized teacher training, and dispersed educational resources, it is essential to advance toward an Environmental Education for the Climate Emergency that goes beyond mere information-sharing to foster critical thinking, collective action, and the empowerment of those on the front lines of climate impacts.

This educational proposition emerges from the integration of three pedagogical approaches: Climate Change Education, Socioconstructivism, and Critical Pedagogy. CCE underscores the importance of students' understanding of key scientific concepts—such as the greenhouse effect, deforestation, and water scarcity—while emphasizing the translation of that knowledge into practical skills and critical capacities for responding to climate challenges. Socioconstructivism strengthens this vision by promoting contextualized learning, in which knowledge is co-constructed collaboratively around the real problems faced by TEBAEV communities. In this model, students cease to be passive recipients and instead become active protagonists of local solutions. Critical Pedagogy contributes an essential transformative dimension by encouraging students to question the structural causes of the socio-environmental crisis and develop a political and ethical consciousness aimed at environmental justice. The convergence of these three approaches yields an Environmental Education for the Climate Emergency that is both comprehensive and situated—one that seeks not only to raise awareness but to effect transformation by empowering youth through critical appropriation of knowledge, collective reflection, and active participation in change processes.

The teaching experiences analyzed in this study confirm that, although interest in integrating environmental activities and local projects within TEBAEV is growing, the dominant focus remains on scientific literacy. Opportunities for fostering genuine environmental awareness and sustained community engagement continue to be limited. This situation underscores the need to rethink teachers' roles, update instructional materials, and strengthen institutional conditions so that CCE ceases to be an ancillary curricular component and instead becomes the organizing axis of educational practice.

In this respect, the combination of CCE, Critical Pedagogy, and Socioconstructivism demonstrates immense potential for crafting pedagogical proposals that respond appropriately to local realities. By interweaving scientific knowledge (ecological awareness and informed action), critical thinking (challenging structural injustices), and community collaboration (co-construction of knowledge from the territory), TEBAEV schools can transform into spaces of resilience, youth agency, and cultural change in the face of the climate emergency.

Nevertheless, structural challenges persist: the lack of contextualized instructional resources, gaps in teacher training, and limited continuity in long-term projects. Overcoming these obstacles demands investments in infrastructure, political support, and partnerships with diverse institutions. To challenge the hegemonic educational model, genuine commitment is required from the entire TEBAEV community—teachers, students, families, and local and state stakeholders—who must be willing to reshape their teaching and learning practices toward climate action and environmental justice.

Consolidating CCE at TEBAEV not only enables adaptation and mitigation of climate-change effects in highly vulnerable areas but also lays the groundwork for emancipatory education capable of empowering youth and communities to confront present and future challenges in a critical, collective manner. This project extends beyond the school itself: it invites a rethinking of the relationship among education, territory, and citizenship, setting a precedent for similar initiatives throughout Mexico and beyond.

References

Anderson, A. (2012). Climate change education for mitigation and adaptation. *Journal of Education for Sustainable Development, 6*(2), 191–206. https://doi.org/10.1177/0973408212475199

Bardin, L. (1996). *Análisis de contenido [Content analysis]* (Vol. 89). Ediciones Akal.

Bello, M., Meira, P., & González, E. (2017). Representaciones sociales sobre cambio climático en dos grupos de estudiantes de educación secundaria de España y bachillerato de México [Social representations of climate change among two groups of secondary education students from Spain and high school students from Mexico]. *Revista Mexicana de Investigación Educativa, 22*(73), 505–532.

Blanco, E. (2021). La desigualdad de oportunidades educativas en México (1958-2010)-Desplazamiento y reconfiguración. *Perfiles educativos, 43*(171), 8–26. https://doi.org/10.22201/iisue.24486167e.2021.171.59387

Blanco, R. (2009). *Los límites de la escuela: Educación, desigualdad y aprendizajes en México [The limits of schooling: Education, inequality, and learning in mexico].* El Colegio de México.

Busch, K. C., Henderson, J. A., & Stevenson, K. T. (2019). Broadening epistemologies and methodologies in climate change education research. *Environmental Education Research, 25*(6), 955–971. https://doi.org/10.1080/13504622.2018.1514588

Bybee, R. W. (1997). *Achieving scientific literacy: From purposes to practices.* Heinemann.

Creswell, J. W., & Creswell, J. D. (2017). *Research design: Qualitative, quantitative, and mixed methods approaches* (5th ed.). Sage Publications.

Espinoza, A. (2014). *Dificultades en lenagua escrita en español de estudiantes de Telebachillerato en context otomí [Difficulties in written spanish among Telebachillerato students in an otomí context]* (Doctoral dissertation). Universidad Veracruzana. https://cdigital.uv.mx/server/api/core/bitstreams/0f27b5d2-112c-44f1-8449-902e923711f6/content

Estrada Ruiz, M. J., & Alejo López, S. J. (2018). Caracterización de los telebachilleratos comunitarios. El caso de Guanajuato [The characterization of telebachilleratos comunitarios. The case in Guanajuato]. Sinéctica, 51, 1–22. doi:https://doi.org/10.31391/S2007-7033(2018)0051-005.

Freire, P. (1970). *Pedagogy of the oppressed.* Continuum.

García-Vinuesa, T., & Meira, P. (2019). Caracterización de la investigación educativa sobre el cambio climático y los estudiantes de educación secundaria [Characterization of Educational

Research on Climate Change among Secondary Students]. *Revista Mexicana de Investigación Educativa, 24*(81), 507–535.
Giroux, H. (2011). *On critical pedagogy*. Bloomsbury.
González, E., & Maldonado, A. (2013). *Los jóvenes universitariosy el cambio climático: Un studio de representaciones sociales [Los jóvenes universitariosy el cambio climático: Un studio de representaciones sociales]*. Universidad Veracruzana. https://doi.org/10.25009/uv.1973.97
González, É., & Meira, P. (2020). Educación para el cambio climático: ¿educar sobre el clima o para el cambio? [Education for climate change: Educating about climate or for change?]. *Perfiles Educativos, 42*(168). https://doi.org/10.22201/iisue.24486167e.2020.168.59464
Instituto Nacional de Ecología y Cambio Climático [INECC]. (2016). *Atlas de vulnerabilidad hídrica en México [Atlas of water vulnerability in Mexico]*. INECC.
Instituto Nacional de Ecología y Cambio Climático [INECC]. (2019). *Compendio de información ambiental de los Estados Unidos Mexicanos [Compendium of environmental information of the United Mexican States]*. INECC.
Instituto Nacional de Estadística, Geografía e Informática [INEGI]. (2020). *Marginalization and geographical coverage statistics of the Telebachillerato del Estado de Vera*cruz *(TEBAEV)*.. Retrieved from https://www.inegi.org.mx/temas/
Instituto Nacional de Estadística, Geografía e Informática [INEGI]. (2021). *Marginalization and geographical coverage statistics of the Telebachillerato del Estado de Veracruz (TEBAEV)*.. Retrieved from https://www.inegi.org.mx/temas/
Intergovernmental Panel on Climate Change [IPCC]. (2022). *Climate change 2022: Impacts, adaptation and vulnerability*. Cambridge University Press.
Lotz-Sisitka, H., Wals, A. E., Kronlid, D., & McGarry, D. (2015). Transformative, transgressive social learning: Rethinking higher education pedagogy in times of systemic global dysfunction. *Current Opinion in Environmental Sustainability, 16*, 73–80. https://doi.org/10.1016/j.cosust.2015.07.018
Mendoza, M. A., & Sandoval, M. (2021). Conocimientos y prácticas locales para el cuidado del entorno social y ecológico a través de procesos educativos situados. *Educación Ecosocial*. https://educacionecosocial.com/2021/10/25/mendoza-y-sandoval-2021-conocimientos-y-practicas-locales-para-el-cuidado-delentorno-social-y-ecologico-a-traves-de-procesos-educativos-situados/
Mezirow, J. (1997). Transformative learning: Theory to practice. *New Directions for Adult and Continuing Education, 74*, 5–12. Retrieved from https://www.ecolas.eu/eng/wp-content/uploads/2015/10/Mezirow-Transformative-Learning.pdf
Monroe, M. C., Plate, R. R., Oxarart, A., Bowers, A., & Chaves, W. A. (2019). Identifying effective climate change education strategies: A systematic review of the research. *Environmental Education Research, 25*(6), 791–812. https://doi.org/10.1080/13504622.2017.1360842
Reid, A. (2019). Climate change education and research: Possibilities and potentials versus problems and perils? *Environmental Education Research, 25*(6), 767–790. https://doi.org/10.1080/13504622.2019.1664075
Robertson, W., & Barbosa, A. (2015). Global climate change and the need for relevant curriculum. *International Journal of Learning, Teaching and Educational Research, 10*(1), 35–44.
Secretaría de Educación de Veracruz. (1999). *Historical background: Directorate General of Baccalaureate*. Secretaría de Educación de Veracruz. Retrieved from https://www.sev.gob.mx/media-superior/bachillerato/antecedentes-historicos/
Subsecretaria de Educación Media Superior y Superior. (2017). *Reference study plans of the common curricular framework for upper secondary education*. Subsecretaria de Educación Media Superior y Superior.
Velázquez-Martínez, B. (2024). *Educación para el cambio climático en el Telebachillerato del Estado de Veracruz: Un acercamiento desde sus materiales curriculares [Education for climate change in the Telebachillerato del Estado de Veracruz: An approach from its curricular materials]*. (Doctoral dissertation,. Universidad Veracruzana. https://www.uv.mx/pdie/files/2024/02/Tesis_Bethzabee-Velazquez_23-feb-2024.pdf

Velázquez-Martínez, B., Maldonado González, A. L., & Vergara Lope Tristán, S. (2023). Unveiling climate change education in Telebachilleratos del Estado de Veracruz (TEBAEV): An exploratory sequential analysis. *Environmental Education Research, 30*(11), 1990–2006. https://doi.org/10.1080/13504622.2023.2273786

Vygotsky, L. S. (1978). *Mind in society: The development of higher psychological processes.* Harvard University Press.

Resistance and Memories of a Brazilian Indigenous Community in the Participatory and Intercultural Confrontation of the Climate Emergency

Hector Barros Gomes, Carmen Lúcia Melges Elias Gattás, and Rosana Louro Ferreira Silva

Abstract This chapter presents the construction of a participatory research conducted with a Guarani Mbya Indigenous community in Brazil, within their territory. A central element highlighted is the alignment of interculturality with an educommunicative perspective, as a tool for socio-environmental transformation. The research involved training indigenous youth in audiovisual production as a practice of engagement regarding socio-environmental issues, documenting the memories and narratives of the community leaders and members. These productions treat the indigenous worldview and the struggle for territory as the foundations for socio-environmental justice. Traditional memory and narratives bring indigenous worldviews to light as an action to confront climate change. Intercultural dialogues fostered through educommunicative practices are fundamental for formative processes that enable social actors to act and engage within their communities. The study highlights both the challenges experienced by the indigenous community and the construction of pathways of resistance through narratives that emphasize traditions, knowledge, and territorial practices as strengths in facing the climate emergency. The results lead us to perceive the existing divergences between dominant worldviews and those of this traditional community.

H. Barros Gomes (✉)
University of São Paulo, São Paulo, Brazil
e-mail: hectorgomes@usp.br

C. L. Melges Elias Gattás
Institute of Biosciences, University of São Paulo—Communication and Education Center of NCE/ECA/USP – CECI Jaraguá, São Paulo, Brazil

R. Louro Ferreira Silva
Institute of Biosciences, Interunit Graduate Program in Science Education, USP Research group on Environmental Education and Teacher Training – GPEAFE, São Paulo, Brazil

J. J. Conde et al. (eds.), *Practices in Education for Climate Emergency*,
https://doi.org/10.1007/978-981-92-2304-6_7

1 Introduction

The choice of didactic proposals based on environmental justice is closely related to the roots of socio-environmental problems in Brazil and worldwide. Ferdinand (2022) explains that violences and dominations provoke catastrophes, just as catastrophes provoke violences and dominations, prompting reflection on relationships between people, countries, rich and poor, men and women, Whites and non-Whites in the political understanding of phenomena by the collectives. According to the author, the environmental issue must be perceived as a decolonial struggle that integrates processes of domination with planetary degradation. From this perspective, environmental education emerges as a worldview-oriented attempt aimed at reclaiming justice, given the violence suffered by pre-Columbian peoples, Black people, *quilombolas*, indigenous peoples, and other oppressed groups.

As national and international references demonstrate, such as the North American Association for Environmental Education (NAAEE, 2018), environmental education transcends specific content and seeks the creation of healthier communities for all, centered around ecological integrity, shared prosperity, and social equity. This project adopts a Critical Environmental Education approach (EE), which understands education as an element of social transformation, based on dialogue, citizenship exercise, and strengthening individuals (Silva & Campina, 2011). Critical Environmental Education demands a dynamic approach in which educational processes are built actively, collectively, cooperatively, continuously, interdisciplinarily, democratically, participatively, and dialogically, pursuing an emancipatory social practice essential to building sustainable societies (Tozoni-Reis, 2006).

It is important to highlight that this perspective is grounded in the *Treaty on Environmental Education for Sustainable Societies and Global Responsibility*, which also states that for the sustainability process to be equitable, communication must be treated as a right and as a privileged channel of education, not only to disseminate information on equal bases, but also to promote the exchange of experiences, methods, and values. Therefore, communication becomes an important tool for critical and emancipatory environmental education.

Accordingly, our educational processes focus on promoting understanding, engagement, and resistance to environmental injustice. Racism, oppression, and environmental degradation are evident in all environments. Lack of potable water, dumpsites, and hazardous areas are problems that predominantly affect marginalized populations. This situation characterizes environmental injustice, defined as: *"the mechanism by which unequal societies assign the greatest burden of environmental harm to low-income populations, racialized groups, and other vulnerable groups"* (Acselrad et al., 2009, p. 41).

In the face of the climate emergency —which is not only an environmental crisis but also a civilizational crisis— it is urgent to recognize that the knowledge sustaining life does not lie in technological solutions or market innovations, but in epistemologies that have never broken ties with the earth, the sacred, and the cycles of life. In Brazil, the inequalities in environmental problems affect ancestral groups such as indigenous communities, which have resisted cultural and environmental degradations since colonization. It is within this context—marked by advancing urbanization, violence, and inequality— that the Guarani Mbya Indigenous territory is situated, and within which the process presented here was developed.

The research involved training indigenous youth in audiovisual production as a practice of engagement and resistance regarding socio-environmental issues, documenting memories and narratives of community leaders and members. This work aims to highlight the challenges faced by the indigenous community and understand the construction of resistance pathways through audiovisual narratives, identifying traditions, knowledge, and territorial practices as strengths to face the climate emergency.

2 Context and Connection with the Youth of the Guarani Community

According to the Socio-Environmental Institute (ISA, 2015), before the colonial period, the Guarani peoples inhabited various regions of what is now Brazilian territory. Currently, Indigenous Lands (ILs) account for only 14% of the country's 851,196,500 hectares (Fig. 1). Most of these areas are located in the Legal Amazon, which concentrates 98.25% of the TIs, representing approximately 23% of the Amazon territory. The remaining 1.75% is spread across the Northeast, Southeast, and South regions, as well as the states of Mato Grosso do Sul and Goiás.

Despite legal protection, the extent of these areas is often questioned by voices claiming there is "too much land for too few Indians." However, the indigenous

Fig. 1 Location of Jaraguá Peak, located within the Jaraguá State Park.

peoples themselves emphasize that they inhabited these lands long before the Portuguese arrival in 1500. Therefore, the demarcation of Indigenous Lands is not merely a legal measure but also a way to restore dignity and ensure the territorial and cultural rights of these traditional communities. Regarding the Guarani peoples, their presence is mainly concentrated in the South, Southeast, and Midwest regions of Brazil.

The Jaraguá Indigenous Land is located around the Jaraguá State Park (PEJ), located in a diverse and complex urban context in the northwest zone of São Paulo. The PEJ preserves the memory of an old farm from the gold cycle, acquired by the state government in 1940. In 1961, the area was designated a protected site and transformed into a State Park, later recognized as a World Heritage Site by the United Nations Educational, Scientific and Cultural Organization (UNESCO) and by the Council for the Defense of Historical, Artistic, Archaeological, and Tourist Heritage (Condephaat). Currently, the PEJ is a fully protected Conservation Unit (UC), dedicated to conserving native vegetation, maintaining ecological relationships, protecting spring areas, and safeguarding local fauna. It represents a vital natural fragment that supports biodiversity in the city of São Paulo.

As shown in Fig. 3, the Jaraguá Indigenous Land (JIL) is located in the Jaraguá district, a region known for the Jaraguá Peak —the highest point in the city of São Paulo, with an altitude of 1135 meters— which is part of the Jaraguá State Park.

The seven villages of the Jaraguá Indigenous Land (JIL) are called *tekoas*, a Guarani term meaning "way of life" or "village." They are: Tekoa Pyau ("new village"); Tekoa Ytu ("waterfall village"); Tekoa Yvy Porã ("fertile land village"); Tekoa Itaendy ("stone of light village"); Tekoa Itakupe ("village behind the stone"); Tekoa Pindó Mirim ("small palm village"); and Tekoa Ita Wera ("shining stone village").

The research was conducted in the Tekoa Pyau village, located opposite the TEJ, where two indigenous schools are situated: the Jaraguá Indigenous Education and Culture Center (CECI-Jaraguá), which serves children from 0 to 6 years old, and the Djekupe Amba Arandy State School, with elementary and high school classes. The CECI was established in 2004 as part of a policy by the São Paulo Municipal Department of Education, addressing the Guarani community demand. In this space, indigenous educators develop an educational approach rooted in Guarani culture, including teaching and the use of the Guarani language (Fig. 2).

The territory forms a mosaic of landscapes that encompasses commerce, public housing, few high-rise buildings, villages, and fragments of Atlantic Forest preserved within the park and nearby villages. Understanding this territory is critical for the research, since the Jaraguá region is characterized by significant contradictions: real estate speculation, heavy vehicle traffic, irregular settlements, and constant urban pressure from urbanization on both the park and the indigenous community. This situation amplifies environmental impacts such as noise and air pollution, improper waste disposal, and criminal fires.

The project was built collaboratively and interculturally, with active participation from indigenous communities in meetings held from the outset (Gattás, 2020). Gradually, we realized that the training provided within public policies and academic

Fig. 2 Representation of Jaraguá State Park, illustrating the locations of villages, main attractions and initiatives related to biodiversity and sociobiodiversity. *Source* Silva et al., 2023.

Fig. 3 Contrast between the city of São Paulo and Jaraguá State Park.

Fig. 4 "Petyngua" photograph taken by a young indigenous participant of the project. The *petyngua* is a sacred pipe used by the Guarani peoples.

practices was insufficient to entirely grasp the depth of forest epistemologies. This communication goes beyond mere information mediation, rooting itself in ancestral knowledge, collectivity, and an ethical commitment to the preservation of life.

In the indigenous context, as in the Guarani Mbya community of Jaraguá, educommunication plays a strategic role in mediating between different rationalities, fostering recognition of the land as a living being and viewing territories as pedagogical, spiritual, and political spaces. Besides spreading traditionally silenced knowledge, the production of audiovisual content by indigenous youth, contributes to building an active collective memory —essential to identity affirmation and the defense of territories. This practice enhances critical environmental education, which is paramount both for those who produce the narratives and for other young people who access them.

3 Environmental Education, Educommunication, and Climate Emergency

Silva et al. (2024) highlight the importance of fostering dialogue and active participation of society in addressing the complexities of contemporary socio-environmental problems. They emphasize the challenges faced by environmental education, such as consolidating more inclusive, critical, and interdisciplinary pedagogical practices capable of shaping conscious, proactive citizens committed to the common good and to confronting socio-environmental injustices. For this purpose, they advocate strengthening collaborative and emancipatory methodologies that value social learning, knowledge exchange, network building, and community-based learning, especially within the context of the Global South, amid challenges such as the climate emergency and biodiversity loss.

So early as the 1990s, the first IPCC assessment report, titled "FAR Climate Change: The IPCC Response Strategies," pointed out that efforts aimed at raising awareness of causes and potential impacts of climate change must consider the social and cultural diversity of nations. This requires educational proposals to meet the specificities of each region.

In order to tackle the issue, particularly in Brazil, it is necessary to understand young people's perspectives about the subject, as well as recognize their relation to social issues. The debate and formative processes must be centered around multiple knowledges and epistemologies that foster counter-hegemonic educational practices and acknowledge the articulation between social and environmental issues, which includes confronting environmental racism. Belmont (2023, p. 16) highlights the urgency of recognizing that there is no climate justice without racial justice and that "without including the debate of traditional peoples, *quilombola* communities, and racial issues, we will continue towards a neocolonial model, a model that prioritizes exploitation of the planet and peoples".

The Oxford Dictionary defines "climate emergency" as "a situation in which urgent action is required to reduce or halt climate change and avoid potentially irreversible environmental damage" (Oxford University Press, n.d.). This concept has been increasingly acknowledged across various situations, as well as in educational processes (Grandisoli et al., 2021). It sustains the need to develop actions aligned with socio-environmental justice, understanding that social inequalities are permanent and structural factors when it comes to the distribution of environmental risks and benefits. Among the most vulnerable groups facing climate change impacts are young people. According to García-Vinuesa et al. (2022), possible solutions to the climate crisis depend on the role played by environmental education.

In this work, critical environmental education intersects with educommunication — a field engaged in political debates connected to citizenship, freedom of expression, and interculturality. Media education is considered as a strategy to promote social justice based on dialogue (Kaplún, 2002), which has advanced towards being a tool of resistance and confrontation of the socio-environmental crisis. In the Latin American context, "educommunication" is understood as a set of participatory and innovative interventions that create new spaces for social participation, inspired by interdisciplinary knowledge. Its theoretical foundations draw from the practices of popular educators and communicators such as Paulo Freire, Jesús Martín-Barbero, and Mário Kaplún, who identified the articulation between communication and education as a vital and transformative process (Soares, 2006). As an emerging field of social intervention, it critically reflects on traditional communication and education models and analyzes their relationships with contemporary social challenges.

The concept of "Socio-environmental Educommunication" has been adopted in official documents, such as the Guidelines for the National Strategy for Environmental Communication and Education in Conservation Units (Brazil/MMA, 2008), expanding its role within socio-environmental public policies. This perspective contributes to science communication and enhances democratic participation.

Educommunication, in the context of this research, constitutes a field of symbolic and political dispute. It allows indigenous youth to reshape their identities, affirm their values, and resist processes of cultural assimilation —transforming media into instruments of care, memory, and envisioning the future (Gomes & Silva, 2024). As Ailton Krenak (2019) emphasizes, the severance from the Earth marks the beginning of the "end of the world" for indigenous peoples. Reconnecting with forest knowledge is paramount to reimagine alternative futures.

Linking educommunicative practices with the principles of climate justice reveals the potential of communication as a territory for struggle, learning, and sense-making. Formative experiences with traditional communities demonstrate that facing the climate emergency requires more than technological solutions: it demands listening to ancestral knowledge, recognizing indigenous cosmoperceptions, and valuing modes of existence rooted in reciprocity, care, and interdependence.

In this horizon, educommunication transcends the mere instrumentalization of media and technologies. It affirms itself as a pedagogy of coexistence and resistance —where communication transforms into a practice of resistance, care, and re-existence. As Catherine Walsh (2009) points out, the power of formative processes aimed at the sustainability of life lies entirely in the ecology of knowledges and critical interculturality.

Constructing educommunicative practices based on participatory methodologies and Southern epistemologies —such as action research and critical ethnography— opens space for community subjects to express their narratives, aspirations, and world projects. As Quijano (2005) highlights, the coloniality of knowledge remains one of the main obstacles to cognitive and climate justice, stressing the importance of recognizing indigenous knowledge as legitimate and essential for building alternatives to the hegemonic development model.

4 Formative Processes in Intercultural and Memory Dialogues

During the research process, constant dialogues were established with indigenous leaders and various members of the Guarani Mbya community through sensitive listening, conversation circles, and collaborative activities at the indigenous school. These encounters, conducted with respect and equality, allowed the research to be built collectively, incorporating multiple perspectives and valuing local knowledge. The mediation of leaders and elders was crucial to understand the Guarani worldview and recognize the centrality of the territory as a sacred, pedagogical, and political space.

The community's experiences and perceptions were documented through audiovisual media from the perspective of the indigenous youth themselves. They expressed their point of view about daily life, territory, and urgent socio-environmental challenges faced by their community. This initiative enabled the youth not only as research subjects but also as authors of memory and active voices in constructing narratives about their world. Their recordings reveal a profound grasping of the relationship between nature, spirituality, and collective life —a unique way of seeing, feeling, and caring for the land, integrating culture, ancestry, and resistance.

This process evidences the intercultural character of the research, where indigenous and academic knowledge dialogue in accordance with Catherine Walsh's (2009) perspective, acknowledging epistemic diversity. In tandem, the participatory construction process reaffirms the importance of collective memory as the foundation for identity and territorial struggles, denouncing the coloniality of knowledge that has historically marginalized and erased indigenous voices (Quijano, 2005). The experience of the Guarani Mbya youth aligns with a decolonial approach and research proposal, in which the territory is lived as a space of living memory and production of futures.

Collective memory is a fundamental concept for understanding how social groups build and sustain their identities over time, linking past, present, and future. Maurice Halbwachs, one of the main theorists on the subject, argues that memory is not exclusively an individual phenomenon but social, structured, and shared within groups and communities that preserve and transmit their traditions, histories, and experiences. In the context of the Guarani Mbya indigenous communities, this collective memory manifests in the preservation of territory, rituals, languages, and ancestral knowledge, which are essential elements for maintaining cultural identity and resisting against external threats. The participatory research with the community's youth, through autonomous recordings and dialogues, contributes to strengthening this living memory by reaffirming the value of their experiences and worldviews.

Malcom Ferdinand complements this perspective by emphasizing the need for policies and epistemologies that incorporate cultural diversity and recognize marginalized groups, such as indigenous peoples. He emphasizes that the struggle for memory is also a struggle for social and environmental justice.

On that account, by considering collective memory from the perspectives of Halbwachs, Quijano, and Ferdinand, we can understand that the participatory research developed in the Guarani Mbya community is not merely documentary recording but a political and cultural act. It is the creation of living memory that resists invisibility and reaffirms interculturality as a pathway for recognizing and valuing multiple forms of existence and knowledge in contemporary Brazil. This approach strengthens community identity, preserves cultural heritage, and contributes to confronting urgent socio-environmental challenges, integrating the ethical and political dimensions of the research.

For this research, the insights of Ailton Krenak (2019), a prominent Brazilian indigenous leader and thinker, were fundamental. Krenak introduces the concept of the "end of the world" to describe a profound environmental, cultural, and existential crisis caused by territorial devastation and the rupture of the ancestral relationship between indigenous peoples and nature. The author warns that the prevailing development model, rooted on the relentless exploitation of natural resources, leads to ecosystem collapse and the loss of ways of life that ensure the survival of various peoples, including indigenous communities.

This "end of the world" symbolizes the exhaustion of the planet and disrespect for forms of life that preserve the earth as a living entity to be cared for and respected. For indigenous peoples, such as the Guarani Mbya, the land is a source of memory, spirituality, and identity. When the territory is destroyed, their histories and knowledge are erased. Krenak's vision underscores the importance of strengthening collective memory and promoting intercultural practices that treasure cultural diversity and connection with the environment —opposing the colonial and capitalist logics that generate this crisis.

5 Theoretical-Methodological Dialogues of the Research Construction Process

In this work, critical ethnography was adopted as the methodology, articulated with participatory research, and approved by the Ethics Committee of the Institute of Biosciences at USP and by the National Research Ethics Commission (No. 9102720.8.0000.5464). This choice is justified as it is a social investigation approach involving the active participation of both researchers and study subjects, enabling a collective construction of knowledge. It is a more engaged practice, contextualized and committed ethically, ecologically, and socially.

This type of research aims to break away from positivist paradigms by incorporating a clear pedagogical and political dimension. It embraces an ideological commitment to social transformation, a foundation in the participants' everyday life experiences, and recognition of the knowledge produced by society as legitimate (Mainardes & Marcondes, 2011).

The connection with participatory methodology requires building a process based on dialogue, active participation, and shared responsibility in decision-making, since it involves the community itself in analyzing its reality through interaction between researchers and subjects in the studied context. Hence, it is understood that there is a commitment to the community along with the responsibility toward the environment, including the ethical action to return the research results to the subjects involved in the study.

In this instance, ethnography developed from a critical perspective is systematized around several fundamental axes: observation and description of the field; analysis of observation data; generation of dialogical data; analysis to establish relationships between individuals, groups, and systems; and the use of relationship systems to explain the research conclusions. As stated by Mainardes and Marcondes (2011,

p. 246): "Research of this kind seeks to understand the relationships between the investigated issues and the broader social system, aiming to grasp the totality of their determinants, as well as to unveil systems of domination, oppression, ideology, and discourses to contribute to changing such situations".

Therefore, this research approach requires time and a thorough description of the situated environment and the observed realities. Following this methodological path, data analysis was based on semi-structured interviews, audiovisual recordings, and informal conversations established with members of the Guarani Mbya community, respecting their timing, modes of expression, and their own forms of organizing knowledge.

As this research concerns a partnership with an indigenous population, it also aligns with the ideas of Martins and Knapp (2020), who propose integrating interculturality as a central element —encompassing diverse aspects such as identity, territory, struggle, and culture— understood as a notion of the social complexity of indigenous peoples. In this sense, data analysis was guided by four main elements: discursive memory, discursive formation, ideology, and silencing, in dialogue with interculturality and environmental education.

6 Linking the Results to the Confrontation of the Climate Emergency

To understand the systematic violence against indigenous peoples, it is essential to dive into the historical, social, cultural, and political roots that underlie the consolidation of colonial thought and oppression of minorities. Colonization was not solely by established coercion; its genealogy is also built through ideological resources and discourse, as tools to manipulate violence and shape rational justifications for its execution.

Memory plays a significant role in indigenous resistance, operating as a cyclical action as it unfolds in itself to remain present. It is a bond with the past and a connection with tradition, securing the preservation of principles inherited from previous generations without renouncing the present.

Educommunicative practices break with stereotypes and stigmatizations frequently reinforced across various media spaces. Furthermore, the development of meetings within the project enabled the social collective discourse to be voiced, creating spaces for the indigenous community to express, produce, and disseminate their knowledge. By integrating educational and communicative practices, this approach facilitates not only the documentation but also the safeguarding of traditional knowledge grounded in the voices and historical experiences of the community itself.

The integrated analysis of audiovisual narratives from different members of the Guarani Mbya community reveals how contemporary indigenous discourses are intrinsically linked to memories of resistance, intercultural education practices, and critiques of colonialism (Fig. 4).

The statements of the interviewed participants highlight accounts of their personal and generational experiences, as well as the ongoing forms of organization. All of which involves the struggle to maintain identity and strategies to resist the pressures of their context.

To prevent indigenous populations —due to their vulnerability— from being erased from processes affecting their territories, it is necessary to reclaim their historically silenced voices and restore political and territorial dignity to those who have been systematically ignored. In an interview with a Guarani Mbya leader and teacher, he addressed his young Guarani disciples and said:

> Memory is very important so that we do not erase part of our history. You need to make a timeline, especially one that shows the struggles and achievements to guarantee our rights. Today, writing and audiovisual media serve us to record this history. What we have achieved was not given by the whites; it was the result of a conquest involving much of our own struggle. (Interview conducted during a history class in Elementary School II at Djekupe Amba Arandy State School)

This statement reveals an understanding of memory as collective heritage and a tool of resistance, aligning with what Maurice Halbwachs (2006) defines as collective memory —a shared construction process that strengthens identities and community bonds. By proposing the elaboration of a timeline, the leader highlights the importance of historicizing struggles and making achievements visible, not only as a record but as a pedagogical and political tool.

The recognition of writing and audiovisual media as means to "record history" demonstrates the strategic appropriation of recording and communication technologies, while remaining grounded in the Guarani people's own epistemologies. This dimension dialogues with the perspective of educommunication as a practice of narrative autonomy (Soares, 2006), placing indigenous subjects themselves as authors and protagonists in their collective memory production.

The final assertion —"what we have achieved was not given by the whites" — explicitly conveys a counter-hegemonic narrative that breaks with official versions that render indigenous agency invisible. Here, the struggle is presented as a structuring element of Guarani Mbya's achievements, situating their history within a continuous process of resistance against the coloniality of power (Quijano, 2005).

In their records, the young members of the Guarani Mbya community reflected on how contemporary lifestyles —based on consumption, exploitation, and appropriation— have eroded deep relationships with nature. The community leaders claim that territorial fragmentation and the imposition of Western values have weakened collective bonds, resulting in ecological and existential violence. During a workshop held during the research, a young Guarani Mbya expressed:

> I can't separate Nhanderu (Guarani God), the forest, and the animals from myself, because everything is interconnected in a single whole. (Interview conducted during a video recording, 2024)

This statement reveals a cosmoperception that challenges the Cartesian dualism between subject and nature, reaffirming a deep sense of belonging to the territory. Such moments demonstrate that formative processes are not merely sustained by content transmission but by creating spaces of listening and expression, where participants become authors of their narratives and agents of their history. According to Munduruku (1999), oral traditions, myths, and rituals passed down through generations carry profound teachings about harmony, interconnection, and balance with the environment.

Another example of active listening occurred during a conversation circle on youth and rights. In this dialogue, concerns about teenage pregnancy emerged, and young community members explained that motherhood from age 15 is common and respected within Guarani Mbya society. For them, this experience is part of the natural cycle of community life and does not carry the "social problem" connotation often attributed by the juruás (non-indigenous people). Such alterity calls attention to the limits of a colonized pedagogical approach. It reinforces the importance of having formative processes that do not impose external values but, instead, build intercultural bridges based on mutual respect and listening. These experiences clearly demonstrate the importance of recognition and dialogue between different cultures.

The Tape Porã (Good Path) trail, built by leaders to educate youth about Guarani culture, fostered dialogue with ancestral knowledge, including traditional hunting techniques, and the artisanal trap-making. This knowledge remains alive in the memory of elders and is orally transmitted as part of the collective identity. The symbolic and pedagogical value of listening at length to the leaders' accounts significantly contributes to the transmission of cultural knowledge. This represents an ancestral technology that expresses a distinct relationship with the environment —centered on attentive observation of the forest, respect for animal cycles, and subsistence ethics rather than exploitation.

The challenges faced by Guarani Mbya youth living in urban villages, such as in Jaraguá, reveal strong tensions between tradition and modernity. The maintenance of the Guarani language, for instance, is threatened by daily contact with Portuguese through social networks, schools, and institutional spaces. As a language preservation strategy, children in these communities learn to speak Guarani first and only begin literacy in Portuguese in Elementary School. The language's survival relies on continuous efforts from leaders since its loss implies not only a rupture in communication but also in spirituality and identity.

Guarani Mbya leaders emphasize the importance of indigenous cultural and spiritual diversity, contrasting with the historical imposition of a single national and religious identity in Brazil. By affirming that "the Guarani have always lived with these other peoples," they celebrate a tradition of respect for plurality, challenging the colonizing logic of homogenization. The ancestral wisdom referenced here is linked not only to knowledge of the Earth but also to a way of life that values coexistence among different peoples and spiritualities.

> Keeping the wisdom that there is one Earth. Various peoples, and the Guarani have always lived with these other peoples. This very old wisdom does not fit with the idea of Brazil and

> a single religiosity as they have always tried to impose, right? (Interview conducted with a Guarani Mbya indigenous leader, 2024)

This cultural fragility is intensified by the lure of the urban capitalist model, marked by consumerism, easy access to drugs and alcohol, and constant exposure to lifestyles that contradict Guarani collective, spiritual, and territorial culture. This context creates fissures in the community fabric and imposes a double sense of belonging on young people —bonded to both their village and the city that permeates them —giving rise to conflicts and inequalities.

The current economic model has intensified socio-environmental problems and the humanitarian crisis. The climate emergency is a result of this crisis of civilization and meaning. In one statement from an indigenous leader, the conflicts become evident:

> We keep seeing the death of leaders. One way or another, this keeps happening. What we hope is that our own relatives, especially younger ones, can seek more knowledge and conditions to defend our people, but never stop resisting. (Interview conducted with a Guarani Mbya indigenous leader, 2024)

In their records, the Guarani Mbya indigenous community demonstrates that the climate emergency cannot be solved solely through technological innovations or international agreements. It rather requires a profound transformation in how relationships between living beings are established. Therefore, the climate crisis cannot be separated from the ethical and political emergency, which demands a re-approach to indigenous values and worldviews. The indigenous way of life rejects the exploitative and utilitarian perspective of the environment, inviting a collective state of consciousness:

> Nhanderekó is resistance and protection. Nhanderekó is our shield. It is our way of being together to walk. Indigenous struggle is a situation we have to face collectively. In the juruá's (non-indigenous) thinking, everything is solved with violence. We do not respond with violence. We become stronger. (Interview conducted with a Guarani Mbya indigenous leader, 2024)

The decolonial approach to the climate emergency questions quick, market-driven solutions that ignore structural roots of this crisis. The discourse always arises from the other(s), the cultural collective. Discursive memories evoke personal histories, taking past and present agents integrated into indigenous culture as reference. These generational stories revisit conflicts from ancestors.

In this conflict zone, the importance of territory emerges as the center of indigenous identity and as an extension of the environment. The word Yvyurpá, used by Guarani Mbya leaders, signifies that this territory is sacred and constantly threatened by the urban growth. For the Guarani Mbya, the effects of climate change manifest concretely in altered ecological cycles, loss of green areas, and contamination of natural resources, endangering cultural and spiritual practices that depend on the territory's integrity. Additionally, environmental fragmentation caused by urban expansion directly affects mobility, rituals, and access to essential elements of traditional life, such as medicinal herbs, prayer sites, and water sources.

Bearing that in mind, it is urgent to recognize that indigenous territories are also territories of climate protection. The preservation and empowering of Yvy Rupá are not only an identity claim but also acts of political and ecological resilience with global implications. Recent studies indicate that areas under indigenous management are the most resistant to deforestation and the most successful in conserving biodiversity, acting as barriers against climate change progression (Makondo & Thomas, 2018).

Therefore, respecting indigenous territorial rights and valuing their ways of life is not only a matter of historical justice but also an essential strategy in resisting the global climate emergency.

7 Final Considerations

Environmental racism is present in everyday life and exposes indigenous peoples to a context of systemic and structural violence. The accounts reveal a complexity of ways of existing rooted in indigenous narratives amid urban pressure. The importance of cultural events such as festivals and community mobilizations come into view as processes of struggle to safeguard identity values.

The audiovisual production process can serve as an ally in digitally disseminating the socio-environmental conflicts experienced by the indigenous community, as well as spreading ancestral knowledge and wisdom aligned with more harmonious and ethical relationships with the territory. Likewise, this collaborative production allows other youth, as well as teachers, to become familiar with different conceptions, which fosters reflection on hegemonic understandings of the relationship between society and the environment.

Educommunication, —as a formative and denunciatory process— contributes both as a pedagogical instrument and a tool for community strengthening. The images and voices documented affirm the continuity of ways of life that resist historical processes of erasure and displacement.

Accordingly, confronting environmental racism involves valuing indigenous voices and building networks of solidarity that serve as sources of strength for indigenous cultures, grounded in collective memory and respect for the land. It constitutes a form of resistance that intersects tradition and innovation.

Finally, recognizing the active role of indigenous communities in environmental preservation demands attentive listening and a commitment to the decolonization of knowledge. The narratives and audiovisual productions created by these communities challenge the dominant development paradigms and point toward alternative futures, where environmental justice intertwines with the right to self-determination of the indigenous peoples.

Acknowledgments This work develops from the intersection of two research projects within the Environmental Education and Educator Training Research Group: *Environmental Education and Management of Conservation Units in the State of São Paulo: Articulation of Knowledge in*

Building Learning Communities (FAPESP Process No. 2019/19528-1) and *Equity and Sustainability in the Articulation Between Curricula, Pedagogical Practices, and Teacher Training in Public Schools in São Paulo* (Process No. 2022/06966-3). The common premises of these projects involve participatory research experience, the construction of learning communities and situated knowledge production, and the development of environmental education processes articulated with social justice.

References

Acselrad, H., Mello, C. C. A., & Bezerra, G. N. (2009). O que é justiça ambiental [What is environmental justice]. Garamond.

Belmont, M. (Ed.). (2023). *Racismo ambiental e emergências climáticas no Brasil [Environmental racism and climate emergencies in Brazil]*. Instituto de Referência Negra Peregum e Oralituras.

Brasil, Ministério do Meio Ambiente. (2008). Educomunicação socioambiental: Comunicação popular e educação [Socio-environmental educommunication: Popular Communication and education]. MMA.

Ferdinand, M., & Mei, L. (2022). *Uma ecologia decolonial: Pensar a partir do mundo caribenho [A decolonial ecology: Thinking from the Caribbean world]*. Ubu Editora.

García-Vinuesa, A., Meira-Cartea, P. A., Caride-Gómez, J. A., & Iglesias da Cunha, M. L. (2022). El cambio climático en la educación secundaria: conocimientos, creencias y percepciones. Enseñanza de las Ciencias, 40(2), 23–48. https://doi.org/10.5565/rev/ensciencias.3526

Gattás, C. (2020). *O Guarani nas ondas do radio [The Guarani on the Radio Waves]*. Magistério/ Secretaria Municipal de Educação. Coordenadoria Pedagógica.

Gomes, H. B., & Silva, R. L. F. (2024). Educação ambiental e patrimonial: Um estudo da produção de vídeos educomunicativos na comunidade indígena Guarani do Brasil. *Revista Eletrônica do Mestrado em Educação Ambiental, 41*(3), 264–282. https://doi.org/10.14295/remea.v41i3.17852

Grandisoli, E., Torres, P. H. C., Jacobi, P. R., Toledo, R. F., Coutinho, S. M. V., & Santos, K. L. (Orgs.). (2021). Novos temas em emergência climática: Para os ensinos fundamental e médio [New themes in climate emergency: For elementary and secondary education]. IEE-USP. https://doi.org/10.11606/9786588109083

Halbwachs, M. (2006). *A memória coletiva [Collective Memory]*. Vértice.

Instituto Socioambiental (ISA). (2015). Guarani Mbya. Povos Indígenas no Brasil. https://pib.socioambiental.org/pt/Povo:Guarani_Mbya

Kaplún, M. (2002). *Uma pedagogia da comunicação (o comunicador popular) [Pedagogy of Communication (the Popular Communicator)]*. Editorial Caminhos.

Krenak, A. (2019). *Ideias para adiar o fim do mundo [Ideas to postpone the end of the world]*. Companhia das Letras.

Mainardes, J., & Marcondes, M. I. (2011). Reflexões sobre a etnografia crítica e suas implicações para a pesquisa em educação [Reflections on critical ethnography and its implications for educational research]. *Educação & Realidade, 36*(2), 425–446.

Makondo, C. C., & Thomas, D. S. G. (2018). Climate change adaptation: Linking indigenous knowledge with western science for effective adaptation. *Environmental Science & Policy, 88*, 83–91. https://doi.org/10.1016/j.envsci.2018.06.014

Martins, D. V., & Knapp, C. (2020). Proposta de uma metodologia intercultural na construção de artigos científicos por estudantes indígenas do PPGET-UFGD [Proposal of an intercultural methodology in the construction of scientific articles by indigenous students of PPGET-UFGD]. *Cadernos CERU, 31*(2), 137–148.

Munduruku, D. (1999). *Coisas de Índio [Indian things]*. Global Editora.

North American Association for Environmental Education. (2018). 2017 annual report: Education we need for the world we want. North American Association for Environmental Education. https://naaee.org/sites/default/files/2022-07/ar_2017.web_.8.21.pdf

Oxford University Press. (n.d.). Climate emergency. In Oxford Languages. Retrieved June 18, 2026, from https://languages.oup.com

Quijano, A. (2005). Colonialidade do poder, eurocentrismo e América Latina [Coloniality of power, eurocentrism, and Latin America]. In E. Lander (Org.), *A colonialidade do saber: Eurocentrismo e ciências sociais* [The coloniality of knowledge: Eurocentrism and the social sciences] (pp. 117–142). CLACSO.

Silva, R. L. F., & Campina, N. N. (2011). Concepções de educação ambiental na mídia e em práticas escolares: Contribuições de uma tipología [Conceptions of environmental education in the media and in school practices: Contributions of a typology]. *Pesquisa em Educação Ambiental, 6*, 29–46.

Silva, R. L. F., Freire, B., da Silva, A., Gattás, C., Thomaz, C., Bacci, D., Pereira, G., Silva, G., Gomes, H., Ghilardi-Lopes, N., da Silva, M., & P., & Veruli, V. (2023). *Educação ambiental em unidades de conservação: Fundamentos e práticas [Environmental education in conservation units: Foundations and practices]*. Diagrama Editorial.

Silva, R. L. F., Grandisoli, E., & Jacobi, P. R. (2024). The role of the environmental education in tackling the climate emergency and biodiversity decline: Articulating Brazilian experiences. *International Journal of Environmental Resilience Research and Science, 6*(2), 1–25. https://doi.org/10.48075/ijerrs.v6i2.33481

Soares, I. O. (2006). *Educomunicação: Interfaces e perspectivas [Educommunication: Interfaces and perspectives]*. Paulinas.

Tozoni-Reis, M. F. C. (2006). Temas ambientais como "temas geradores": contribuições para uma metodologia educativa ambiental crítica, transformadora e emancipatória. *Educar em Revista, 27*, 93–110. https://revistas.ufpr.br/educar/article/view/6467

Walsh, C. (2009). Interculturalidad crítica y pedagogía de-colonial: Apuestas (des)de el Sur [Critical interculturality and decolonial pedagogy: Commitments from the South]. In C. Walsh (Org.), *Interculturalidad, descolonización del Estado y del conocimiento* [Interculturality, decolonization of the state and of knowledge]. Ediciones Abya-Yala.

Educating for Action From Schools: Innovative Pedagogies, Art and Communication

School Climate Assemblies in the Socio-Critical Paradigm: Methodological Approach and Key Processes

Núria Monterde-Miralles, Gisela Cebrián-Bernat, and Mercè Junyent-Pubill

Abstract This chapter presents the methodological approach of school climate assemblies developed within the EDU4CLIM project. Grounded in the socio-critical paradigm, this approach aims to promote transformative learning for sustainability. School climate assemblies are introduced as an innovative educational strategy that fosters authentic student participation in deliberative democratic processes to address local socio-environmental issues. The chapter outlines the four main phases of the assembly process—learning, deliberation, decision-making, and communication—and connects each to key concepts and processes of the socio-critical paradigm, including ideology critique, reflective practice, communicative action, deliberative democracy, active participation, and empowerment. Through the assembly sessions, students reflect on environmental challenges, deliberate on possible solutions, engage in dialogue with peers and political representatives, and share their proposals at a final Fair on Sustainability Ideas. This process helps develop core sustainability competencies such as systems thinking, critical thinking, futures thinking, action competence and eco-citizenship, enabling students not only to learn about sustainability but also to actively participate in shaping a more just and sustainable future for their communities. Educational initiatives, such as school climate assemblies, are essential for creating learning environments that support deliberative and participatory democracy within the school context thereby proposing a concrete educational strategy and pathway for fostering active and empowered citizens from an early age.

N. Monterde-Miralles (✉) · G. Cebrián-Bernat
Research Group EDIT – Educational Transformation, Leadership and Sustainability, Department of Pedagogy, Universitat Rovira i Virgili, Tarragona, Spain
e-mail: nuria.monterde@urv.cat

G. Cebrián-Bernat
e-mail: gisela.cebrian@urv.cat

M. Junyent-Pubill
Department of Teaching of Mathematics and Experimental Sciences, Universitat Autònoma de Barcelona, Bellaterra, Spain
e-mail: merce.junyent@uab.cat

J. J. Conde et al. (eds.), *Practices in Education for Climate Emergency*,
https://doi.org/10.1007/978-981-92-2304-6_8

1 Introduction

Today's societies must face various complex global challenges such as the climate crisis, droughts, biodiversity loss, extreme weather events and increasing social inequalities. To address these challenges, it is necessary to promote quality education for everyone, as well as a critical and committed citizenship capable of reflecting on and questioning its own visions and actions. Moreover this self-awareness should lead to more sustainable individual and collective practices (UNESCO, 2017). Education for Sustainability from a critical approach seeks to transform society by promoting values of justice, tolerance, equity and care of the environment, as well as responsibility and respect for others, for the environment and for the resources of the planet we inhabit. This education applies active and participatory learning, as well as innovative teaching and learning practices for developing sustainability competencies (Macintyre et al., 2025). It is based on a transformative and action-oriented pedagogy, on autonomous learning and orientation towards problem solving and critical reflection, and also on collaboration and dialogue processes that involve the entire system and stakeholders, promoting active participation (UNESCO, 2017).

Active citizen participation in climate action is essential for individuals and communities to be able to engage in community and transformative learning processes on sustainability that improve wellbeing, personal and collective fulfilment and that lead to individuals adopting more sustainable lifestyles (Cebrián et al., 2025). These processes allow individuals to be more autonomous and better informed concerning current societal challenges. People are therefore able to make decisions in favour of sustainability in their daily lives (Alvarado & García, 2008).

The real participation of citizens in social processes is key, just as the participation of students in the educational context is essential; however, in many cases it ends up being a symbolic participation due to adults not understanding the particular interests of younger people (Hart, 1997). With the aim of bridging this gap, school climate assemblies have been proposed in secondary education as an innovative educational strategy positioned in the socio-critical paradigm that seeks to empower young people in climate action (Cebrián et al., 2025). The school climate assemblies promote authentic participation, reflection and inclusive dialogue for students through deliberative democratic processes for identifying and agreeing on informed solutions and responses for mitigation and adaptation to climate change in the local environment. The assemblies offer a safe and inclusive space for students to share, exchange, reflect and discuss collectively and freely on issues related to sustainability and climate change. Students can propose improvement actions at a local level that are later shared with the educational, social and political community of the area. Furthermore, the assemblies encourage the development of sustainability competencies such as problem solving, critical thinking, interpersonal competence, systemic thinking, future thinking, action competence, and strategic and normative competencies (Bianchi et al., 2022).

In this chapter we introduce the methodological approach to school climate assemblies, presenting their phases, sessions and activities, as well as their connection with the key processes of the socio-critical paradigm. First, we outline the key processes associated with this paradigm. Second, we present Education for Sustainability in relation to the socio-critical paradigm. Third, we detail the process followed in the school climate assemblies organised as part of the EDU4CLIM project. Finally, we link the different phases, sessions and activities of the organised school climate assemblies with the key processes of the socio-critical paradigm, explaining how these are put into practice in the assemblies and showing the sustainability competencies that are promoted among students.

2 The Socio-Critical Paradigm: Key Processes

The socio-critical paradigm, originating in the critical theory that emerged from the Frankfurt School and presented in the work of Carr and Kemmis (1986), symbolizes change and liberation that alternates criticism with ideology (Bisquerra, 2019). It is based on self-reflection, rational autonomy and the internal knowledge of the individual in relation to their interests and needs, all of which enable the citizen to actively participate and generate transformative changes in society (Alvarado & García, 2008). According to Freire's pedagogy of the oppressed (1985), this capacity is called critical consciousness. It is a process that requires transformative and emancipatory actions so that the individual becomes aware in order to reflect on and challenge their own worldviews, perspectives and beliefs, which have been developed uncritically. They can therefore become active subjects of their own actions and lives (Lucio-Villegas, 2015, 2021). Individuals are co-creators of their realities, they actively participate through their experience, intuition, creativity, thoughts and actions, which allow them to build their conception of society both individually and collectively (González Morales, 2017). Within the socio-critical paradigm, schools have a fundamental power in defining the social reality because they constantly seek to transform both existing structures and educational content that perpetuates social, economic and cultural inequalities (Bisquerra, 2019). This paradigm proposes creating a more democratic, plural and holistic education to ensure an egalitarian educational practice. Unlike other paradigms, the socio-critical paradigm tries to unmask the existing ideologies, promoting political commitment and supporting the idea that knowledge is a way of liberating individuals and making them the protagonists and active agents in the education community (González Morales, 2017).

The socio-critical paradigm in the educational field seeks to promote social transformation and foster an education for liberation (Muñoz-Gaviria, 2017). According to Freire (1985) and Muñoz-Gaviria (2017), social transformation is based on questioning dominant ideologies and the historical and sociocultural contexts in which we live, to create new social practices and alternatives. Thus, new fairer and more egalitarian forms of social life are nurtured through social transformation. On the other hand, education for liberation aims to change social conditions and free people from situations of domination and oppression so they become active subjects in their own society with autonomy and freedom (Freire, 1985). This education promotes a humanizing political practice that helps us generate more socially just and equal communities through the exchange of ideas and collective reflection (Lucio-Villegas, 2015).

Placing ourselves in the socio-critical paradigm to address social transformation through education means considering and putting into practice a set of key processes:

- **Ideology critique:** This is based on the ability to reflect on reality and social, cultural and political contexts. When people acquire this consciousness, they can build their own worldviews and culture in a more conscious and active way. Education should allow people to understand and value their culture critically in order to build new visions and ways of perceiving it, freeing themselves from the social pressures imposed by oppressive structures (Lucio-Villegas, 2021). This ideology aims to show the self-conceptions that are implicitly found in individuals' distorted ideas and to propose a way to overcome the contradictions of the individual's current beliefs (Carr & Kemmis, 1986).
- **Reflective practice:** reflection is not just a mental act, it involves a conscious and transformative action. According to Freire (1985), education is not just a matter of transmitting knowledge and abilities, it is a reflective process. The aim of education should be to help people develop a critical consciousness that allows them to position themselves and act in the world as citizens with rights. This implies that educators should not only be transmitters of knowledge, they should also actively reflect on their own educational practices and on how their actions can contribute to the liberation of students (Lucio-Villegas, 2015).
- **Communicative action:** this is the engine for individual, collective and social transformation. The dialogic relationship refers to an open and respectful communication between people, where the focus of criticism is placed on the social, cultural, political and existential possibilities that arise from this mutual exchange with others. Dialogue not only serves to share ideas, but also to recognize and value the other as a subject with the capacity for thought and action, as well as helping to verify the claims of veracity of opinions (Carr & Kemmis, 1986). Dialogue breaks with merely reproductive practices, which occur when people are treated as objects or are forced to follow a single way of thinking (Muñoz-Gaviria, 2017).
- **Deliberative democracy:** the educational project must train democratic subjects, that is, people capable of actively participating in political and social life, who can free themselves from the limitations imposed by systems based on oppression and violence (Freire, 1985). These oppressive systems make it difficult for people to actively participate in democracy, since they impose limitations to maintain control and social exclusion. Therefore, the objective is to help people free themselves from these limitations and have an active and equal participation in democracy to discuss and decide on the issues that affect their society in an open and inclusive way (Lucio-Villegas, 2015).
- **Active participation:** one of the goals of education should be that it is not only a process of passive absorption of information, but rather it should also train people to express themselves and intervene in society in a reflective and conscious way

through active participation. Active participation requires learning and dialogue, from which knowledge is built with the commitment of the people involved (Carr & Kemmis, 1986). Therefore, according to Freire (1985), people involved in the educational process must play an active role in defining and constructing the topics that are addressed. Education must be a space in which participants can actively identify and reflect on the problems that affect them, with the aim of transforming individual and social realities and practices (Lucio-Villegas, 2015).

- **Empowerment:** According to Freire (1985), the central objective is not only to teach basic abilities, but also to help people achieve a critical awareness that allows them to position themselves and act in the world from the perspective of a citizenship with rights. This implies that education must be a reflective process, in which students not only learn passively, but can also transform their social reality. That is, Freire's critical consciousness process is key for the empowerment of the individual because it is not only about theoretically understanding reality, but also about acting on it and transforming it.

These key processes promote the development of a critical consciousness that allows the individual to situate themselves and act in the world as a citizen with rights, to reflect on reality and social contexts, to communicate openly and respectfully with other people and to play a part in transforming society in a reflective and conscious way through equal and democratic participation (Carr & Kemmis, 1986; Freire, 1985).

3 Education for Sustainability in the Socio-Critical Paradigm

Education for Sustainability is based on learning processes such as collaboration and dialogue, systemic thinking, constant innovation of the educational curriculum and active and participatory learning, as well as learning linked to asking critical questions, questioning one's own values, visualising future scenarios, responding in an informed manner and exploring the dialectic between tradition and innovation (Macintyre et al., 2025; UNESCO, 2017). Education for Sustainability seeks to develop competencies that empower students by allowing them to understand the impact of their actions and make informed decisions about them, taking into account the effects of these decisions on the social, cultural, economic and environmental spheres, both present and future, from a local and global perspective, at all educational stages (Cebrián et al., 2020). In a globalised and uncertain society, being competent in matters of sustainability in the educational context is an ongoing task. Students must attain and put into action a series of multidisciplinary and transversal competencies that enable them to understand complex systems and take more sustainable action and practices (Bianchi et al., 2022). Sustainability competencies are conceived as a combination of knowledge, practical abilities, values and ethical attitudes that empower individuals and communities to contribute to sustainability through their actions and in their daily lives. In recent decades, different conceptual frameworks of sustainability competencies have been proposed and developed. These frameworks coincide in the main key competencies required to deal with sustainability challenges (Bianchi et al., 2022; Cebrián et al., 2020; UNESCO, 2017):

- **Systems thinking:** This involves the ability to identify and understand relationships within systems, analyse complex systems, reflect on how these systems are interconnected across domains and scales, and navigate uncertainty effectively.
- **Future-thinking competence:** This encompasses understanding and evaluating multiple future scenarios, including possible, probable and desirable futures, developing personal visions for the future, applying the precautionary principle, dealing with uncertainty, assessing the impact of actions and managing change and risk.
- **Normative competence:** This involves understanding and reflecting on the underlying ethical norms and values that guide personal actions and behaviours, negotiating values, principles, objectives and sustainability goals, supporting equity and justice between generations, and recognizing the rights of all species.
- **Collaboration competence:** This involves learning from others, respecting and empathizing with their needs and actions, managing conflicts within a group, and facilitating cooperative problem solving.
- **Critical thinking:** This involves evaluating information and different viewpoints, as well as the ability to question existing norms and practices, reflect on personal values, perceptions and actions and actively participate in debates on sustainability.
- **Action competence:** This involves understanding the political system, demanding responsibility and accountability for unsustainable practices, and promoting effective policies for sustainability. It also requires collaborating with other actors to drive change as well as recognizing one's own potential to actively contribute to improving the community and the planet.

All educational institutions, from early childhood to higher education, formal or informal, have the responsibility of offering tools and learning opportunities so that people can address issues related to sustainability by promoting the development of these competencies (UNESCO, 2017). Education for Sustainability in schools should train students to build explanatory models about world phenomena, allowing them to ask reflective and critical questions about the phenomena they encounter, understand the processes that develop, plan actions and predict their consequences by imagining preferable and sustainable futures, as well as clarifying sustainability values to identify whether their lifestyles are aligned with them or not (Bianchi et al., 2022; Macintyre et al., 2025). There is a broad consensus within the academic community on the most effective teaching-learning methodologies for promoting the development of sustainability competencies. These methodologies include learner-centred approaches such as problem- and project-based learning, service-learning, action learning, systemic thinking, agile methodologies, and collaborative and participatory decision-making processes (Cebrián et al., 2020).

Therefore, the principles of Education for Sustainability are closely linked to the principles of the socio-critical paradigm: both are pedagogical, empirical, civic and moral frameworks, with a transformative, holistic and democratic vision, and they both promote critical reflection to make informed decisions through dialogue and collective action, considering the emotional dimension (Cebrián et al., 2012). Research in the field of Education for Sustainability in relation to a socio-critical paradigm is necessary for transforming mental and action models in favour of sustainability since this paradigm manifests values such as freedom, equality and justice in relation to the environment.

Considering the transformative potential offered by the socio-critical paradigm, it is worth highlighting its importance in research that can be transformative by promoting individual and collective autonomy and responsibility, as well as its capacity to generate changes to promote more just, democratic and equitable societies (Alvarado & García, 2008). In the socio-critical paradigm, there are three types of associated research: action research, collaborative research and participatory research. All three take the individual as an active protagonist of the research within society, to transform reality through critical reflection on the behaviours of this reality, which implies its redirection (Cebrián et al., 2012; González Morales, 2017). In educational research, based on the basic assumption that education is neither neutral nor individual, the researcher establishes horizontal links with teachers and students as facilitators of the processes generated in the research, such as data collection. The researcher develops attitudes of complicity and synergy with these groups who are conceived as active subjects and co-researchers, establishing an effective commitment by being part of the same transformation of their realities through processes of collaboration, learning and critical reflection (Bisquerra, 2019; Lucio-Villegas, 2015).

In this sense, school climate assemblies are the context for the EDU4CLIM applied research project, which has a participatory and pragmatic approach. The general objective of this project is to develop, test and implement school climate assemblies in secondary schools for the co-creation of climate change mitigation and adaptation strategies and the development of sustainability competencies in students.

4 The School Climate Assemblies of the Edu4clim Project

It is necessary to develop new practices that enable young people to understand and make sense of climate change, so they can act accordingly and actively contribute to real sustainability challenges in their communities (McGimpsey et al., 2023). Citizen climate assemblies are one of the increasingly popular deliberative democratic processes in the environmental field. These are forums for citizen participation to share knowledge, as well as to identify, deliberate, make decisions and develop action proposals and policy recommendations for tackling the climate emergency and sustainability (Cebrián et al., 2025; Elstub et al., 2021). In recent years, school and youth climate assemblies have emerged at national and international levels as a youth-based process for giving a voice and empowering young people in relation to climate change, such as the Nottingham Youth Climate Assembly (NYCA) and the Children's Parliament part of Scotland's Climate Assembly (Reid, 2024; Wilson et al., 2024).

Within the framework of the R&D&I project *School climate assemblies for developing sustainability competencies in Secondary Education* (EDU4CLIM), school climate assemblies have been designed and implemented in eight secondary schools in the province of Tarragona, with the aim of exploring their influence on the development of sustainability competencies and the co-creation of climate change mitigation and adaptation strategies at the local regional level. This project is based on Education for Sustainability within a socio-critical paradigm, proposing school climate assemblies as an innovative educational strategy that promotes the development of sustainability competencies associated with critical thinking, action competency and democratic participation, fostering inclusive dialogues, deliberation processes, consensus and decision-making (Bianchi et al., 2022). The assemblies are conceived as an educational process focused on empowering children and young people in the search for solutions to address climate change (Cebrián et al., 2025). They are based on the methodological approach and phases of the citizen climate assemblies that have emerged in recent years at the national level in countries such as Ireland, the United Kingdom, Scotland, Denmark, France and Spain (Elstub et al., 2021). The four phases of the school climate assemblies implemented are:

1) **Learning phase:** this includes materials, preparatory activities and the participation of climate change and sustainability experts through presentations, talks, workshops, group discussions and/or expert panels to raise awareness and provide a deeper understanding of the climate change and sustainability challenges and solutions addressed in the assemblies.
2) **Deliberation phase:** This involves facilitating dialogue and debate among participants on the environmental issues discussed in the assemblies, through collaborative and cooperative activities, in order to reflect, share and further develop the participants' opinions with their peers, through active learning and work in small groups, role-playing games and future visualisation exercises.
3) **Decision-making phase:** this focuses on facilitating the development of joint and consensual conclusions based on the participants' reflections and discussions by voting and prioritizing the actions that emerged from the assembly to produce action plans with political recommendations.
4) **Communication phase:** the actions proposed in the assemblies and their results are disseminated and communicated with different political representatives, agents of the local and political community and the general public through the organization of events and activities. The actions and results are also disseminated through different channels, the media and social networks.

Below we present the school climate assembly process followed in the EDU4CLIM project. We also describe the five sessions organised, all supported and guided by expert facilitators who are members of the project team, as well as the activities and dynamics used. In the process, four assembly sessions are organised per class group participating in the project and the process concludes with a fifth session in which all class groups present and share their action proposals for promoting sustainability at a final Fair on Sustainability Ideas. Each session is linked to one of the phases of the school climate assemblies.

- **Session 1 (*Learning phase*):** a one-hour training workshop is held, which includes dynamic and participatory activities related to environmental topics associated with sustainability (water, energy, biodiversity, mobility, waste), with the aim of reflecting on and delving deeper into these topics and their main environmental problems, their causes and consequences. In the second part of the session each class group focuses on one of the environmental issues that students have previously chosen. Specifically, this session begins with a brief presentation and explanation of the EDU4CLIM project, its phases and the objectives of the different assembly sessions. Three dynamic and participatory activities are then carried out:
  - The game *"Green or Red: Climate Emergency"* is played, in which students learn about the climate emergency and its relationship with aspects such as waste, mobility, biodiversity, energy and water, through a "True or False" game based on concepts and misconceptions related to climate change.
  - The game *The Dictionary* is played. The objective of this game is to collaboratively write a definition of the environmental issue that each class group focuses on, which could be biodiversity, air, mobility, waste, etc.
  - The game *The Relay* is played. This game consists of presenting various real socio-environmental problematic situations on the environmental issue that the class group is working on. In groups, participants have to analyse the origin and the actions necessary to address these problematic situations, while doing a team relay race.

Finally, the session is closed with a group reflection and final joint conclusion, including the aspects that have been learned, as well as those that have attracted the most attention. In addition, students are asked to observe their surroundings between this first session and the second session and to identify socio-environmental situations in which they would like to take action in their local area, in order to be able to share them in the second session.

- **Session 2 (*Deliberation phase*):** Different deliberative and participatory activities are carried out in this two-hour session. The class group is organised into small working groups to work on the selected environmental issue. The existing challenges are identified, and actions are proposed to address them, which are then shared with the rest of the class. Deliberations are held within each working group to reach a consensus on the most relevant and priority actions to promote sustainability in their local area. This session is divided into three main activities:
  - The assembly facilitator shows both positive and negative images on the environmental issue selected by the class group to inspire the students and

encourage debate. Then, in small working groups, the students create a conceptual map showing the positive and negative aspects of the environmental issue, as well as possible opportunities or actions for solving the problems associated with the topic at an individual level, at the level of the school, neighbourhood or town/city.

- Once each group has made its conceptual map, they debate as a group and agree on possible actions for addressing the identified problems. Each group must agree on five action proposals to be carried out in their local area, which are presented to the rest of the class.
- Once each small group has presented and shared their proposals with the class, a vote is held individually. Each student votes on the three actions that considers most relevant and a priority of the five actions proposed by each team. Afterwards, a count is made, a group reflection is held and the four action proposals that received the most votes from the entire class are presented and evaluated jointly with political representatives in the next session.

- **Session 3 (*Decision-making phase*):** this session consists of a meeting between students and local council technicians and policy representatives to present and share the initial four action proposals that received the most votes from students in the second session. This third session is an opportunity for students to ask political agents questions to delve deeper into the environmental problem addressed, as well as to learn about the local policy and strategy for addressing it, and the actions or initiatives that are already being implemented or planned at a local level. This session allows students to obtain information about the problem in its context, to subsequently choose and specify an action that is relevant and feasible at a local level.
- **Session 4 (*Decision-making phase*):** after the meeting with local council technicians and political representatives, a one-and-a-half-hour session is held in which the class decides and agrees on an action to be developed. This action is further researched and contextualised to be presented to the educational community, political agents, local stakeholders and citizens at a Fair on Sustainability Ideas (fifth session). To do this, in small groups, students must work on contextualizing the action based on a series of guiding questions: *1) What do we need for our local area? 2) What do we need to know/research? 3) What do we propose?* Between this session and the next, students and teachers create a final product for presenting their action at the event, such as creating a model, making a video, writing a song or performing a skit, among others.
- **Session 5 (*Communication phase*)**: a Fair on Sustainability Ideas is organised in the local area. At this event, all the class groups from the participating schools present the selected actions for promoting sustainability at a local or regional level to the rest of the students, groups and schools in the area, as well as to political representatives, social and environmental entities, and the general public. The aim of this event is to disseminate the action proposals, share perspectives and receive feedback from the educational, political and social stakeholders in the area.

Once the Fair on Sustainability Ideas is finished, the final report resulting from the school climate assemblies is prepared and distributed, which includes the actions proposed by the students, as well as political recommendations based on them. This report is distributed and disseminated to the different stakeholders, including other education communities and schools outside the project, as well as to the political representatives of the area, environmental entities and associations, through different channels and means of communication such as press releases, email and dissemination through social networks and the website.

5 School Climate Assemblies and Key Processes of the Socio-Critical Paradigm

In this section we present how the different phases, sessions and activities of the school climate assemblies organised are linked to the key processes of the socio-critical paradigm, explaining how these are brought into practice through the assemblies: ideology critique, reflective practice, communicative action, deliberative democracy, active participation and the empowerment of individuals (Carr & Kemmis, 1986; Freire, 1985). We also discuss how the phases and activities of the school climate assemblies contribute to developing sustainability competencies, such as systemic thinking, critical thinking, futures thinking, eco-citizen competence and action competence (Bianchi et al., 2022). At a general level, school climate assemblies pursue a common goal: to promote an education for sustainability that cultivates the empowerment of individuals and social transformation towards sustainability.

Key processes of the socio-critical paradigm such as communicative action and active participation (Carr & Kemmis, 1986) are promoted and developed throughout the different sessions of the assemblies, in which an active role is given to students and working groups, creating spaces for group dialogue and reflection, open and respectful communication and student participation in solving environmental problems in their local area. In Fig. 1 we present a visual diagram of the phases and sessions of school climate assemblies, and their link with the key processes of the socio-critical paradigm.

During the *learning phase*, the questioning of assumptions and mental models related to climate change is activated through the dynamic and participatory activities proposed by experts in climate change and sustainability during the training workshop (Session 1). Ideology critique and reflective practice are promoted through the exchange of opinions and points of view on the origin of socio-environmental problems and their nuances (Freire, 1985; González Morales, 2017). Activities such as the game "Green or Red: Climate Emergency", the Dictionary game or the Relay game promote these processes, since they help participants reflect on their previous conceptions about climate change, as well as possible origins, impacts and solutions associated with real socio-environmental situations and problems. This phase encourages the development of different sustainability competencies, such as systems

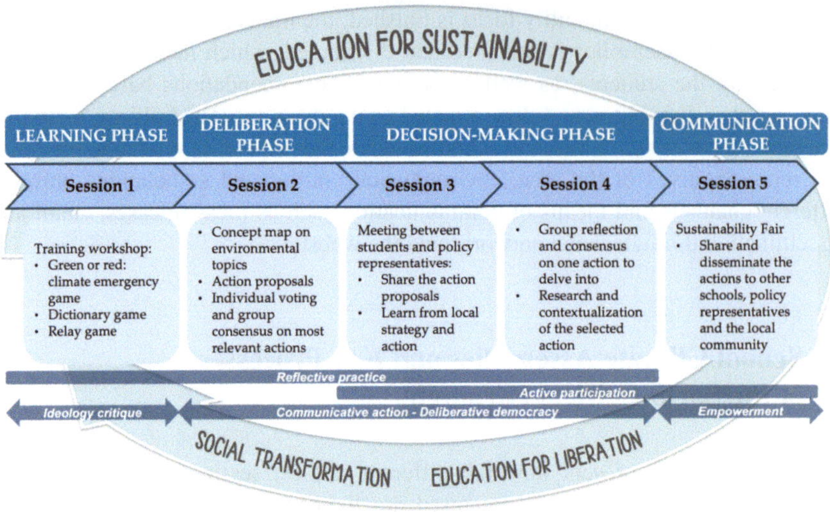

Fig. 1 The process of school climate assemblies in relation to the socio-critical paradigm.

thinking and normative competence, since environmental situations are analysed from a complex perspective, taking into account their different dimensions (social, environmental and economic), and participants reflect on the ethical norms and values related to the principles of sustainability (Bianchi et al., 2022; UNESCO, 2017).

In the *deliberation phase*, students reflect, debate and deliberate on the environmental problems that exist in their local area and possible solutions to improve their area in environmental issues such as water consumption, energy production, care of the biodiversity and the promotion of sustainable transport by creating a conceptual map, proposing actions, voting and reaching a group consensus (Session 2). This phase promotes the reflective practice and communicative action through collective reflection, dialogue and group deliberation (Carr & Kemmis, 1986; Freire, 1985). It is based on implementing deliberative democratic processes, training students for active participation in environmental problems that affect their area, through debate and proposing actions. In turn, these processes contribute to developing sustainability competencies such as systems thinking, a cognitive framework necessary to explain, argue and understand problems related to climate change and the interconnections between environmental, social and economic dimensions (Clark et al., 2017). Competencies such as critical thinking are also promoted in students: it not only enables them to reflect on issues related to sustainability, but also helps them to evaluate them objectively, form judgments, justify them, propose actions and make informed and consensual decisions cooperatively (Bianchi et al., 2022; Rudd et al., 2020).

In the *decision-making phase,* key processes such as reflective practice, communicative action and deliberative democracy are activated by the students meeting with council technicians and political representatives of the local area (Session 3), where they reflect and debate on feasible actions to address the sustainability challenges of the territory. This phase allows students to develop competencies linked to critical and committed action to address sustainability issues in an informed and successful way, as well as to become agents of active change in their own community (Pauw et al., 2015). This process enables students to develop the action competence for sustainability, which is fundamental, since it includes the sense of both individual and collective responsibility, the ability to influence and actively contribute to sustainability, a concern for the common good, democracy, active participation and a commitment to improving the community (Bianchi et al., 2022; Sauvé, 2017). After discussing and agreeing on the action to be presented at the Fair on Sustainability Ideas, following the visit with local council technicians and political representatives, participants do further research into the action selected, analysing which specific improvement proposals are necessary to carry it out (Session 4). In addition, a decision is made on what final product will be created to present the action at the Fair on Sustainability Ideas. This research process activates the sustainability competence of future-thinking, since it encourages the individual and the group to question and visualise different possible futures related to environmental problems in their local area and to make informed decisions for the future (Lucio-Villegas, 2015, 2021; UNESCO, 2017). Addressing the competence of future-thinking in students involves helping them recognise that they have influence over the future in relation to external problems, allowing them to make conscious decisions that take into account the impact that these decisions can have on their neighbourhood or local community (Bianchi et al., 2022).

In the *communication phase,* other key processes of the socio-critical paradigm are also put in practice such as the active participation and empowerment of participants. These processes are developed through the organisation and participation in the Fair on Sustainability Ideas (Session 5) in which students present and share the proposed actions with other students and schools, political representatives, entities of the area, local stakeholders and the public. The active participation of students is encouraged, and they are empowered by being able to present, discuss and receive feedback on their actions from political representatives and local and regional stakeholders, who visit, listen to and consider the proposals for action in sustainability put forward by students from the different schools. Moreover, participants, who share a common environmental reality, are encouraged to develop interpersonal relationships and social connections. This phase activates sustainability competencies such as eco-citizenship and action competence for sustainability, which involve understanding the political system, demanding accountability and acting individually and collectively to promote effective policies in terms of sustainability (Sauvé, 2017). These competencies seek to promote a conscious and critical citizenship committed to the relationships between society and nature, training students to participate in public debates and in influencing change in sustainability policies and practices that affect their community.

6 Conclusions

In this chapter we present the methodological approach of the school climate assemblies organised in the EDU4CLIM project and in relation to the socio-critical paradigm, analysing how the different phases, sessions and activities of the assemblies are linked to the key processes of this paradigm. The four phases of the assemblies —learning, deliberation, decision-making and communication— facilitate processes such as ideology critique, reflective practice, communicative action and deliberative democracy (Carr & Kemmis, 1986; Freire, 1985). Ideology critique, which involves the ability to reflect on reality and challenge existing worldviews to build new ways of perceiving the world and acting in it, is activated in the learning phase through a training workshop with experts in climate change and sustainability. Regarding reflective practice, which seeks to develop critical awareness in students so that they become active citizens in matters of sustainability, is mainly encouraged in the learning, deliberation and decision-making phases. The communicative action or dialogue, understood as a process for sharing ideas and recognising the other as a thinking subject capable of acting, is developed through sessions aimed at identifying and specifying action proposals and meeting with local council technicians and political representatives, in the deliberation and decision-making phases. Likewise, deliberative democracy is encouraged in these phases, leading to group debate and decision-making that goes beyond listening to the participants' opinions (Elstub et al., 2021). Active participation, which involves engaging participants to play an active role in the construction of the topics discussed in the assemblies, is mainly promoted in the decision-making and communication phases (Wilson et al., 2024). Participants are empowered to rethink, to act and to transform the social and environmental reality of their local area, through consensus building, contextualisation and dissemination of actions to promote sustainability in their local area.

School climate assemblies also promote the development of sustainability competencies, particularly those related to critical thinking, by allowing students to learn about the problems that affect their immediate environment and understand their origins, impacts and consequences (Bianchi et al., 2022; Cebrián et al., 2025). The assemblies encourage systems thinking, as they allow local socio-environmental issues to be understood from a holistic perspective, as well as future-thinking, by encouraging students to visualise alternative long-term scenarios by reflecting on what action to research to improve the sustainability of the local area (UNESCO, 2017). In addition, they promote eco-citizen and action competencies, as they involve not only students in researching actions to be carried out in the local area, but also political agents and local stakeholders, promoting a common collective commitment (Sauvé, 2017).

To foster active and empowered citizens in sustainability matters, it is necessary to create action-oriented learning spaces that are open, inclusive and safe. Children and young people must be able to experience sustainability and participate actively and democratically in decision-making processes in their schools and in their local community from an early age. Educational initiatives such as school climate assemblies are essential for creating structures and spaces for deliberative and participatory democracy in the school context, in which students can become agents of active change in response to environmental problems in their local areas.

Acknowledgments Grant *CNS2023–143709* funded by MICIU/AEI/10.13039/501100011033 and by European Union NextGenerationEU/PRTR.

References

Alvarado, L., & García, M. (2008). Características más relevantes del paradigma socio-crítico: Su aplicación en investigaciones de educación ambiental y de enseñanza de las ciencias realizadas en el Doctorado de Educación del Instituto Pedagógico de Caracas. *Sapiens. Revista Universitaria de Investigación, 9*(2), 187–202. https://www.redalyc.org/pdf/410/41011837011.pdf

Bianchi, G., Pisiotis, U., & Cabrera Giraldez, M. (2022). In M. Bacigalupo & Y. Punie (Eds.), *GreenComp. The European sustainability competence framework.* Publications Office of the European Union. https://doi.org/10.2760/13286

Bisquerra, R. (2019). *Metodología de la Investigación Educativa.* Editorial La Muralla.

Carr, W., & Kemmis, S. (1986). *Becoming critical: Education, knowledge and action research.* Deakin University Press.

Cebrián, G., Boqué, A., Olano, J. X., & Prieto, J. (2025). School climate assemblies: An educational tool for empowering pupils and youth to take climate and sustainability action. *Sustainability Science, 20*, 135–153. https://doi.org/10.1007/s11625-024-01583-6

Cebrián, G., Grace, M., & Humphris, D. (2012). Developing people and transforming the curriculum: Action research as a method to foster professional and curriculum development in Education for Sustainable Development in Higher Education. In W. L. Filho (Ed.), *Sustainable Development at Universities: New Horizons* (pp. 273–284). Peter Lang Scientific Publishers.

Cebrián, G., Junyent, M., & Mulà, I. (2020). Competencies in education for sustainable development: Emerging teaching and research developments. *Sustainability, 12*, 579–587. https://doi.org/10.3390/su12020579

Clark, S., Petersen, J. E., Frantz, C. M., Roose, D., Ginn, J., & Rosenberg Daneri, D. (2017). Teaching systems thinking to fourth and fifth graders using environmental dashboard display technology. *PLOS ONE, 12*(4), e0176322. https://doi.org/10.1371/journal.pone.0176322

Elstub, S., Carrick, J., Farrell, D. M., & Mockler, P. (2021). The scope of climate assemblies: Lessons from the climate assembly UK. *Sustainability, 13*(20), 11,272. https://doi.org/10.3390/su132011272

Freire, P. (1985). *Pedagogía del oprimido.* Siglo XXI.

González Morales, A. (2017). Los paradigmas de investigación en las ciencias sociales. *ISLAS, 138*, 125–135. https://islas.uclv.edu.cu/index.php/islas/article/view/572

Hart, R. A. (1997). *Children's participation: The theory and practice of involving young citizens in community development and environmental care.* Routledge.

Lucio-Villegas, E. (2015). Paulo Freire. La Educación como Instrumento para la Justicia Social. *Revista Internacional de Educación para la Justicia Social, 4*(1), 9–20. https://doi.org/10.15366/riejs2015.4.1.001

Lucio-Villegas, E. (2021). Concientización y criticismo. Aportaciones de Raymond Williams y Paulo Freire a la educación de personas adultas. *Tendencias Pedagógicas, 38*, 57–67. https://doi.org/10.15366/tp2021.38.006

Macintyre, T., Tilbury, D., & Wals, A. (2025). *Education and learning for sustainable futures: 50 years of learning for environment and change.* Routledge.

McGimpsey, I., Rousell, D., & Howard, F. (2023). A double bind: Youth activism, climate change, and education. *Educational Review, 75*(1), 1–8. https://doi.org/10.1080/00131911.2022.2119021

Muñoz-Gaviria, D. A. (2017). La educación como práctica de la libertad: una lectura antropológico pedagógica al pensamiento de Paulo Freire. *Kavilando, 9*(1), 26–41. https://kavilando.org/revista/index.php/kavilando/article/view/193/167

Pauw, J., Gericke, N., Olsson, D., & Berglund, T. (2015). The effectiveness of education for sustainable development. *Sustainability, 7*(11), 15,693–15,717. https://doi.org/10.3390/su71115693

Reid, K. (2024). *Children and young people's participation in climate assemblies.* KNOCA – Knowledge Network on Climate Assemblies. https://www.knoca.eu/guidances-documents/children-and-young-peoples-participation-in-climate-assemblies

Rudd, J. A., Horry, R., & Skains, R. L. (2020). You and CO2: A public engagement study to engage secondary school students with the issue of climate change. *Journal of Science Education and Technology, 29*(2), 230–241. https://doi.org/10.1007/s10956-019-09808-5

Sauvé, L. (2017). Educación Ambiental y Ecociudadanía: un proyecto ontogénico y político. *REMEA- Revista Electrônica do Mestrado em Educação Ambiental*, 261–278. https://doi.org/10.14295/remea.v0i0.7306

UNESCO. (2017). *Education for Sustainable Development Goals. Learning Objectives.* UNESCO. http://unesdoc.unesco.org/images/0024/002474/247444e.pdf

Wilson, R., Keddie, K., Arya, D., & Henn, M. (2024). Climate policy, youth voice and intergenerational justice: Learning from Nottingham youth climate assembly. *Children's Geographies.* https://doi.org/10.1080/14733285.2024.2352368

The Co-Creation of Nature-Based Schoolyards as a Climate Change Education Initiative: An Analysis of Sustainability Competences

Andrea Corres, Isabel Ruiz-Mallén, and Filka Sekulova

Abstract The implementation of nature-based solutions in school environments for climate adaptation can be a powerful climate change education initiative for community members when guided by an inclusive approach in both design and maintenance. In this chapter we explore how the co-creation of nature-based schoolyards promotes stakeholders' knowledge, skills and attitudes to make and keep their school environment more sustainable and climate adapted while disentangling perceived barriers to build these competences. We relied on the GreenComp approach in sustainability education as a conceptual and analytical framework, and used the Barcelona pioneering programs on schoolyards climate adaptation and naturalization as a case study. We discuss how the co-creation of nature-based schoolyards can be understood and applied as a disruptive educational intervention, boosting climate change education based on reflection and action at schools while promoting sustainability competences in teachers and other stakeholders. Finally, we also make recommendations on how these sustainability competences could be enhanced and promoted among users in future school greening projects.

A. Corres (✉) · I. Ruiz-Mallén (✉)
Faculty of Psychology and Education Sciences, Universitat Oberta de Catalunya, Barcelona, Spain
e-mail: acorres@uoc.edu

I. Ruiz-Mallén
e-mail: iruiz_mallen@uoc.edu

TURBA Lab, Interdisciplinary Research Centre on Social and Cultural Transformations (TRÀNSIC), Universitat Oberta de Catalunya, Barcelona, Spain

F. Sekulova
Department of Economic History, Institutions, Politics and World Economy, Universitat de Barcelona, Barcelona, Spain

J. J. Conde et al. (eds.), *Practices in Education for Climate Emergency*,
https://doi.org/10.1007/978-981-92-2304-6_9

1 Introduction

The multiple challenges around climate change entails rethinking traditional pedagogies and therefore the ultimate education purposes. Critical environmental education scholars advocate that Education for Sustainable Development or Sustainability Education needs grounding in disruptive and wild pedagogies to become climate action-oriented and transformational (Jickling, 2018). These are reflective and active pedagogies, responding to real problems and challenging education paradigms that reproduce inequalities of the capitalist society, while trying to overcome their limitations (Jickling, 2017). On the ground, a first step for this endeavour entails tackling the rigidity of the educational system by redefining restrictive norms, growing creativity, and being open to conceiving nature as a co-teacher (Jickling, 2018). The naturalization of school environments, particularly schoolyards, can be an innovative strategy to navigate this transition towards more transformational approaches of climate change education in formal education settings (Ruiz-Mallén et al., 2023).

The Green Education Partnership (GEP), an international initiative on climate change adaptation and education, is bringing a new vision on the relationship between the school and nature by embracing a *whole school approach* (WSA). The WSA engages all members of the school community in climate education, overcomes single-subject approaches and places education outside of built spaces to reconnect with nature, representing an opportunity to develop novel and disruptive pedagogies in climate education (Global Education Monitoring Report Team, 2024). Relatedly, Nature-based Solutions (NbS) are increasingly promoted and implemented in school environments to mitigate and adapt to climate change, as these increase the green areas, provide shade, replace pavement with permeable soil and eventually decrease heat island effects, while retaining rainwater (Antoniadis et al., 2016, 2018). At the same time, evidence shows how NbS in schoolyards broadens children's learning opportunities and empathy for the natural world, expands social interactions, play opportunities and physical activities during recreation and formal education processes, while reducing stress and improving physical and mental health (Van Den Bogerd et al., 2023; van Dijk-Wesselius et al., 2018).

Furthermore, when nature-based schoolyards are co-created, this is done through a participative process of interaction, collaboration and co-production to support their design and implementation by engaging a diversity of actors in a social learning process (European Commission, 2023; Sekulova & Ruiz-Mallén, 2024). Hence, the co-creation of nature-based schoolyards can be a powerful educational initiative by itself for its potential impact on stakeholders' sustainability competences (García-Serrano & García, 2023). Nonetheless, there is lack of evidence analysing such impact on learning in terms of how sustainability competences are put into practice and translated into promoted skills, knowledge and attitudes among the different stakeholders involved in the schoolyards' co-creation. This study put the focus on the ways the co-creation of nature-based schoolyards can contribute to the development of sustainability competences among teachers, students and other actors involved (e.g., families, architects, municipal technicians). Our study uses the GreenComp approach (Bianchi et al., 2022) on sustainability competences as a conceptual and analytical framework.

2 The Greencomp as a Reference Framework

The competence-based approach is relevant for climate change education as it has the potential to respond to relevant socio-environmental issues by mobilizing cognitive, affective, volitional and motivational elements (Corres et al., 2020; Díaz-Barriga, 2019; Rieckmann, 2012). The recent European sustainability competence framework named as GreenComp proposes a novel angle to both sustainability and competence-based approach on education. The GreenComp moves forward towards a sustainability discourse that emphasizes agency development through thinking, planning and acting for a common goal while respecting the balance between nature and human beings. While it overcomes the focus of sustainability competences on a specific educational level and age, proposing one that is relevant to all ages and to all educational levels and forms (Bianchi et al., 2022). Such novel approach makes the GreenComp framework suitable for our study to analyse the way sustainability competences, including knowledge, attitudes and skills, are promoted across the different typologies of actors involved in the process of co-creating nature-based schoolyards. Results from such analysis will thus elucidate whether and how the co-creation of nature-based schoolyards entails a disruptive initiative in climate change education.

We adapted the GreenComp competences to our study aim focusing on the co-creation of nature-based schoolyards (Table 1).

3 Methods

3.1 Context and Sample

The co-creation of nature-based schoolyards in Barcelona (see Fig. 1) has been put in place by two main programs launched by the Barcelona City Council. One is the 3-year pilot program "Refugis Climatics" (Climate Shelters) that got European financial support through an Urban Innovation Action (UIA) in 2019 aiming to adapt 11 school buildings and schoolyards to climate change. The other is the ongoing "Transformem els Patis" ("We Transform the Schoolyards") program that started in 2020 and focused exclusively on the transformation of schoolyards into more inclusive, gender-balanced, playful, climate-adapted and greener spaces. As of 2024, the initiative reached and transformed 71 schoolyards. Both municipal programs are implemented by the Barcelona Education Consortium (hereafter Consortium), a co-management body formed by the Barcelona City Council and the Catalan government

Table 1 The GreenComp framework' areas, competences and descriptors (extracted from Bianchi et al., 2022, pp. 14–28) and examples of learning outcomes when applied to the co-creation of NbS in schoolyards

Area	Competence	Descriptor	Examples of Knowledge (**K**), Skills (**S**) and Attitudes (**A**) applied to the co-creation of nature-based schoolyards
1. Embodying sustainability values	1.1 Valuing sustainability	To reflect on personal values; identify and explain how values vary among people and over time, while critically evaluating how they align with sustainability values	**S:** can articulate and negotiate sustainability values in the co-creation process, principles and objectives while recognising different viewpoints; **A:** is prone to acting in line with values and principles for sustainability during the co-creation
	1.2 Supporting fairness	To support equity and justice for current and future generations and learn from previous generations for sustainability	**A:** is committed to address and respect the interests of future generations during the co-creation
	1.3. Promoting nature	To acknowledge that humans are part of nature; and to respect the needs and rights of other species and of nature itself in order to restore and regenerate healthy and resilient ecosystems	**A:** considers the need for a harmonious relationship between nature and humans when participating in the co-creation of the schoolyard
2. Embracing complexity in sustainability	2.1 Systems thinking	To approach a sustainability problem from all sides; to consider time, space and context in order to understand how elements interact within and between systems	**K:** knows that the co-creation of the nature-based schoolyard has environmental, social and cultural impacts; **A:** is concerned about the short- and long-term impacts of the co-creation

(continued)

Table 1 (continued)

Area	Competence	Descriptor	Examples of Knowledge (**K**), Skills (**S**) and Attitudes (**A**) applied to the co-creation of nature-based schoolyards
	2.2 Critical thinking	To assess information and arguments, identify assumptions, challenge the status quo, and reflect on how personal, social and cultural backgrounds influence thinking and conclusions	**S**: can analyse and assess arguments, ideas, actions and scenarios to determine whether they are in line with evidence and values in terms of the sustainability of the co-created nature-based schoolyard
	2.3 Problem framing	To formulate current or potential challenges as a sustainability problem in terms of difficulty, people involved, time and geographical scope, in order to identify suitable approaches to anticipating and preventing problems, and to mitigating and adapting to already existing problems	**K:** knows that to identify fair and inclusive actions, it is necessary to look at sustainability problems in the schoolyard from multiple stakeholder perspectives
3. Envisioning sustainable futures	3.1 Futures literacy	To envision alternative sustainable futures by imagining and developing alternative scenarios and identifying the steps needed to achieve a preferred sustainable future	**S:** can envisage alternative futures for the schoolyard that are grounded in science, creativity and values for sustainability;
	3.2 Adaptability	To manage transitions and challenges in complex sustainability situations and make decisions related to the future in the face of uncertainty, ambiguity and risk	**S:** can take into account local circumstances, or unplanned/unexpected outcomes, during the NbS co-creation; **A:** is willing to discontinue unsustainable practices and try alternative solutions during the NbS co-creation

(continued)

Table 1 (continued)

Area	Competence	Descriptor	Examples of Knowledge (**K**), Skills (**S**) and Attitudes (**A**) applied to the co-creation of nature-based schoolyards
	3.3 Exploratory thinking	To adopt a relational way of thinking by exploring and linking different disciplines, using creativity and experimentation with novel ideas or methods	**K:** knows that sustainability problems related to the NbS must be tackled by combining different disciplines, knowledge cultures and divergent views to initiate systemic change; **A:** is committed to considering sustainability challenges and opportunities in the schoolyard transformation from different angles
4. Acting for sustainability	4.1 Political agency	To navigate the political system, identify political responsibility and accountability for unsustainable behaviour, and demand effective policies for sustainability	**S:** can identify relevant social, political and economic stakeholders in one's own community and region to engage with the NbS co-creation; **A:** demands accountability for unsustainable/non-responsible behaviour or actions related to the co-creation process

(continued)

Table 1 (continued)

Area	Competence	Descriptor	Examples of Knowledge (**K**), Skills (**S**) and Attitudes (**A**) applied to the co-creation of nature-based schoolyards
	4.2 Collective action	To act for change in collaboration with others	**K:** knows how to work with diverse participants to create inclusive visions for a more sustainable schoolyard; **S:** can create transparent, inclusive and community-driven co-design process;
	4.3 Individual initiative	To identify own potential for sustainability and to actively contribute to improving prospects for the community and the planet	**S:** can act promptly, even in the face of uncertainty and unforeseen events, keeping in mind the precautionary principle related to the co-creation of the schoolyard;

that is responsible for managing and maintaining schools in the city. The Consortium also defines the criteria for the public tenders associated with naturalizing schoolyards. Besides, a few nature-based schoolyards in the city have been co-created either through other municipal funding grants such as the Barcelona Pla Clima (2018–2030) providing small amounts for climate adaptation at schools (in these cases architects were hired by the school) or with the construction of the new school buildings. In selecting the schools of our sample we opted for a diversity of co-creation methods and even geographical spread.

3.2 Data Collection and Analysis

We conducted 29 semi-structured interviews and one focus group covering overall 9 primary schools in Barcelona, representing a total reach of approximately 3320 students. The centres were located in the districts of Sant Martí, Sant Andreu, Ciutat Vella, Sants-Montjuïc, Gràcia, Horta-Guinardó and Sarrià-Sant Gervasi. This sample covers from working-class neighbourhoods and multi-ethnic environments to upper-class areas and areas with gentrification dynamics. Among the participants we interviewed teachers, educators and headmasters working at the schools (n = 17), family members (n = 4), architects (n = 3) and governmental officials (n = 5). Three stakeholders in each school community were generally interviewed. This involved, a

Fig. 1 Nature-based schoolyard co-created in Barcelona (Author: I. Ruiz-Mallén)

director, a teacher and a parent, so as to allow for triangulation across stakeholders. Yet, in some schools we could only interview one or two actors. The rest of interviewees were stakeholders with major involvement in the schoolyard transformation process, including representatives of the Barcelona City Council and the Consortium, and architect companies that have contributed to the design of the respective programs. We followed a purposeful sampling of interviewees, who were then asked to refer to other relevant stakeholders, reaching 29 adult informants, 22 female and 7 males. In addition, five children (2 female and 3 male) from School 4 participated in a focus group. The interviews took place mostly in person, over the year 2022 and 2023. All interviews were conducted with interviewees informed consent, and when these were children – with the written consent of their parents.

Participants were asked about the history of the particular intervention or program they were involved with, and their motivation, experience with the participative and implementation stages of the schoolyard transformation and view on its impacts. One of the questions furthermore pursued interviewee's interest in, and experience with, outdoor schooling, prior and after the transformation.

Once data were anonymized we analysed the interviews using the Atlas.ti software and the GreenComp categories See Table 1). This first implied identifying which competences from the GreenComp participants put into practice, or refereed that other stakeholder hold, prior, during and after the co-design of the schoolyard. Second, we identified barriers or challenges as described by participants and analysed their relationship with the promoted competences from the GreenComp. This said, our study is *not* an assessment of stakeholders' competences. Rather, it is an interpretative, yet limited, analysis of the sustainability competences stakeholders put into practice during the co-creation of a natured-based schoolyard as revealed by participants during the interviews. In what follows, we present these results by competence area.

4 Results: Sustainability Competences Identified in the Co-Creation of Nature-Based Schoolyards

4.1 Acting for Sustainability

Before engaging in the co-creation of nature-based schoolyards, many interviewed teachers reported that they were already used to collaborative work with families and students around greening schoolyards so by that time they demonstrated *Collective action* knowledge and skills:

> I remember a couple of years before, when I was starting as headmaster, I asked all the children through a questionnaire around which thing they were missing in their school (…) and one of the things we could do was around greening, so we started to act on this need. (T5S3)

Furthermore, these teachers demonstrated competence in *Individual initiative* skills before participating in the co-creation, because they already anticipated and influenced sustainable changes around implementing actions for greening their schoolyard. In some schools, they organized small green schoolyard interventions with students and architects, and in others there were sustainability school committees setting up science weeks and other sustainability education interventions.

During the co-design meetings, many teachers gave a central role to students in discussions and decisions on schoolyard greening in order to empower them in the long term:

> One of the things we have learned over the years is that if we grownups do something it doesn't last. But if we do it in collaboration with the kids, it will last longer. Because they need to be part of it, so that they took care of it differently. (T12S8)

Interviewed students in one of the schools also referred to their active engagement at this stage of the transformation. In their own voice:

> Before (co-creation), this quiet space did not exist and almost everything was a paved area, until we had the idea of making a more sustainable schoolyard. (Ch1S4)

These students then shared the passion and care they developed for the greened schoolyard with younger schoolmates. In some schools (S2, S5), however, interviewees perceived that children's voices were not sufficiently heard due to a lack of time or poor facilitation, deterring students from reinforcing their competence in *Collective action.*

Teachers involved in the co-design process also demonstrated *Individual initiative skills* as they stepped forward in asking for funding to naturalise the schoolyard, looking for collaboration with architects, getting involved voluntarily, participating in trainings and advising others on how to naturalize the schoolyard based on the needs and limitations identified. In some cases (S1, S8), they even rejected the design proposed by the architects when it did not respond to the schools' needs.

Interviewees also referred to families' *Collective action* and *Individual initiative* skills and attitudes during the co-design particularly in those schools from middle class. In schools located in economically unprivileged neighbourhoods, informants explained the absence of this competence mostly due to low parental involvement associated with language/cultural barriers or economic precarity.

Interestingly, interviewees' perceptions on the way architects and public administration showed the *Collective action* competence varied across schools and type of interventions. In general, in those schools where the construction work was led by architects who had won the public tender for transforming schoolyards in Barcelona, interviewees perceived that the municipality was strongly promoting collaborative work and architects were willing to engage proactively. However, in one of these schools a teacher commented that architects did not seem to have a collaborative attitude:

> We (teachers) have made participatory sessions with students, families and all of us said the same and yet they (architects) did not listen to us, it was strange and very frustrating. (T4S3)

On the contrary, in the school (S4) where the architects were chosen by the school, they were described by teachers as mastering a wider range of skills to create inclusive design processes. These architects supported teachers in the write-up of a funding application for transforming the schoolyard and provided professional expertise to a schoolyard greening design.

Despite the efforts of the municipal administration to use inclusive participatory approaches and methods around schoolyards' transformation, especially in schools located in economically unprivileged areas, collaborative action was sometimes hampered by bureaucratic norms. Some of these schools were nonetheless able to find ways to overcome structural barriers through *Collective action*:

> Once the architects [chosen by the school] started to work, they started to have contact with the people at the Consortium. These people made changes in our plan insisting in keeping the paved court because they said we needed to follow the regulated measurements. Even they called someone in charge of the public-school sports from that entity, and also said the court had to be in that way, even though we do not have any federated sport clubs (…) at the end of the day the green space was so beautifully transformed that it is called a garden instead of a patio. (F3S4)

After the naturalization of the schoolyards in most of the sampled schools, teachers, hand-in-hand with students and families continued to work together in the maintenance of the new green areas, showing *Collective action* and *Individual initiative* skills. Teachers from schools where the transformation was a school' initiative, or where green areas were already part of the school, such competences were also more emphasized in maintaining and expanding the naturalized spaces. In one of them

(S3) for instance, teachers were already involved in watering the plants, and after the transformation they actively looked for more funding to hire a gardener. In another school (S4), teachers, students and families decided upon the norms for taking care of the nature-based schoolyard.

When it comes to maintenance, interviewees acknowledged that collective and individual action is fundamental. The schools that expected that the Consortium would maintain the newly planted vegetation saw much of their plants wither or deteriorate. Whenever the school community (teachers and directors) lacked time and interest in taking care of the new green areas, or in using these spaces for educational purposes, either as part of the curriculum or not, the vegetation in the newly transformed schoolgrounds was not thriving.

4.2 Envisioning Sustainable Futures

Some of the interviewed teachers manifested *Exploratory thinking* skills, knowledge and attitudes by combining different disciplines and angles, and using creativity and novel methods before joining the co-creation of the nature-based schoolyard. For instance, some envisioned and used the schoolyard as an outdoor learning environment and a pedagogic scenario to tackle sustainability challenges through project-based or service-learning methodologies and art-science techniques. One of them explained:

> In our school, we've always conceived that there shouldn't be a separation between the inside and outside, that the outside is a learning space too, it isn't a space where children go out for half an hour, take out their energy and then go back inside, but it needs to be connected. (T13S8)

Teachers made different efforts to integrate the nature-based schoolyard into the curriculum both during the co-creation and afterwards. Some encouraged students to make models of the future schoolyard, calculate the material needed for the construction, participate in debates around the best way to proceed and do research on the soil types present in their schoolyard. Others were willing to use the naturalized schoolyard as a "classroom", acknowledging how the greening broadened outdoor education possibilities and actively searched for training courses to use outdoor learning spaces. In the nature-based schoolyard, some teachers organized debates, discussions, presentations, storytelling, scenic representations and nature observations to work on different subjects:

> We have added the schoolyard as a learning space so that people are aware that you can use it for a particular activity; in doing this we want to give the schoolyard the same importance as any other conventional classroom. (T3S2)

Some teachers achieved to integrate the schoolyard maintenance in the curriculum and learnt how to do it effectively through trial and error:

> In other years, all the classes were doing it [using the schoolyard as a classroom], but this year, specific groups are on it. Because it was very difficult to share the tasks [among all classes]. For instance, to take care of a plant pot between three different groups was almost impossible because some watered 1 day, and then the other shouldn't. And now it works better. (T9S5)

Other interviewees identified limitations associated with using the schoolyard as a learning environment. These were mainly associated with the lack of teachers' training in outdoor education, with teachers' and families' perception about students' attention dispersal outdoors, or the lack of budget to contract more personnel to support teachers:

> If you go to the schoolyard to make a class in which you need that students are focused, that is difficult because there is a lot of stimuli and there is always someone who, unfortunately, relates being outside with recreation, and unconsciously, that is what they do. So, as a teacher, I have to make an extra effort to help them to regulate themselves. (T3S2)

During the co-design teachers manifested skills and attitudes associated with *Exploratory thinking* and *Futures literacy* competences around imagining alternative designs for the schoolyard grounded in values for sustainability or when embracing creative methods and a diversity of disciplines (architecture, ecology, education, arts) to imagine diverse, inclusive and nature-based designs. Notably, the architects hired by the S4 community used non-verbal methods with students to explore their environment in the schoolyard that activated all five senses so that kids could connect with their bodies and express in diverse ways their needs around the desired schoolyard:

> Instead of asking questions, we [architects from S4] became observers to later on reflect that on our schoolyard needs' map. We have a battery of materials to explore the space in a body-driven fashion and through the game we activate all senses in students. With the sight we propose a game telling them for instance: <Look for a place nobody could see you>. And, as they did not find any shaded corner, all ended up in the only box they could found (…). When we activated the touch, we blindfolded them and make them touch different materials and asked them if they like those, and they would say things like <No, this is plastic, no this is too cold> So, the information we gathered emerged from their whole perception, not only their verbal descriptions. (Aq2S4)

A*daptability* skills were also used and applied by members of the City Council in guiding teachers and children to prioritize the elements of the new green schoolyard, for instance, by highlighting the importance of reflecting on the size of the area that needs to be de-paved. This said, not all schools agreed on achieving substantial depavement or physical transformation of their schoolyards. One school community (S1), for example, opted for keeping most of the pavement during the transformation process as it was used for the hockey afterschool activities.

After the transformation, most teachers manifested *Futures literacy* skills when the final design did not match the expectations derived during the co-creation phase. They were able to envision further improvements, such as incrementing the green areas and improving their educational uses.

4.3 Embracing Complexity in Sustainability

Problem framing and *Critical thinking* skills were crucial for the schoolyard co-design. This involved analysing ideas and proposals for the naturalization of the schoolyard, identifying challenges, limitations and opportunities around the co-creation process. The size of the areas that could be de-paved was one of the core issues that were subjects of critical discussion. Yet reducing concrete was subject to regulation-based, budget-based, architectural/technical, and use-related limitations. Safety and the socio-cultural background of the neighbourhood were other concerns that typically emerged in discussions around the barriers to schoolyard greening in the schools. Biodiversity enhancement, diverse play, gender justice and thermal comfort were then put forth as opportunities that could help overcoming some of the limitations listed above in critical discussion around transforming schoolyard during co-design workshops. As narrated by a teacher:

> We are getting to the conclusion that it is a lot around budget and maintenance, which are interrelated, and then also around regulations that you need to have to comply with in terms of security (…) We wanted this [point a pergola] to become green putting at the bottom some plant pots. But they [municipal staff] told us we couldn't do it because kids could have an accident, as we cannot put anything that kids could jump high from. So, you have this kind of limitations that you ended up restraining ideas, because maybe nothing would happen to the kids, but if it will, it is our responsibility. (T3S2)

Students also put the *Problem framing* competence into practice when looking at and thinking about the conflict between those who wanted to keep the pavement to play football and those who preferred greener schoolyards and have more play options by considering and understanding both perspectives:

> The yard used to be just a big football court with basketball baskets, it was like you could only choose between playing or soccer or basketball (…) it was super difficult for children who wanted to walk, talk, or relax or who don't want to play football (…) and there were a lot of problems with how to divide the yard (…) I remember those from 6th grade made a pool about who used the paved court to play football and it was less than a quarter of the students (…) we complained a lot because there was a lot of room to play soccer and there were a lot of people who didn't like soccer and didn't want to play (…) now (after the greening) we have more space to play other things, a schedule to play football specific days and hours, and they (children who like to play football) have half of the court with pavement, but they still complaint. (Ch1S4)

4.4 Embodying Sustainability Values

Interviews clearly show how stakeholders' sustainability values underpin schoolyard transformation actions. Teachers interested in incorporating environmental awareness and biodiversity conservation activities in their daily educational practice revealed that their action was led by their values around the role and meaning of nature that they already hold before the transformation. Some had created a vertical garden in the school, or were active participants of the Barcelona Sustainability Schools network (a collaborative scheme led by the municipality that promote sustainability and green education at schools). Nature-protection values were further manifested by the interviewed municipal representatives, who were continuously striving to encounter responses to existing eco-social emergencies through urbanistic innovations like the "Super blocks" promoting urban greening and pedestrianization in dense areas of the city (GT3G1).

These and other participants also showed skills and attitudes around promoting and taking care of nature during the co-design of the schoolyard. Notably, the architects who undertook the schoolyard transformation in school S4 made a special effort in fostering this competence to students, highlighting the relevance of experiencing natural elements textures, smells and sounds. These architects used hands-on and sensory techniques based on experiential learning with students when analysing children's' needs and preferences for the envisioned schoolyard:

> We built a temporary landscape to potentiate the contact with nature which at that time was completely absent in the schoolyard; so we produced a circuit where they could enter as behaving like animals and walking on four legs. We covered the pavement with different materials: potting soil, fall leaves, stones, tree bark and a roof made of aromatic plats, including vegetation touching their face. It was very powerful since at the end they had to make a drawing and one kid saw himself as a wolf: "I have felt like a wolf". And several kids said the same (…) so later on we included a cabin with natural elements in the final design. (Aq2S4)

This team of architects understood the project, in their words, "from a bio-philia point of view", that is, embracing nature throughout the co-design and construction process so as students *"learn to love the wild, appreciate its beauty and realize the value of these natural elements beyond aesthetical reasons, for instance in the way they contribute to urban biodiversity* (Aq2S4)".

Nonetheless, such approach and competences were rarely observed among the rest of interviewed architects who participated in the design of the other schoolyards. Some family members and teachers that embodied nature-related values, played an important role in fostering the way nature was perceived and placed in these schoolyards. They often defended the reduction of pavements and its replacement by permeable terrains and vegetation, while facing the resistance of specific architect companies that worked with the Consortium and stemmed from a conventional, or catalogue-driven approach to schoolyard transformation Resistance to the reduction of pavements was also found on some occasions within the school community, particularly parents, for example where hockey or ball-based games were a popular afterschool activity (T2S1).

Overall, learning from and in nature continued after the transformation of most schoolyards, whenever teachers undertook efforts to enhance *Promoting nature* competences among students by encouraging them to take care of, and maintain the new green areas:

> We first introduced to students the norms, and we told them: here we are going to work on the respect for all beings, the animals, people, plants, as they give us so much. We work part of the curriculum this way, they plant a seed (…) they see how much care this process needs. By doing it, they enjoy and they are aware. (T1S1)

In a school where students were deeply engaged with the all stages of the nature-based transformation process children showed stewardship attitudes:

> There are kids from 4th and 5th grade that want pull out the mint plants and we tell them not to do so as there will not be none left. (Ch1S4)

This said, in many schools the balance between the children's need to interact with the plants (by touching, or picking them), which may involve vandalism at times, versus the strict protection of all plants so that these could thrive, was thin and something that rested on the very interpretation of the sustainability values competence embodied at the level of each educational space. Such competences were fostered whenever teachers and family members manifested a willingness to continue working towards a greener and more sustainable schoolyard and found joy in collaborating in these new spaces with the kids.

Finally, a few interviewees engaged with matters of *Intergenerational fairness* or manifested the associated competence by explaining how the transformed schoolyard could be beneficial for future students' generations (Aq2S4, T9S5).

5 Discussion and Conclusions

By using the GreenComp as a conceptual and analytical framework, we found a set of sustainability competences that are more emphasized among stakeholders when they are engaged in the co-creation of nature-based schoolyards (Fig. 2).

Competences from the Acting for Sustainability area were the most recalled. In particular, the *Collective action* and *Individual initiative* competences stood

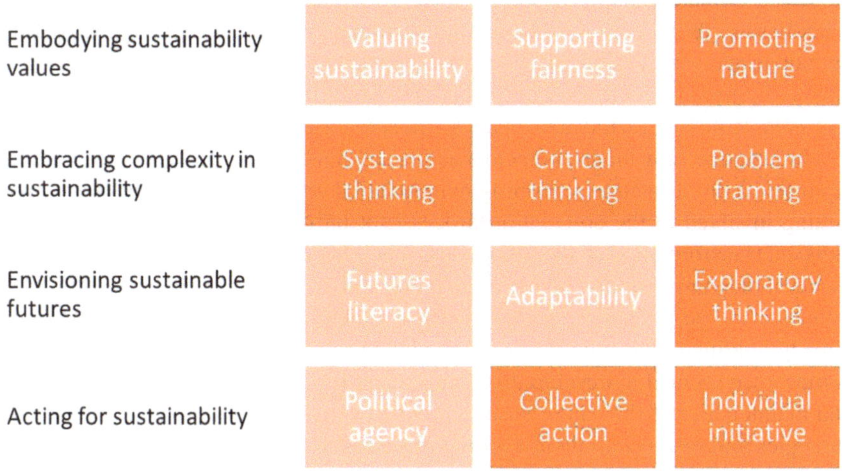

Fig. 2 GreenComp sustainability competences mainly promoted in the process of co-creating nature-based schoolyards (highlighted in orange colour)

out when referring to the skills, knowledge and attitudes that teachers, students, families, architects, municipal technicians put into practice or acquire to mobilize action throughout the co-design of the naturalized schoolyard and its maintenance. Within the Envisioning sustainability futures area, interviewees mainly referred to the *Exploratory thinking* competence mastered by many teachers and some architects, such as being able to combine different disciplines and angles, and using creativity and novel methods during the co-design and maintenance. Regarding the Embracing complexity in sustainability area, interviewees referred to *Systems thinking, Critical thinking* and *Problem framing* competences in an intertwined fashion during the co-design. This puts in evidence, on the one hand, the central role such competences played for stakeholders to understand the schoolyard as a system and consider its elements when participating in the co-creation process; on the other hand, it shows how these competences are challenging to neatly discern between each other in practice. Finally, within the Embodying sustainability values area, the *Promoting nature* competence was highlighted among municipal technicians, teachers, students and many families, as well as some architects. Nonetheless, sustainability competences on *Political agency, Futures literacy, Adaptability, Valuing sustainability* and *Supporting fairness* competences were less evident. Bureaucratic and time limitations, other structural factors (i.e., impossibility to grow trees in the schoolyard due to the existence of an underground gym), lack of actors' previous training and know-how and their resistance to achieve more radical transformations of the schoolyards were the most cited barriers in promoting these sustainability competences as we will discuss here.

Our findings suggest that the process of co-creating nature-based and climate adapted schoolyards can enhance and promote relevant sustainability competences among the different stakeholders engaged in it, and thus facilitating new knowledge production, skills development and values identification on their environment, including climate change. Based on this, we argue that co-creating nature-based schoolyards can be considered a disruptive initiative in climate change education, but not necessarily feasible in all school contexts and not necessarily influencing all involved actors mostly because its potential depends on a series of social, educational and even economic factors and process characteristics.

For instance, how much competent actors are in mastering creativity and interdisciplinary skills, as well as their pro-environmental attitudes and judgements, before getting involved in the co-creation process not only shapes the final result of the physical transformation of the schoolyard. Their ways of acting, and overall assessments of the transformative power of greening and diversifying schools, significantly influence the sustainability competences of others. In this regard, those teachers already holding skills, attitudes and knowledge related to the *Exploratory thinking* competence seemed to develop this competence further as a result of the co-design and maintenance of the nature-based schoolyard by incorporating new pedagogical strategies and means into their teaching practices in a creative and holistic fashion. Teachers in one school included the construction work as part of the objectives of a subject like mathematics. In other schools, teachers increased the use of schoolyards as a classroom setting after the naturalization, taking advantage of the schoolyard as a green and climate adapted area to increase students' contact with nature. This also corresponds with previous evidence indicating how nature-based environments in schools could expand formal education processes (Bohnert et al., 2022).

However, our findings also show that teachers' *Exploratory thinking* competence and other sustainability competences need to be reinforced when they do not feel competent enough or resist using the new green areas for educational purposes, particularly for climate change education. Enhancing these sustainability competences among teachers, along with critical thinking perspectives, is fundamental for guaranteeing and fostering students' sustainability learning and competences (Corres et al., 2020), and particularly those associated with holistic and creative skills as well as managing uncertainty (Corres et al., 2024).

Architects' use of holistic skills associated with the *Exploratory thinking* competence and teachers, families and students' *Collective action* skills, knowledge and attitudes are another precondition for enabling the co-creation of nature-based schoolyards as a disruptive educational initiative. When this happened, students' learning was promoted by using active-oriented competences since they were given more protagonist roles in deciding the future transformation and maintenance of the schoolyard into a greener and more climate-adapted learning space.

Yet, as identified in previous research (Sekulova & Ruiz-Mallén, 2024) bureaucratic and time limitations throughout the co-cocreation process and power unbalances in co-design workshops can deter participants from deciding for more transformational and ambitious interventions, meaning ones that require profound changes in the frames of reference and behaviours related to sustainability and climate change. This, however, ultimate translates in fewer transformative learning opportunities (Mezirow, 1997), particularly related to competences from both Acting for sustainability and Embodying sustainability values areas. As previous studies on the take-up of green schoolyards have reported (Giezen & Pellerey, 2021), the bureaucratic barriers behind the (collective) production of nature-based schoolyards need to be simplified.

Interestingly, our results also suggest that in the schools where teachers, students and families already mastered these sustainability skills, knowledge and attitudes, such barriers forced them to look for alternative ways to green, diversify and transform their schoolyards triggering the use of these competences. Some teachers struggling with the conventional way that schoolyard greening or climate adaptation was enacted, meaning aesthetic and consultative, rather than inclusive, approaches to design, actually enhanced their sustainability competences. This said, it is essential to provide architects with training on issues like children engagement, participation, landscape-based approaches to planning, and the promotion of horizontal and participative processes that rely on creativity as a key sustainability skill. In that way, architects can accompany and guide school communities to carry out the transformation and climate adaptation as a collective process that changes not only the physical conditions but also the social and eco-pedagogical aspects of the school environment. Future WSA implementation should therefore consider how the greening of school environments takes into account the development of sustainability competences among all actors, in order to expand their learning opportunities (Global

Education Monitoring Report Team, 2024). This also aligns with the GreenComp competence approach, designed to promote sustainability learning at all ages and beyond formal education (Bianchi et al., 2022).

As per the *Promotion of nature* competence, when nature-laden values underpinned co-design, final designs and eventual implementation were ambitious and transformative (e.g., emphasizing the reduction of pavement and augmenting wilderness). As a result of that, after the transformation students built upon the knowledge and attitudes associated with this competence, and exhibited care and respect for the rights of nature. Importantly, care-based interaction with nature implies, and enables conceiving "nature as a co-teacher" (Jickling, 2017), which is particularly important for physical and mental health and stress relief (Van Den Bogerd et al., 2023; van Dijk-Wesselius et al., 2018).

Learning opportunities that emerge beyond classroom settings are those that can enable positive disruption in pedagogy, especially when these provide active response to real problems like climate change. These are also capable of creatively reshaping traditional pedagogies and enhancing creativity (Jickling, 2018). In this sense, future efforts in applying the WSA should incorporate the co-creation of nature-based, climate adapted schoolyards as an educational initiative into the curriculum due to its potential in boosting key sustainability competences. In doing this, it is important to overcome the focus on individual responsibility and lifestyle change as primary solutions of the climate crisis as it has been done in many green schools, to further analyse the role of social and economic systems on it and develop political capabilities at the collective level to challenge and transform these systems (Tannock, 2021). Without that critical and transformative focus, the co-creation of nature-based schoolyard cannot become a disruptive educational initiative.

Acknowledgement This work builds on the COOLSCHOOLS project (http://coolschools.eu) and has received funding from the European Union's Horizon 2020 research and innovation programme under grant agreement No 101003758 (NextGenerationEU"/PRTR) and the Spanish Research Agency (PCI2022-132958; MCIN/AEI/10.13039/501100011033).

References

Antoniadis, D., Katsoulas, N., & Kittas, C. (2018). Simulation of schoolyard's microclimate and human thermal comfort under Mediterranean climate conditions: Effects of trees and green structures. *International Journal of Biometeorology, 62*(11), 2025–2036. https://doi.org/10.1007/s00484-018-1612-5

Antoniadis, D., Katsoulas, N., Papanastasiou, D., Christidou, V., & Kittas, C. (2016). Evaluation of thermal perception in schoolyards under Mediterranean climate conditions. *International Journal of Biometeorology, 60*(3). https://doi.org/10.1007/s00484-015-1027-5

Bianchi, G., Pisiotis, U., & Cabrera, G. M. (2022). GreenComp The European sustainability competence framework. *Publications Office of the European Union.* https://doi.org/10.2760/570517

Bohnert, A. M., Nicholson, L. M., Mertz, L., Bates, C. R., & Gerstein, D. E. (2022). Green schoolyard renovations in low-income urban neighborhoods: Benefits to students, schools, and the surrounding community. *American Journal of Community Psychology, 69*(3–4), 463–473. https://doi.org/10.1002/ajcp.12559

Corres, A., Rieckmann, M., Espasa, A., & Ruiz-Mallén, I. (2020). Educators competences in sustainability education: A systematic review of frameworks. *Sustainability, 12*(23), 9858. https://doi.org/10.3390/su12239858

Corres, A., Ruiz-Mallén, I., & Rieckmann, M. (2024). Educators' competences, motivations and teaching challenges faced in education for sustainable development: What are the interlinkages? *Cogent Education, 11*(1), 1–16. https://doi.org/10.1080/2331186X.2024.2302408

Díaz-Barriga, F. (2019). Competencies assessment in higher education: Experiences in the Mexican context. *Revista Iberoamericana de Evaluación Educativa, 12*(2), 49–66. https://doi.org/10.15366/riee2019.12.2.003

European Commission: Directorate-General for Research and Innovation, Andersson, I., Ferreira, I., Arlati, A., Bradley, S., Buijs, A., Caitana, B., Garcia-Mateo, M., Hilding-Hamann, K., Lupp, G., Mahmoud, I., Mateus, A., Canto Moniz, G., Nunes, N., Sejdullahu, I., & Van der Jagt, S. (2023). *Guidelines for co-creation and co-governance of nature-based solutions: Insights form EU-funded projects.* http://op.europa.eu/en/publication-detail/-/publication/dd7b9f43-9a33-11ee-b164-01aa75ed71a1/language-en. Retrieved, November 10th, 2024.

García-Serrano, P., & García, J. A. (2023). Evaluación del impacto del "Modelo heydi-patios que educan" sobre la comunidad escolar del ceip Ramón María Del Valle Inclán. In J. Cáceres Iglesias & K. Gajardo Espinoza (Eds.), *Soñar grande es soñar juntas. En busca de una educación crítica e inclusiva* (pp. 1075–1092).

Giezen, M., & Pellerey, V. (2021). Renaturing the city: Factors contributing to upscaling green schoolyards in Amsterdam and The Hague. *Urban Forestry & Urban Greening, 63*, 127190. https://doi.org/10.1016/j.ufug.2021.127190

Global Education Monitoring Report Team. (2024). *Education and climate change: Learning to act for people and planet.* UNESCO. https://unesdoc.unesco.org/ark:/48223/pf0000389801?posInSet=29&queryId=0959bc36-fbed-496b-9e9b-76e59fa0b464

Jickling, B. (2017). Education revisited: Creating educational experiences that are held, felt, and disruptive. In S. S. B. Jickling (Ed.), *Post-sustainability and environmental education* (pp. 15–30). Palgrave Macmillan.

Jickling, B. (2018). On education. The Crex Crex collective. In B. Jickling, S. Blenkinsop, N. Timmerman, & M. De Danann Sitka-Sage (Eds.), *Wild pedagogies* (pp. 63–76). Palgrave Macmillan. https://doi.org/10.1007/978-3-319-90176-3_4

Mezirow, J. (1997). Transformative learning: Theory to practice. *New Directions for Adult and Continuing Education, 74*, 5–12. https://doi.org/10.1002/ace.7401

Rieckmann, M. (2012). Future-oriented higher education: Which key competencies should be fostered through university teaching and learning? *Futures, 44*(2), 127–135. https://doi.org/10.1016/j.futures.2011.09.005

Ruiz-Mallén, I., Baró, F., Satorras, M., Atun, F., Blanc, N., Bortolamiol, S., Casas, L., Clauzel, C., Gallez, E., Grand-Meyer, E., Gras-Velázquez, A., Ivanova, I., Nawrot, T., Presser, P., Reckien, D., & Sekulova, F. (2023). Nature-based solutions for climate adaptation in school environments: An interdisciplinary assessment framework. In Z. llam (Ed.), *Sustainable urban transitions. Urban sustainability* (pp. 87–105). Springer. https://doi.org/10.1007/978-981-99-2695-4_6

Sekulova, F., & Ruiz-Mallén, I. (2024). The governance configurations of green schoolyards. *Environmental Science & Policy, 156*, 103752. https://doi.org/10.1016/j.envsci.2024.103752

Tannock, S. (2021). *Educating for radical social transformation in the climate crisis.* Palgrave Macmillan. https://doi.org/10.1007/978-3-030-83000-7

Van Den Bogerd, N., Hovinga, D., Hiemstra, J. A., & Maas, J. (2023). The potential of green schoolyards for healthy child development: A conceptual framework. *Forests, 14*(4), 660. https://doi.org/10.3390/f14040660

van Dijk-Wesselius, J. E., Maas, J., Hovinga, D., van Vugt, M., & van den Berg, A. E. (2018). The impact of greening schoolyards on the appreciation, and physical, cognitive and social-emotional well-being of schoolchildren: A prospective intervention study. *Landscape and Urban Planning, 180*, 15–26. https://doi.org/10.1016/j.landurbplan.2018.08.003

ESenRED, Organizing Schools Towards Education for Climate Emergency

José Manuel Gutiérrez Bastida and José Ignacio de Guzmán Alonso

Abstract ESenRED (Escuelas hacia la Sostenibilidad en Red/Schools towards Sustainability Network) was created in 2011 as a network of networks of non-university educational institutions in Spain that works to make environmental education and sustainability the cornerstones of their pedagogical projects. This network unites networks of initiatives of public administrations (Autonomous Communities, City Councils, Provincial Councils, etc.) and it is organised as a community of practices based on cooperative networking to promote the culture of sustainability in educational centres, enhancing the leading role of students via different actions and programmes. Its main actions are the seminar of technical staff of the networks, the teachers' symposium and the Let's Take Care of the Planet Youth Conference (Confint). The main keys to these actions are: working with real and local eco-social problems linked to the climate emergency; the leading role of students in the teaching-learning process and their eco-social empowerment; the implementation of participatory research-action projects; and real and direct intervention in eco-social situations in the immediate school or local surroundings, especially for climate change mitigation, adaptation or communication actions.

1 Introduction

The urgent response to the climate emergency, the transition towards socially just and ecologically sustainable horizons, requires people who are trained to provide the necessary reply. This transition can only come from people who are ethically and culturally educated in an eco-social approach which means human beens who deeply value life and consider it in every decision, who are aware of their dependence in

J. M. Gutiérrez Bastida (✉) · J. I. de Guzmán Alonso
ESenRED, Segovia, Spain
e-mail: josemanu.gutierrez@gmail.com

J. I. de Guzmán Alonso
e-mail: jigdeguzman@gmail.com

J. J. Conde et al. (eds.), *Practices in Education for Climate Emergency*,
https://doi.org/10.1007/978-981-92-2304-6_10

human relations, both with other forms of life and with the inanimate environment, thus promoting a sustainable future.

It is in this context that ESenRED (Escuelas hacia la Sostenibilidad en Red/ Schools towards Sustainability Network) is organized as a network of networks of schools towards sustainability that works innovatively in eco-social education with the aim of empowering students to freely become active agents of social change.

In the following sub-sections, we describe some of the most important activities of ESenRED's annual plan – the seminar and the symposium – while dedicating a separate section, due to its relevance in the context of this publication, to The International Youth Conference: Let's Take Care of the Planet.

2 ESenRED, the Network of Networks

In 2011, ESenRED was created as a network of networks that integrate non-university educational institutions that choose to make environmental education the core of the school's educational project. These programmes are created and promoted in a decentralised way by public administrations with competence at different geographical scales. Thus, we find networks coordinated by town councils, provincial councils and autonomous communities, which are the competent authorities in the field of formal education.

ESenRED currently brings together 17 networks (from the autonomous communities of Andalusia, Asturias, Castilla-León, Catalonia -which brings together 20 municipalities-, Extremadura, Balearic Islands, Canary Islands, La Rioja, Madrid, Murcia, Valencia, Basque Country; the autonomous city of Ceuta, and the provinces of Albacete and Palencia) in whose projects a total of 4200 centres and more than 1,500,000 students participate (Fig. 1).

ESenRED is presented as a community of practice that generates cooperative work in networks to develop a culture of sustainability in school communities from an eco-social perspective, which emphasizes the role of students. All the networks that constitute ESenRED assume to act and take responsibility in terms of equality through collaborative actions, reflection and permanent practice, in order to improve the programmes, which has led to significant progress in their educational approaches (CENEAM, 2019).

Environmental education programmes are configured as the educational response that public administrations offer to the demand that society formulates in different ways due to the socio-ecological challenges currently facing the world in its globality and citizenship in their locality. The ESenRED networks consider that this educational response involves empowering young people so that, as citizens of both the present and the future, they assume responsibilities with regard to the social and ecological crisis affecting the planet. To this purpose, it is necessary to develop school structures and eco-socially meaningful teaching-learning proposals that, collaboratively, help the climate transition of their schools and their local communities towards socially fairer and ecologically more sustainable societies.

Fig. 1 Puzzle of the different networks that make up ESenRED. Photograph by the authors

Four are the objectives of the network of networks (CENEAM, 2019):

1. To facilitate the meeting, exchange, collaboration and dissemination between the different networks of actions, resources, materials and ideas.
2. To promote reflection, evaluation and innovation on one's own practice in order to collaboratively build knowledge in reference models.
3. To develop common or shared projects by the networks that seek the permanent improvement of the students' competence learning, through their participative protagonism, as well as the permanent improvement of the teachers' professional competence.
4. Establish links, relationships and common projects with other international networks of schools towards sustainability.

The fundamental contribution of ESenRED, a consequence of the systematic work on reflection, innovation and evaluation of practices aimed to build reference models, has been to promote the evolution of the basic approach of the programmes of the different networks towards the ecosocial paradigm.

3 Commitment for Eco-Social Education in ESenRED

ESenRED declares that human societies are part of the natural environment in which they operate and can in no way be understood outside of it. Furthermore, the ecosocial vision situates life, its care and protection, as the centre around which humanity must perform its activities. The ecosocial paradigm integrates the awareness of people's essential eco-dependence on the environment in which they live. Life is also understood as interdependent. This interdependence leads human beings to assess that the care received and given by others is fundamental for the development of the life of each individual person.

Educationally, ESenRED fosters the integration of thinking and living the complexity, of admitting and managing the uncertainty and foreseeing future scenarios. It stimulates the feeling of indignation with social injustices, climate denialism and ecological depletion and, with the same intensity, intragenerational and intergenerational solidarity. To this aim, ESenRED encourages educational processes aimed to assume responsibility, both in reflection and in action, in order to collaborate in improving or overcoming the crisis. This approach has the students as the protagonist of their learning, which is, placed in community-relevant situations, enterprising and collaborative. It also responds to the challenges of educative communication.

4 Impact of the Activity of the ESenRED

ESenRED has been configured as a space for meeting, exchange and diffusion that has allowed the evolution – for the better, we believe – of the different programmes it integrates. The networks have evolved within the constantly changing educational and environmental world. As networks with programmes for formal education, ESenRED has experienced the regulatory changes derived from the successive national educational laws (LOE, 2006; LOMCE, 2013; LOMLOE, 2020), in which the transformation of the European educational framework and the formulation of the GreenComp (Bianchi et al., 2022) have been transcended, and as environmental education proposals it has been in harmony with the programmes of the United Nations, the Sustainable Development Goals and Agenda 2030.

Before ESenRED, the programmes of the networks presented serious differences in their approaches. They ranged from being aimed at promoting changes in individual habits, to those dedicated to the understanding and conservation of natural spaces, as well as those based on the paradigm of sustainability, in dialogue and critical contrast with the canon of sustainable development.

In their evolution, the programmes have recognised the central role of students, who have gone from participating in the educational proposal to acquiring a leading role in it. They have also encouraged the creation of local networks as promoters of educational action. These networks included, in the early days, schools whose students began to work collaboratively and, subsequently, have gradually incorporated various social agents. This structure has made it possible for students to take on responsibilities beyond the school walls and to become involved in dialogue and cooperation with municipal political authorities. The concepts of global citizenship, climate emergency and competence for sustainability have found spaces for development in these educational actions and projects.

The concern about the educational coherence of schools, which have become real spaces for pupils to take responsibility about, has led the programmes to give a significant role to the structures for the participation of students and to systematise student decision-making in educational projects and in the environmental management of the schools themselves. This systematic openness to pupil participation in the projects, but also to other social actors, has made it necessary to study in depth the characteristics of the functioning of the structures of participation and, consequently, to provide those who drive and participate in these structures with tools suitable to their psycho-social development.

In this way, participation and the needs of the teaching-learning process have revealed the inexcusable need to strengthen the role of educommunication as a learning tool through interaction between those who participate in school projects and the communities on which they are based. Likewise, the centrality of the students as subjects of their education has led to a deeper understanding of the task of peer mediation in the learning sequences, especially those in which students are called upon to make decisions in the projects in which they participate.

Subsequently, the partner schools, aware of the educational value of the transformations in their organization, in the didactic approaches and in the learning results from the environmental projects, demanded a social recognition of the quality of these projects. The construction of this award, of its criteria and evidence, has become a powerful tool both for the consolidation of the projects in the schools, as well as for the motivation to participate for other educational institutions. The educational quality recognition awards have made it possible to form nodes in local networks that contribute, through peer-to-peer learning, to improving the quality of educational projects.

In another order of things, the continuous improvement of teaching competence in environmental and climate education has also been a constant concern of ESenRED. To meet this objective, collaborative strategies have been developed between the different networks for the design of training itineraries oriented to their application in the realization of school projects. Within these approaches, the proposals articulated around peer training stand out, where significant experiences are shared that concretely embody the pedagogical propositions of ecosocial education. The number of teachers involve in the training activities has reached very significant levels. Consequently, the need to answer to this request has improved the level of preparation of the staff leading the programs.

ESenRED is a space for cooperation between networks that, based on the knowledge of the possibilities of collaborative work, seeks to establish contacts, relationships and common projects with other entities and networks dedicated to education for sustainability. Since the beginning, ESenRED has had the support of the National

Centre for Environmental Education (CENEAM), dependent on the Ministry for Ecological Transition and the Demographic Challenge, which offers its spaces for the coordination and materialization of lines of work. Likewise, collaboration is maintained with the General Subdirectorate of Territorial Cooperation and Educational Innovation of the Ministry of Education, Vocational Training and Sports. A notable aspect of the relationship with the Ministry of Education has been the contribution to the curricular development decrees of the LOMLOE (2020), which includes elements of ecosocial education, both in basic knowledge and in the evaluation criteria of the areas.

4.1 ESenRED Seminar

Within the framework of the Working Groups supported by CENEAM (Valsaín, Segovia), the School Networks for Sustainability seminar has been held since 2010. It is the meeting space in which, in 2011 and at the proposal of the recently created Xarxa d'Escoles per a la Sostenibilitat de Catalunya (XESC), ESenRED was founded, defined its composition, objectives and first annual plan. From 2012, this meeting place is called the ESenRED Seminar.

This seminar is a collaborative work space where the technical staffs in charge of promoting the environmental education programs of their respective public entities, in busy working days, coordinate the joint progress of ESenRED. The seminar also becomes a formal and informal welcoming opportunity for those who join from new networks or for new people who join the coordination of ESenRED.

The seminar has consolidated a defined structure:

- Updating information on participating networks, which serves to learn about and become familiar with their characteristics, circumstances and latest progress.
- Evaluation of the previous year's action plan, including lessons learned and proposals for improvement.
- Reflection and training on current challenges, opportunities in sight or events in the near future that arise in the field of formal education or in environmental education with an eco-social approach.
- Drafting of the annual ESenRED work plan, which seeks to achieve the four objectives through actions ad hoc.

The responsibility for specific coordination and collaboration in the development of the activities of the annual plan is voluntarily assumed by the participating networks.

4.2 ESenRED Teachers Symposium

This is one of the most important formative event in the annual plan. In the CENEAM facilities, with a collaborative methodology and eco-social approach, a space of training, meeting, reflection and exchange of ideas and experiences is built by the teachers joined in ESenRED. This activity allows them to improve daily practice, their teaching competence and to foster their proposals in environmental education. It is also intended that the formative space can facilitate the generation of new educational projects between different schools.

At the same time, the Symposium serves to learn more about initiatives, innovative projects and new theoretical frameworks that bring to a deeper understanding of eco-social education projects.

The Symposium is organized in the following way:

1. Plenary sessions: In addition to an initial welcome from ESenRED and the presentation of its main objectives and projects, there is a main presentation, which marks the axis of the symposium, in the format of a dialogue with experts in environmental and climate education to improve the understanding of ecosocial education. In plenary sessions too, the main ideas and conclusions of the exchange sessions and workshops carried out are also shared, as well as their assessment.
2. Exchange sessions: where the teachers of ESenRED share the specific experience that they contribute to the symposium about the environmental project of their educational institution. These sessions are articulated by significant axes (creativity and communication, care for life in the school institution, community and participation in change, ecodependence and interdependence…) and are moderated by facilitators who direct them to make explicit the most notable elements of each presentation.
3. Workshops: active spaces in order to provide tools to implement environmental education proposals: design of learning situations, development of educommunication strategies, service-learning projects, response to climate change, food sovereignty, youth leadership and participation….
4. Informal meetings: unstructured situations of exchanges and interrelation between teachers, aimed at group consolidation and possible inter-school projects.

The experiences and materials presented in the symposium remain open available to all interested persons (Fig. 2).

The assessment and demand of ESenRED teachers, in addition to its significant training potential, has led to this action being organized annually, although it was initially planned to be held every 2 years.

5 The International Youth Conference: Let's Take Care of the Planet

In the eco-social curricular praxis of ESenRED, Confint, the International Youth Conference: Let's take care of the planet, deserves a special place. A process that arose in the spirit of the environmental education program *Vamos a cuidar do Brasil*

Fig. 2 ESenRED 2022 teachers symposium. Exchange of experiences. Photograph by the authors

(Let's take care of Brazil), in the implementation of which the School Conference of Youth for the Environment had a special relevance. A proposal for a dialogic, co-educational and intergenerational approach that embraces the diversity of natural life, culture, social justice, ethnicity and plurality. These were the first steps that later included the dimensions of learning and policies for sustainability.

In 2003, the I National Conference of Youth for the Environment, promoted by the Ministry of the Environment in partnership with the Brazilian Ministry of Education, transformed thousands of schools all over the country into spaces for ecosocial mobilization that gave young people, teachers and educational communities the opportunity to meet, express their opinions and prioritize their suggestions and actions on how to take care of Brazil (Trajber, 2004).

5.1 A Bit of History

One day in 2003, Marina Silva, Brazil's Minister of the Environment, was at home studying the project of the first Environmental Conference. Her daughter Moara wanted to know what it was that was stealing her attention so much. Marina explained that it was a Conference where people spoke and contributed their ideas to influence public policies to improve the environment. Moara is said to have asked, "And girls like me, do we get a say too?". Sensing that youth would have a lot to contribute, the Ministry of Environment launched a creative idea to regenerate the environment: the action of young people.

As a result of this transformative initiative, the environmental education program *Vamos a cuidar do Brasil* was implemented and three national conferences were held to give a voice to Brazil's students (2003, 2006 and 2009), involving more than 20,000 schools and more than 8,000,000 students.

From 2003 to 2018, more than 20 million people participated in the five editions of the conference. Among them, students from 11 to 14 years old (as representatives); young people from 16 to 29 years old (as motivators, facilitators and workshop leaders); teachers, school communities and public authorities in the areas of education and environment.

During the second Brazilian National Conference, in 2006, international observers – among them Edith Sizoo (coordinator of the Charter of Human Responsibilities network) – pointed out to the organizing team that Brazil should share this experience with the world. In view of this success and international pressure, in 2009, the Brazilian Ministries of Education and Environment issued an invitation to all countries to participate in the International Children and Youth Conference: Let's take care of the planet (Confint), to be held in Brasilia from June 5 to 10, 2010 (Gutiérrez-Bastida, 2012).

In attendance to this invitation, over 400 students from 53 countries met in Brasilia. Each of them had been selected by their schoolmates in the earlier stages of the process, which the organization itself structures in successive levels: school, regional, national, and continental, leading up to the international encounter. These young people had previously completed their corresponding school, national and continental Confint. In Spain, 57 schools from Aragon, Galicia, the Basque Country and Catalonia participated in different encounters. Finally, a delegation of 8 boys and girls represented them in Brasilia, where they had the opportunity to exchange experiences and renew their responsibility for sustainability. They also appealed to the commitment of public representatives to take measures in this regard and to encourage the creation of participatory forums for young people to show that they are also active agents of change and social transformation. Furthermore, they took part in the drafting of the Charter of Youth Responsibilities, which they composed under the slogan "If not now, when? If not us, then who?

Based on the experience in Brazil, the Confint was successfully integrated into the programs of different Spanish networks and was the spark that generated ESenRED. To date, there have been thousands of school conferences, dozens of regional conferences, six state Confint (Vitoria-Gasteiz, 2012; Barcelona, 2014; Logroño, 2016; Alcaraz, 2018; Guadarrama, 2021; Granada, 2023) and three European conferences (Brussels Committee of the Regions, 2014 and 2016 and Lisbon 2018).

Rachel Trajber was the coordinator of Confint in Brasilia. In a recent paper, she stated:

> At the turning of the decade, Brazil has experienced a sad situation. There have been years of total destruction and I doubt about the reversibility of rights, of laws, of everything that has been razed to the ground. In this period there have been no school conferences. In 2023, President Lula was democratically re-elected. The country has begun to rebuild its path towards sustainability and resilience. The Minister of Environment and Climate Change,

> Marina Silva, has laid the groundwork for the 6th National Children and Youth Conference for the Environment to be held in 2024. (Trajber, 2024)

In 2025, the conference will mobilise children and young people to act for climate justice, improve the resilience of schools and communities to climate change and contribute to a better quality of life in schools. The event is to coincide with Brazil's preparation for COP30, to be held in Belém (PA) in November 2025.

5.2 *Why Confint?*

The first reason is the global ecosocial crisis itself, which threatens many forms of life, including human life.

The second justification could be the idea raised by Moara, Marina's daughter, today a Brazilian reference in the struggle for gender equality. The curiosity and interest of young people to participate in 'adult debates' on the environment requires a space in which they can interact from the very beginning, where they can develop and learn from their own particular vision and needs.

And thirdly, because different legal agreements respect and encourage children's participation in everyday life.

Every year on 20 November, the 1959 Declaration of the Rights of the Child is commemorated, as well as the 1989 Convention on the Rights of the Child, treaties which recognise, among other rights, the right of children to express their opinion and to be heard, and to participate freely in cultural life. The aforementioned Convention is of particular relevance, as it gives full rights to all persons under 18 years of age. For the first time and in a global manner, the full citizenship of children was recognised.

On its part, the EU Strategy on the Rights of the Child (European Commision, 2021), in its thematic area on EU measures to empower children to be citizens and active members of democratic societies, specifies: 'Children [...] shall be able to express their views freely. Their views shall be taken into account in matters affecting them, in accordance with their age and maturity'. And the ecological and social crisis situation affects them fully.

To this end, the Strategy, among other aspects, proposes a series of specific actions distributed in six priority thematic areas:

- participation in political and democratic life.
- socio-economic inclusion, health and education;
- combating violence against children and ensuring child protection;
- child-friendly justice;
- digital and information society;
- global dimension: supporting, protecting and empowering children around the world, including during crises and conflicts.

And last but not least, the very importance of child and youth participation, which is now a key element of the ecological, social and educational environment. Although, as mentioned above, it is built on a solid legislative basis that recognises and promotes it, it is essential that it ceases to be a well-intentioned statement and becomes part of everyday life, enabling students to exercise their civil rights with maximum autonomy. Participation is a process, not a single event, and must satisfy several conditions and requirements: transparent and informed, voluntary, respectful, appropriate, child-friendly and inclusive. Participation is learned through participation in the community.

Participation is a formative process and, therefore, it must be accompanied by people who share their experiences and opinions, and by others who promote and care for the fulfilment of the necessary conditions. Participation turns learning scenarios into educational and transformative ones where the learner grows in autonomy and control (Novella et al., 2022). Hart (1997, p. 16) adds:

> We need fewer clichéd examples of children saying or singing that they are the future or that they understand global environmental problems better than anyone else, and more models that truly recognise their capacity to play a meaningful role in community-based sustainable development, especially through collaboration with adults. (…) In addition, children are more receptive to change and less integrated into the existing economic system and social order.

Confint is an authentic model that not only takes participation as a sine qua non of the process but is a platform for the empowerment of students.

5.3 *What is Confint and What Are the Reasons for its Success?*

Confint is a multidimensional educational process that:

- With regard to the eco-social crisis, invites youth to reflect and act to improve the eco-social matters of the planet, to generate commitment and responsibility.
- It links the environmental policy dimension to formal education, as it mobilises and engages young people to analyse and debate, with the rest of the educational community, the main current global and local ecological and social challenges.
- From a social point of view, it offers an interactive process carried out by young people from different localities, regions, countries and continents, who learn and act together for a common goal: to take care of the planet.
- From an educational approach, Confint is a process of learning construction in which young people meet, deliberate about important eco-social problems, assume responsibilities, develop actions to face the selected problems and elect representatives who will communicate the agreed ideas in different spaces of exchange.

In short, Confint is a participatory action-research process that calls for personal and collective action by students, locally and globally, in the fight against the eco-social crisis. Confint is founded on four methodological and ethical principles:

1. Environmental education and planetary ethics. The aim is to constitute an environmental education process based on ethical reflection and action.
2. Responsibility. This is the key word of the process, and it is based on its etymological origin, *respondere*, which means to promise, deserve or pay, or *responsum*, who is obliged to answer for something or someone. According to this other approach, individual responsibility appears from the moment the person becomes aware of the link between his or her way of life and its consequences on nature or on the concrete living conditions of other human beings (notion of space) or over future generations (notion of time). The recognition of individual and collective responsibilities or commitments in the face of the global problem is the key to the proposal. In any case, it is important to bear in mind that the responsibility of each citizen is differentiated: those who are jointly responsible for the global eco-social crisis are responsible for resolving it within the framework of their competence and possibilities of access to information and power.
3. Coordinated action with youth. It is the principle that recognises the role of youth as a social subject that lives, acts and intervenes already in the present, and not the one that will be able to intervene in the future. It has some special characteristics:

   - Youth chooses youth. In the Conference, young people are the protagonists; they become the centre of decision making, taken by the students themselves and not by third parties' indications.
   - Young people educate young people. Older young people teach younger young people. They prepare the material and guide the dynamics and discussions. This mobilisation process is built up by the students, respecting and trusting their capacity to take on commitments for transforming actions.
   - One generation learns with another. The Conference promotes the alliance between the different generations involved. In environmental education, this characteristic becomes especially important, as it involves innovative concepts that students transmit to their families, immediate environment and teachers. While children and young people easily appropriate transformative ideas, it is up to adults to provide and guarantee the conditions for the necessary changes to take place.

4. Local action on global issues. A common topic was discussed in all the schools during the first Confint: Climate Change. The pupils of each participating school had to choose a responsibility, a commitment in response to global warming, and carry out an action in each school or in the surrounding area.

5.4 *What for?*

The ultimate goals of these long processes acquire global dimensions in the framework of shared collective learning:

- Empowering learners to participate socially in the necessary and urgent eco-social transformations to face the climate emergency.

- Give a voice to millions of young people who have the right to participate in building a sustainable future for their communities and for the planet.
- Create and strengthen in schools spaces to debate the social and environmental problems that affect the community, and to perceive how these issues are interconnected with the rest of the world.
- Promote the exchange between local, regional, national and international experiences that contribute with lines of action for the improvement of serious global eco-social issues, through education and the participation of society.
- Enable as many young people, teachers and communities as possible to become locally involved in planetary commitments, taking responsibility for sustainable societies to disseminate and deepen fundamental issues for the acceptance of diversity, the culture of peace and planetary survival.
- Engage local institutions to work with youth groups to improve the quality of life in their communities.

5.5 *How is it Developed?*

Confint proposes to implement a constructivist process in which people meet, discuss the issues raised and elect representatives to present the ideas they agree and present them to other institutions. This initiative respects and values the opinions of students and promotes awareness that they can take individual and collective responsibility for improving the quality of life both locally and globally. Such projects provide young people with knowledge and skills that enable them to understand the complexity of sustainability challenges, reflect on their responsibilities and take concrete actions.

This approach transforms each school into an active space that debates and enquires the habits of knowledge construction, while allowing for democratic engagement and debate about the global and the local. In this way, the school is conceived as a 'republican space' (based on the idea of *res publica*, a public affair), of lifelong education and for all. As the Conference progresses, the space becomes a Greek agora or even a 'circle of culture', in the manner imagined by the educator Paulo Freire, whose reflections enlighten the process. The simplicity of the proposal stimulates and strengthens public participation in discussions about issues that are normally confined to specialised research or public policy decision-making centres. This action respects and values young people's opinions and promotes the understanding that we can take individual and collective responsibility for improving the quality of life at local and global levels (Gutiérrez-Bastida, 2014).

The process takes place at school, local/regional, national, European and international levels:

5.5.1 School Level

1. Motivation and organisation. A 'core group' is created, composed of the pupils who will lead the project in the school. It can be developed with part of a class group or with the environmental committee or similar, formed by students from several different classes. It can be carried out with motivated or invited students. It should not go beyond 6 or 8 members.
2. Research. This involves analysing the global and local reality with an eco-social vision. The core group detects what problems are present in their surroundings or there is a thematic proposal from the national organisation of the Confint. What causes them? What are their consequences? What is the relationship between these local issues and global ecosocial problems?
3. Planning the Conference. This will involve, on the one hand, the dates, spaces, resources, etc., to be negotiated with the managers of the educational institution. On the other hand, the preparation of the Conference: prior communication, dynamics to be proposed, information to be presented, etc.
4. Lead the Conference, with several details to be taken into account:

   (a) Introduction. Welcoming the participants and explaining the Conference: what it is, what it is for, how it is going to take place, among other aspects.
   (b) Description of the problem. Based on the information gathered by the core group.
   (c) Group dynamics with the purpose of reflection. Different inter-level groups are proposed.

   Where do we want to go? What changes are proposed? How do we imagine the result of our actions? What will allow us to know that we have achieved the change we propose?

   (d) Proposals for action. The proposals of the different groups are presented at the Conference and actions to bring about change are decided democratically.
   (e) We plan and seek alliances. The execution of the actions is planned: who, how and when. Who can help in this process? Which actors (organisations, administrations, other schools…) can be allies and help to achieve the proposed change?
   (f) Election of representatives. Following the premise 'young people choose young people', the delegates who will participate in the regional Conference are democratically elected.
   (g) Evaluation and educommunication. All the necessary aspects of the development of the school Confint are evaluated in order to improve the successive editions. Both the process and the results are communicated via the usual channels of the educational centre with productions from the pupils.

5. Action. Following the Conference, actions are implemented according to the foreseen planning.

6. Evaluation. The results and the process are assessed: How did the course of action go? How was it experienced? Was the intended objective achieved? What can be improved?

5.5.2 Regional, National, European Level…

In this moment of the process the organization corresponds to the people responsible at each level within each regional, national or European administration. The Confint of this scale hosts the representatives elected at the previous level.

Basically, the sequencing of activities is similar, except that now there is the figure of facilitators, young people up to 30 years of age, who dynamise and facilitate the development of group dynamics and decision-making.

These Confints usually have a time for the presentation of participants, another time for the exchange of experiences, for the debate of ideas and suggestions, for the election of proposals and representatives, for educommunication and for the evaluation of the Conference itself.

These Conferences culminate in participation in political decision-making processes on eco-social issues. Confint draws up a document presenting an analysis of the issue, the commitments made and the proposed actions to be taken by the responsible actors. The Confint representatives submit the work carried out, the responsibilities acquired and the proposals to the relevant authorities at the administrative level, and a response is requested.

The different representatives give their experience back to those who have chosen them, to share how far their ideas have gone, how they have been modified, what response has been obtained. Schools make educommunication proposals to inform their educational communities. This learning practice can lead to modifications of the planned actions or even the introduction of new actions in the planned ones (Fig. 3).

5.5.3 Potential for Improvement

Obviously, this splendid process is not without problems. Taking González et al. (2021) as a reference, weaknesses, obstacles or elements for improvement can also be observed:

1. As it is a proposal in which young people from many different schools, localities or countries participate, it is difficult to ensure a similar prior process for all participants. This often complicates an adequate development of the subsequent levels.
2. School conferences require active participation in order to hear the opinions of the largest number of students in the school. The proposed methodology to achieve this sometimes does not fit in with the organization, ideology and functioning of the school.
3. While a representation of young people from all schools participates in the regional conferences, only a representation of them can participate in the national

Fig. 3 Educommunication at the Confint. (Participants' faces have been blurred to protect their privacy) Photograph by the authors

conferences and, therefore, some schools are not represented at the national conference. Thus, the return of the representatives must be well managed so that it reaches all the schools, with or without a representative at the state conference.

4. The different authorities, in their own learning process in relation to participatory processes such as Confint, could become more involved in the educational cause, i.e. the experience for a boy or girl of presenting their ideas to them could become an educational practice in itself: being welcomed in public institutions, seeing their opinions and contributions respected, feeling a commitment and receiving justification for their responses, etc.
5. Greater involvement, collaboration and support between the different departments of Education and Environment or Sustainability would be very welcome

5.5.4 Impact

Nevertheless, the impact of the Confint process is really huge. We can summarise it in these ideas:

1. Confint has given young people a voice and offered them the possibility to participate in building a sustainable future for their schools, communities, cities, regions, countries and for the planet.
2. The experience is imprinted in the memory, most profoundly of course, of those participating at various levels who have had the opportunity to meet young people

from all over the world who share their commitment to walk together towards a more sustainable world. It is a process they remember fondly as one of the first times they have been asked in a serious way what they think about local and global eco-social problems, that they see that their ideas are heard, that they feel they can change some things, etc.

3. Confint empowers youth. Over the years, we have seen how former students that are now more mature young people and feel very much shaped by the experience. It is common to find participants who are now studying degrees related to the ecological crisis, who are activists in social movements, or who are even running for local elections with eco-social proposals. The process makes them aware that each person has to take responsibility, individually or collectively, about the challenges affecting life on the planet and makes them active agents of change.
4. School Confint transforms the educational processes of many schools. The loss of the teacher's protagonism and 'power' has given way to confidence in the students, to developing their autonomy, to learning among equals and unequals, etc. Confint, despite the difficulties of school organization, has been able to change obsolete dynamics thanks to the initiative of its students. Confint has created and strengthened spaces for debate on social and ecological challenges in many educational institutions and has promoted the acquisition of responsibilities and the awareness that the answers to these challenges must be local, with a global vision.
5. Local, regional, state or European administrations have become accustomed to (realy, they have lost their fear of) dealing with students who take on responsibilities, expose their proposals and ask for answers. Many processes have led to changes in municipal budgets and many ideas have been taken into account in other administrations

6 Evaluation and a Glance to the Future

In summary and to conclude, in this context of global ecosocial crisis, there are experiences of environmental education with an ecosocial focus that give the young people of ESenRED a leading role, through the different programmes of each network, by participatory research-action processes that offer spaces for the development of students' competences, as well as for individual and collective, local and global, transforming ecosocial action. However, based on Gutiérrez-Bastida (2022), there is ample room for improvement:

1. Integration. Today, 2025, we find ourselves with a network that lacks autonomous communities, although efforts are still being made in this sense to complete the territorial map.
2. Vulnerability. ESenRED has grown a lot in terms of the number of schools and the complexity of the actions it proposes. This requires a stable organizational structure (technical secretariat) to provide a rapid response to the more bureaucratic

issues related to the network. But on the one hand, the organization basically relies on the voluntarism and commitment of the networks' technical staff and, on the other hand, it depends on administrative dynamics, sometimes linked to the precariousness of some technical jobs and sometimes linked to political and budgetary decisions regarding the different programmes.
3. Polarization. The social and political scenario of growing denialism, hoaxes and vexation of eco-social struggles transforms the former space of security and universal agreements (human rights, SDGs, climate change…) into an insecure and vulnerable territory for environmental education and for those who work in it.
4. Lack of institutionality. ESenRED has no legal entity or budget, which makes it dependent on the different administrations that promote each of the networks. The political ups and downs are risks for ESenRED, as well as the difficulties to reach important and binding agreements.
5. Visibility. Despite its trajectory and achievements, ESenRED's activity is largely unknown in the educational field.

On the positive side, ESenRED's journey has left behind some major milestones:

1. Exchange. The exchange of experiences and learning among peers throughout these years has been immense and has borne great fruit.
2. Union. The fusion of the experience accumulated in each and every one of the networks has generated new spaces for development, growth and deepening of the different projects.
3. Synergy. The symbiosis of the networks is significant for the realization of common projects, hopes, strategies, and the feeling of belonging to a collective with a common objective that supports those who begin.
4. Empowerment. The ability of ESenRED to give voice, offer real decision-making scenarios and provide tools to students, for their protagonism and Freirean empowerment and for their collective transformative action, stands out.
5. Educational impact: The experience of participating in ESenRED transforms both students and schools, promoting a change in the traditional dynamics towards a more participatory and autonomous approach.
6. Glocal action. ESenRED has encouraged each school to have its own project, contextualized at the local level, but with a shared fundamental axis and common lines of action at the global level.
7. Diversity. The diversity of types of networks, approaches and projects is another strong point.
8. Partnerships. Projects shared with ministries and collaborations with other entities (Sense – within the European eTwinning program-, Let's take care of the planet – organization of the European Confint-, #EA26, REDEEA, Unesco…) are evidence of its good work.

The gratitude of teachers and students for the work done so far (training, advice, proposals, etc.) is perhaps the greatest reward that the network of networks can receive. Even so, the recognition of ESenRED's work also comes from multiple activities. On the one hand, ESenRED, since its creation, has participated and has been in demand in national and international environmental education forums, consolidating its work in environmental and ecosocial education. Just to cite a few examples, the seventh World Environmental Education Congress (WEEC, Marrakech, 2013), the III Pan-European Conference on Environmental Education towards Sustainability (Barcelona, 2015), various CONAMA (National Environmental Congress), among others. On the other hand, it has received direct recognitions such as the NAOS (Nutrition, Physical Activity and Obesity Prevention) Strategy Award for special mention 2023 (AESAN, 2023), the III *#EducaciónAmbienta*l #EA26 award (#EA26, 2020), and has been candidate for the UNESCO-Japan Prize for Education for Sustainable Development (SGCTIE, 2021), as some examples.

In addition, young people elected in the different national Confints have been called upon to represent the youth of the networks in various international events such as the reception with Mr. H. Van Rompuy, President of the European Council, Mrs. Mette Plauborg, member of the Climate Action Cabinet of the European Commission, in Brussels, or Mr. Van Ypersele, vice-president of the IPCC (Gutiérrez-Bastida, 2010); at the Rio + 20 Summit (ESenRED, 2012); in several Youth Conference (YouthCore) within the Environmental Conference of the Regions of Europe (ENCORE) (2014, 2016…), among others.

In summary, and also looking to the future, we believe that it is necessary to move towards the integration of the environmental education programmes of the different cities, provinces and autonomous regions of Spain, consolidate a stable structure and move towards its own legal entity, address the situation of social and political polarization, strengthen and deepen the methodologies of student empowerment, and increase visibility and social impact. These ideas and challenges would allow ESenRED, based on its successes, to consolidate itself as an innovative, empowering and resilient educational network, with greater educational and eco-social impact on a local and global scale.

References

#EA26. (2020). *#EA26 Educación Ambiental.* (retreived 8 June 2026). https://educacionambiental26.com/2022/01/27/esenred-iii-premio-educacion-ambiental-ea26/

AESAN. (2023). *XVII* Premios *Estrategia NAOS, edición 2023.* (retreived 8 June 2026). https://www.aesan.gob.es/AECOSAN/docs/documentos/nutricion/premios/2023/ACTA_XVII_PREMIOS_NAOS_web.pdf

Bianchi, G., Pisiotis, U., & Cabrera Giraldez, M. (2022). In M. Bacigalupo & Y. Punie (Eds.), *GreenComp—The European sustainability competence framework.* Publications Office of the European Union. https://joint-research-centre.ec.europa.eu/greencomp-european-sustainability-competence-framework_es

CENEAM. (2019). Qué es ESenRED. Ministerio para la Transición Ecológica y el Reto Demográfico. https://www.miteco.gob.es/es/ceneam/recursos/mini-portales-tematicos/esenred/que-es-esenred.html

European Commision. (2021). EU strategy on the rights of the child and the European child guarantee. Communication from the commission to the European Parliament, the Council, the European Economic and Social Commettee and the Committee of the Regions. EU strategy on the rights of the child. COM (2021) 142. Bruselas. (retreived 8 June 2026). https://eur-lex.europa.eu/legal-content/EN/TXT/?uri=CELEX:52021DC0142

ESenRED. (2012, junio 29). La voz de los y las jóvenes de Europa en Rio + 20. ESenRED. https://confint-esp.blogspot.com/2012/06/la-voz-de-los-jovenes-en-rio-20.html

González, C., Vilar, M., & Pérez, P. (2021). *Conferencia Internacional de Jóvenes: Cuidemos el Planeta (CONFINT).* Ministerio de Educación y Formación Profesional.

Gutiérrez-Bastida, J. M. (2010, enero 10). *La Confint en Bruselas: UE e IPCC.* Ingurugela Montrealen. https://ingurugelamontrealen.blogspot.com/2011/01/la-confint-en-bruselas-ue-e-ipcc.html

Gutiérrez-Bastida, J. M. (2012). La Confint, una conferencia internacional de jóvenes para cuidar el planeta. Boletín del CENEAM, enero.

Gutiérrez-Bastida, J. M. (2014). La Confint y ESenRED: Crisoles de aprendizaje hacia la eco-ciudadania. Boletín del CENEAM, noviembre.

Gutiérrez-Bastida, J. M. (2022). *Escuelas hacia la Sostenibilidad en Red (ESenRED).* Secretaría General Técnica. Centro de Publicaciones. Ministerio de Educación y Formación Profesional. https://esenred.blogspot.com/2019/12/posicionamiento-de-esenred-despues-de.html

Hart, R. A. (1997). Children's participation: The theory and practice of involving young citizens in community development and environmental care. Earthscan.

Novella, A., Ferrús, L., Noguera, E., & Buxarrais, M. R. (2022). *La participación, derecho indiscutible de la infancia.* Fundación La Caixa.

Subdirección General de Cooperación Territorial e Innovación Educativa-SGCTIE. (2021). *Candidatura 2021 de ESenRED al Premio UNESCO-Japón de Educación para el Desarrollo Sostenible.* Subdirección General de Cooperación Territorial e Innovación Educativa-SGCTIE. https://www.educacionfpydeportes.gob.es/mc/sgctie/comunicacion/noticias/2021/junio2021/esenred.html

Trajber, R. (Coord.). (2004). *Formando Com-Vida Comissão do Meio Ambiente e Qualidade de Vida na Escola: construindo Agenda 21 na Escola.* Ministério da Educação, Ministério do Meio Ambiente do Brasil.

Trajber, R. (2024). Al aula… desde la experiencia brasileña: Rachel Trajber. En J. M. Gutiérrez-Bastida y J. I. De Guzmán-Alonso (Coords.), *Educación ecosocial a la luz de los siete saberes de Edgar Morin.*

Lifting Voices and Sharing Promising Practices in Fire Education and Awareness Communications

M. Conceição Colaço, Natasha Caverley, Amy Cardinal Christianson, Briony Towers, and Leigh Johnson

Abstract Alongside air, land, water and space, fire exists in nature. Climate change, socioeconomics, land abandonment and wildfire management are exacerbating the frequency and intensity of wildfires; annually affecting millions of people. One of the first fire education and awareness campaigns was "The Smokey Bear" campaign in the USA which focused on wildfire risk behaviours. Adapted by many countries, this line of thinking resulted in "good fire" (a source of fire stewardship) being banned with strict law enforcement. "The Smokey Bear"-styled campaigns were typically designed by state-led agencies who rarely recognized Indigenous-led fire practices, rural traditional practices, geographic context and socio-economic priorities. While advances in ecology, forestry and risk analysis have shown that this line of thought decreases the number of ignitions, conditions were then created for larger and more extreme wildfires to occur due to fuel build up on the landscape. A shift in fire education is occurring from schools using interdisciplinary methods that engage neighbouring communities to grassroots-led knowledge resource development. This chapter presents five case studies: Indigenous fire stewardship in Canada; wildfire education in Australian schools; Portugal's citizen-led fire prevention exhibition; a European student-teacher training course (Portugal, Spain, Ireland); and FireExplorer, a European platform promoting fire education through cultural, practical, and community-based approaches.

N. Caverley and A. Cardinal Christianson are of Indigenous heritage.

M. C. Colaço (✉)
University of Lisbon, School of Agriculture, CEABN-InBIO, Lisbon, Portugal
e-mail: ccolaco@isa.ulisboa.pt

N. Caverley
Turtle Island Consulting Services Inc., North Saanich, Canada

A. Cardinal Christianson
Indigenous Leadership Initiative, Mashteuiatsh, Canada

B. Towers
Learning Ecologies for Action on DRR and Resilience (Leadrrr), Melbourne, Australia

L. Johnson
Harkaway Primary School, Harkaway, Australia

J. J. Conde et al. (eds.), *Practices in Education for Climate Emergency*,
https://doi.org/10.1007/978-981-92-2304-6_11

1 Introduction

1.1 Fire: A Short Historical Context

"Traditional, rural, native or indigenous fire has coexisted with lightning-caused fire since the origin of Humanity" (de Oliveira et al., 2023). This fire has assisted societies to evolve worldwide through a strong role in land management, allowing humans to protect themselves; care for animals; and plant, grow and harvest foods. At present, the same type of fire, rural and Indigenous, continues to exist by helping to manage the land through fuel load and vegetation reduction and mitigation of wildfire impacts. However, cultural burning and the traditional use of fire undertaken by particular groups (for example, rural community members, Indigenous Peoples) is misunderstood – leading to law enforcement that often criminalizes the actions and actors (for example, Poland, Italy) or limiting the burning with permits (for example, Portugal, Spain, Canada).

Cultural burning has been carried out since time immemorial. In many Indigenous cultures in Canada and abroad, fire is a sacred and powerful element that can help on landscapes and in ceremonies. Cultural burning consists of Indigenous-led fire practices that involve the use of fire on the land to achieve cultural and holistic community objectives; for example, sustaining diverse animal life and plants that serve as food or medicine. Common uses of fire on the land include, but are not limited to, cultural and language revitalization (for example, use of traditional knowledge), fuel mitigation/fuel hazard reduction, food and medicinal plant gathering and access, habitat enhancement (for example, hunting and trapping), forest product access (for example, firewood collection), trail clearing and access, and pest reduction (for example, reduce wood ticks). Cultural burning is community-driven, follows community-specific governance structures and natural law, and involves comprehensive engagement and guidance from Elders or Indigenous fire practitioners, at times, in partnership with interagency collaborators (such as wildfire management agencies and local structural fire departments) (Greenler et al., 2024; Hoffman et al., 2021).

In Europe, the community-driven sacred dimension of fire is present in some village traditions that merge the "pagan with religious" practices like "Ndocciata" in Italy or "Madeiro" in Portugal both on the 24th of December, among many others throughout the year. However, when considering the traditional use of fire that allowed communities to manage the land on a traditional rural system, there has been a gradual loss of this knowledge because of the reduction in knowledge transfer. Nonetheless, there are still some remnants of this past fire use, mainly in pastoral burning and slash and burn of the agriculture and forestry leftovers (de Oliveira et al., 2023).

Indigenous and rural societies often created their own rules and/or were guided by natural laws for living with and alongside fire – passing this knowledge from one generation to another. Studies have shown that with the development/growth of societies and changes in land ownership, conflicts due to the use of fire increased. As a result, fire regulations and related rules (like fire exclusion policies) were created (de Oliveira et al., 2025). For example, in the Iberian Peninsula in Europe, the Roman Empire (from first century BCE to the fourth century CE) created laws that if damage to property occurred due to a wildfire, the perpetrator would be responsible and would have to pay with money, goods or physical punishments. Throughout the centuries, rules and laws evolved embracing not only damage due to fire, but also limitations to fire use. These rules and laws over time created a paradigm of suppressing wildfires, regulating wildland firefighting and promoting wildfire prevention.

Limitations to the use of fire were also brought out through colonialism from Europe to countries on the American (Vinyeta, 2021), African and Australian (Mariani et al., 2022) continents. Indigenous-led fire practices were banned and the idea that all fire should be suppressed prevailed in most of the world until nearly the end of the twentieth century (Donovan & Brown, 2007).

This vision of zero fires was strengthened by strong fire awareness and education campaigns. One of the first fire education and awareness campaigns was "The Smokey Bear" campaign in the USA which focused on wildfire risk behaviours (Vinyeta & Bacon, 2024). Adapted by many countries, this line of thinking resulted in "good fire" (a source of fire stewardship) being banned with strict law enforcement. "The Smokey Bear"-styled campaigns were typically designed by state-led agencies who rarely recognized Indigenous-led fire practices, rural traditional practices, geographic context and socio-economic priorities. European campaigns such as Sylvain in France or "Portugal sem fogos (Portugal without fires)" are further examples of this style of fire awareness and education campaign. Advances in ecology, forestry and risk analysis have shown that the line of thought regarding "zero fires" decreases the number of ignitions. However, conditions are then created for larger and more extreme wildfires to occur due to fuel load build up on the landscape.

1.2 Fire and Global Changes

According to the United Nations Environment Program – UNEP (2022) report "Spreading like Wildfire: The Rising Threat of Extraordinary Landscape Fires", the number of wildfires will rise by 50% by the year 2100, increasing risks even in regions previously not affected by wildfires. Climate change across the globe is increasing the risk of wildfires, which in turn are becoming significant contributors to climate change due to their carbon emissions and the reduction of carbon sink capacity.

Worldwide, the ecological and socio-economic losses and damages due to wildfires are incredibly high and can last for several years. The number of people directly and indirectly affected has exponentially grown in the last three decades (UNDRR, 2023). The 2017 Portuguese wildfires showed how a combination of meteorological events can create a disaster with the largest burned area and the highest number of fatalities ever recorded due to wildfires in Portugal (Turco et al., 2019). In Canada, wildland fire evacuations alone cost CAD $4.6 billion in productivity losses over a 42-year period (Christianson et al., 2024).

Although the reduction of fuel on the landscape can help in wildfire mitigation, studies have shown that this success only occurs if fire weather conditions are not extreme. All these extreme situations are occurring since "fire is changing because we are changing the conditions in which it occurs." (UNEP, 2022). As a consequence, "wildfires are burning longer and hotter in places they have always occurred, and are flaring up in unexpected places too, in drying peatlands and on thawing permafrost." (UNEP, 2022).

A "business as usual" status quo approach to fire education and awareness campaigns is not enough. Therefore, a shift in fire education and awareness communications is occurring using interdisciplinary methods that engage grassroots-led knowledge resource development. These methods seek to change the narrative and paradigms associated with risk education and environmental education with a focus on recognizing that the answers and knowledge are also in the community.

1.3 Creating a Shared Path to Wildfire: Risk and Environmental Education

While there is a shared focus on learning, wildfire risk education and environmental education do not have the same objectives and backgrounds. They are based on different scientific fields, namely disaster risk reduction and forest sciences, respectively.

Risk education aims to increase awareness and preparedness, helping people understand how to respond to and mitigate risks. This can include strategies for disaster preparedness, safety protocols, and understanding emergency responses (Johnson et al., 2014). In contrast, environmental education is centred on understanding the environment, ecosystems, and the relationships between humans and the natural world. It emphasizes long-term ecological understanding and sustainability, promoting environmental actions on a given situation or issue (Sauvé, 2002).

Focusing on wildfires, both approaches are complementary and have important roles to play. Communities in the wildfire interface need to be prepared in the case of wildfires. The preparedness measures to protect houses and livelihoods, protocols for safety evacuation, understanding risk alerts and knowing what to do in case of an extreme wildfire is the central focus of risk education. On the other hand, creating knowledge on the relation of nature-humans; understanding landscapes and ecosystems specificities; co-creating resilient landscapes and the methods to achieve them are in the field of environmental education. Encouraging communities to better understand wildfire by co-developing local collaborative-based solutions can promote actions towards safer and more resilient societies to wildfires.

In the following sections of this book chapter, five different cases are presented from three continents: North America, Europe, and Australia. Each case study will provide a narrative on promising practices and related reflections on fire education and awareness communications.

2 Case Study #1: Canada – Blazing the Trail: IFS and Knowledge Resource Development in Canada

2.1 Background

Indigenous Peoples across the world are diverse in terms of geography, languages and cultural practices (for example, uses of fire on the land). However, there are universal connections that transcend Indigenous worldviews, unite identities and are sources of strength such as: Indigenous title and inherent rights; relationships between land, water, animals, place and people; truth and reconciliation (recognition of the living history of Indigenous Peoples in relation to settler-colonialism); wholeness and interconnectedness (mind, body, spirit and emotions); and kinship, the role of families and communities.

In many Indigenous cultures across Canada, fire is a sacred and powerful element that can help in ceremony and on landscapes. But fire in the wrong place at the wrong time on the land can be destructive to communities. As Canada continues to experience more intense wildfire seasons each year, there is a disproportionate number of Indigenous communities over the last decades that have been impacted by wildfires. Of note, 80% of Indigenous communities in Canada are in forested regions and therefore are at risk from wildfires (McGee & Christianson, 2021). While it is not uncommon for wildfire agencies to prepare wildfire risk communication resources for Indigenous populations, Indigenous Peoples and communities are rarely involved in authentic engagement and collaboration in fire education and awareness communications from design concept to resource development. There is an increased urgency to recognize the unique voices, perspectives and priorities of Indigenous communities in reducing wildfire risks while healing the land. This includes the recognition of Indigenous fire stewardship (IFS).

Indigenous fire stewardship is a strengths-based, rights-based and complementary pathway to recognize Indigenous self-determination and empowerment in uses of fire on the land. Indigenous fire stewardship (IFS) supports Indigenous self-determination by using fire for ecological, cultural, and community benefits. It involves sharing traditional knowledge and practices to improve the environment, enhance safety and resilience, and strengthen cultural connections across generations. IFS helps maintain healthy, fire-resilient ecosystems in Canada and other countries, using Indigenous ecological knowledge.

FireSmart[1] Canada has produced various publications for segments of the population at risk to wildland fire, including municipalities, communities, homeowners, farm and acreage owners, and the oil and gas industry. However, prior to the launch of the *Blazing the Trail* book (2020), there were no FireSmart Canada publications developed by and for Indigenous communities.

[1] FireSmart consists of evidence-based guidelines and practices that are intended to protect homeowners and communities from wildfires.

From 2018–2020, FireSmart Canada embarked on their first ever IFS knowledge product called, *Blazing the Trail: Celebrating Indigenous Fire Stewardship,*[2] a 50+ page book. Utilizing principles from the "Two-Eyed Seeing" (Bartlett et al., 2012) approach, FireSmart Canada facilitated IFS knowledge resource development which recognizes Indigenous wildfire prevention, risk reduction approaches and wise practices alongside western FireSmart principles. Numerous Indigenous Peoples and groups co-wrote, peer-reviewed and designed *Blazing the Trail.* Key themes explored in *Blazing the Trail* include multiple uses of fire in Indigenous communities in support of "good fires" and dealing with "bad fires," current fire risks and the roles of community leaders, and IFS. *Blazing the Trail* is comprised of detailed, first-person stories that provide insight and context about Indigenous Peoples and fire stewardship, and the narrative encourages celebration and sharing of IFS wise practices and lessons learned across generations. The publication acknowledges that Indigenous communities have been leading, in many ways, fire mitigation and prevention. In 2023, the *Blazing the Trail* book was translated into French. In 2024, *Blazing the Trail* was made accessible online.

2.2 Knowing Your Audiences

As an IFS knowledge resource, the central objective of *Blazing the Trail* was to develop a cohesive, integrated and culturally safe book highlighting Indigenous-informed (First Nations, Métis and Inuit) wise practices and related narratives about IFS. The key audiences for this book are leaders at all levels in Indigenous communities across Canada, from local Youth to Elders, Fire Keepers, Fire Knowledge Holders/Carriers, elected officials, fire chiefs, and senior community managers/administrators. In the spirit of allyship, non-Indigenous agencies that work alongside Indigenous communities across Canada also benefit from *Blazing the Trail.*

2.3 Creating the Writing Circle

A technical writing team was recruited and selected to work together in writing the initial draft of *Blazing the Trail* in a collaborative process, revising the book (based on four rounds of peer review feedback), and preparing a final version of the publication. The technical writing team consisted of 11 people from across Canada – with the team possessing expertise in Indigenous relations and wildland firefighting. The purpose of the technical writing team was to provide cultural and technical expertise in co-developing an IFS book for Indigenous community leaders. Throughout the

---

[2] https://firesmartcanada.ca/wp-content/uploads/2024/06/Booklet_Blazing-the-Trail_Web-version.pdf; accessed June 12, 2026 and https://intellifeucanada.ca/wp-content/uploads/2025/10/Livre_Tracer-la-voie_Web-Version.pdf; accessed June 12, 2026.

writing journey, the technical writing team utilized story(telling), situated themselves and cultures, and engaged in meaning-making pertaining to (re)introducing IFS in Canada.

2.4 *Learning Modalities and the Role of Art in IFS Education and Awareness Communications*

People learn in many different ways. Therefore, *Blazing the Trail* incorporated artwork, photos, technical writing, creative writing and case studies throughout the book acknowledging the significance of storytelling and expressive arts as ways of sharing knowledge about the roles and uses of fire in Indigenous communities across Canada. Expressive arts are just as important as written and oral forms of knowledge sharing. Indigenous illustration and graphic design advisory services were retained to ensure relevant artwork and culturally congruent designs were utilized in *Blazing the Trail* that describes wise practices and related narratives pertaining to IFS.

During the creation of *Blazing the Trail*, FireSmart Canada held a national art contest for Indigenous artists to be featured in the book. Six pieces of art were chosen to be featured, with each artist receiving compensation. Indigenous artists could submit illustrations on various themes related to Indigenous Peoples and wildfire. This national call celebrated the creativity of Indigenous artists from across Canada. As part of ethical and respectful engagement, Indigenous artists provided their biographies, brief narratives about their illustrations and their artistic methods utilized which were recognized in *Blazing the Trail*.

2.5 *Peer Review in IFS Knowledge Resource Development*

A peer review process was developed to obtain broader feedback on various draft versions of *Blazing the Trail* – seeking value-added feedback from existing organizational networks with representatives in identified Indigenous communities and organizations, orders of government (local, provincial, federal), and third parties (for example, not-for-profits, academic institutions) to review the core content and design ("look and feel") of the book – ensuring that the information is current, relevant, accessible, and culturally-informed in nature. Fourteen peer reviewers from across Canada were identified based on respective leadership roles in FireSmart, Indigenous wildfire prevention and risk reduction wise practices, and corresponding skills and subject matter expertise related to IFS. Four rounds of peer review were carried out during the IFS knowledge resource development journey.

Key outcomes from the *Blazing the Trail* knowledge resource journey were promoting greater awareness and sharing of IFS; and highlighting wise practices, case studies, approaches and lessons learned that highlighted Indigenous communities' efforts in IFS at the local and regional levels. To date, there remains high interest to have more IFS knowledge resources available across Canada which use a similar Indigenous-led and informed design and review process as the *Blazing the Trail* publication.

2.6 Wise Practices in IFS Knowledge Resource Development

Embrace Indigenous-led and informed IFS knowledge resource development. Indigenous-led and informed IFS knowledge resources are designed by and for Indigenous Peoples to be tools to inform, reclaim, reconcile, (re)connect and apply Indigenous fire knowledge to the land; educate on the benefits and practices of IFS across Canada; and recognize and celebrate the connections between fire, socio-cultural values and ceremonies in Indigenous communities.

Non-Indigenous wildfire agencies and related organizations have roles in developing IFS knowledge resources. However, rather than leading IFS knowledge resource development, they need to partner with and support Indigenous Knowledge Carriers/Keepers[3] and regional Indigenous organizations in developing IFS knowledge resources and housing IFS knowledge resources with local, regional and national Indigenous organizations.

Authentically know your audiences. At the outset of IFS knowledge resource development, it is important to identify the primary audiences for IFS knowledge resources. Taking a "nothing about us without us" approach, it is respectful and ethical to include Indigenous voices throughout the duration of the IFS knowledge resource development journey. Of particular note, multigenerational perspectives and kinship relations of Indigenous families and communities from Indigenous Children and Youth to Elders are leading and guiding this journey – both in terms of IFS content and the journey (process) itself.

Embrace multi-modal approaches to sharing knowledge. Building upon the above wise practices, individuals and groups involved in IFS knowledge resource development greatly benefit from taking a multi-modal approach to knowledge sharing. This can range from developing hands-on practice and demonstration activities (such as mentoring and experiential learning; learning directly from recognized Indigenous Fire Keepers; on-the-ground training and live exercises); producing audio-visual material (such as videos, podcasts, webinars, online curriculum delivery for all ages); developing children/youth activities (such as colouring books, activity books and FireSmart games) to creating tools/methods (includes a virtual portal of IFS promising practices, lessons learned, research papers and resources). The use of

[3] Indigenous (First Nations, Inuit or Métis) individuals in what is now called Canada who are recognized by their respective communities for their culturally significant knowledge and Indigenous worldviews. Knowledge Carriers/Keepers hold important roles in sharing traditional values and sacred teachings about various topics such as the uses of fire.

visual, written and auditory learning modes extends the depth and breadth of stories about IFS that can be shared.

Take a program-based approach. A program-based approach to knowledge resources development ensures continuous growth and improvement in IFS. This includes developing IFS knowledge resource programs that (re)connect Indigenous Youth and Elders in meaningful ways on the land that preserve Indigenous knowledge and language systems. It is important to note that, in general, funding for Indigenous-led and informed knowledge resource development and related extension has been limited and project-based in nature. As such, there is interest expressed by Indigenous Peoples, communities and organizations to have timely access to sustainable funding support that facilitates opportunities to share their knowledge and practices in IFS.

In summary, lifting voices and sharing wise practices in IFS education and awareness communications reinforces the "nothing about us without us" approach – honouring Indigenous ways of knowing and perspectives as recognized science in advancing IFS knowledge resource development. *Blazing the Trail* is an example of such an approach that creates a safer space to transform and share knowledge.

3 Case Study #2: Australia – The Harkaway Bushfire Safety Manifesto: Promoting Student Voice, Agency and Influence in School-Based Education for Risk Reduction and Resilience

The Harkaway Bushfire Safety Manifesto emerged from a 2-year action research project which sought to establish how problem-based learning (PBL) can be used to promote student voice and agency in bushfire education and support children's genuine participation in bushfire risk management in their homes, schools and communities (Harkaway Primary School Bushfire Safety Committee et al., 2020; HPSBSC, 2021).

The research was conducted in 2018–2020 with two classes of Grade 5/6 students (10–12 years old) at Harkaway Primary School on Melbourne's peri-urban fringe. Consistent with good practice in PBL (Reed et al., 2021), the process began with a broad exploration of the various factors that contribute to bushfire risk and the students were challenged to ask a multitude of questions. In small groups, the students then identified a local problem of concern, planned their inquiries, collected and compiled information, reflected on their findings, designed a solution, and shared that solution with a broader audience of vested interest.

Throughout the PBL process, a suite of ethnographic research methods, including focus groups, semi-structured and informal interviews, and participant observation, were employed to gather data on the learning process from the perspectives of students and their teachers. Upon completion of the process, key findings were shared and validated with the students in a participatory workshop. Using methods of Socratic discussion,[4] the students were encouraged to critically reflect upon and critique the key findings and expand upon key themes. The outcome of this discussion was the production of a 4-point manifesto which articulates key principles for good practice in school-based bushfire education from the perspectives of students themselves.

Since the Manifesto was ratified by the school in early 2020, it has served as a touchstone for the school's approach to bushfire risk management. In this case study, we present each point in the Manifesto (Harkaway Primary School Bushfire Safety Committee et al., 2020), as articulated by the Harkaway students, and we reflect on its meaning and significance for school-based bushfire education. We also consider the influence and impact of the Manifesto on policy and practice beyond the gates of Harkaway Primary School.

#1. We need to learn about bushfire safety. *People need to have a good understanding of bushfires and how to survive. Otherwise, when there is a bushfire, there will be more casualties and more deaths. It will be more catastrophic. Kids also need to know what to do to be safe and they need to learn it from a young age. When kids learn about bushfire safety from a young age, they can grow up knowing about their environment. Most people don't know about bushfires because they didn't learn about it when they were kids. If a bushfire is happening, kids need to know what they can do so they can stay calm and help their parents, instead of the parents doing everything and the kids just going "arrrghhh". When kids know what to do, there will be more hands on deck and everyone will be safer.*

At Harkaway Primary School, problem-based learning for bushfire risk management creates opportunities for students to plan, implement, and evaluate their own activities and initiatives in response to authentic challenges. This moves the students beyond the realm of knowledge acquisition and opinion formation. They are not merely learning about bushfire safety; they are learning to take informed action for bushfire safety, both individually and collectively. Through this process, students are positioned as genuine partners in bushfire risk management and agents of intergenerational change. They are not viewed as 'human becomings' who might be able to contribute to bushfire risk management at some point in the future, but as 'human beings' who can make valuable contributions now.

By positioning students as genuine partners in bushfire risk management, the Harkaway manifesto challenges the deeply entrenched assumption that children are passive victims of disaster who are entirely reliant on adults for their safety and protection (Ralls et al., 2022). While this assumption has been challenged many times before, including in the UN Sendai Framework for Disaster Risk Reduction 2015–2030, it remains implicit in many of Australia's bushfire risk management policies and programs, from the local to the State level. In this context, the manifesto provides the Harkaway staff and students with a framework for identifying and creating opportunities for authentic learning and action. It also provides a framework through which they can advocate for school-based education as a valid and legitimate adaptation pathway for their local community.

[4] https://tilt.colostate.edu/the-socratic-method/. Accessed 12 june 2026.

#2. We want to make decisions about our own learning and action. *Kids have really good ideas and sometimes we see problems and risks that adults don't see. We also know what kind of action is possible for us to achieve at home, at school and in our community. If we can make decisions about our own learning and action, we can solve real problems that matter to us. We all learn differently and we have our own special interests and talents. When we can make decisions about our own learning and action, we can achieve more than if we are just told what to do. We also work harder to succeed because we are doing what we are good at and what we enjoy. We need independence. We need to figure out how to learn and discover stuff on our own. We just need the adults there to guide us, put us on the right track and help us if we get stuck. We can do the rest.*

Historically, school-based bushfire education in Australia has been dominated by top-down approaches that prioritise knowledge transmission and position students as passive recipients of standardised information and instructions (Ralls et al., 2022; Towers, 2019). However, as research has shown, these approaches are not only ineffective, they can leave students feeling confused, disempowered, anxious and fearful (Towers et al., 2018). This is because top-down approaches fail to acknowledge and accommodate the social, cultural, economic, political and environmental factors that influence how students understand and engage with bushfire risk management in the context of their own everyday lives (Ralls et al., 2022). Top-down models also deny students opportunities to draw on their own unique capacities for identifying and solving high-priority problems in their households, schools and communities, thereby undermining their potential as agents of change (Towers, 2019).

As the Harkaway manifesto attests, if bushfire education is to empower students to take meaningful and purposeful action, it must afford them voice and agency in decision-making about what they learn and what they do with that learning. Affording students voice and agency in decision-making recognises that they have unique perspectives on teaching and learning and they should have the opportunity to actively shape their own education (Mitra, 2018). It also recognises that students have the ability and the will to set goals, reflect, and act responsibility to positively influence their own lives and the world around them. While this call for student voice and agency in decision-making represents a direct challenge to the top-down approach that characterises many school-based bushfire education programs, it is entirely consistent with the government policies and practice frameworks that guide teaching and learning in Australian schools (e.g., Victorian Department of Education and Training, 2019).

#3. We need access to trustworthy information from experts. *There is a lot of information about bushfires on the internet, but sometimes the information we are looking for isn't available or it's not explained in ways that we can understand. If we can talk to experts, we can get the information we need and it can be explained to us in ways that make sense. We really want to learn from Aboriginal people too because they have been here for a really long time and they know so much. If we have learned from experts, people will trust us and take us more seriously. Learning from experts also gives us opportunities to improve our communication skills. Making phone calls, writing emails, and doing research interviews are good skills for us to* *learn. These are skills we can use to learn more about bushfires, and other important topics, throughout our lives.*

While the Harkaway manifesto firmly positions students as agents of change who should be afforded voice and agency in decision-making about their own learning, it does not overlook or discount the importance of established knowledge in fire management. On the contrary, it explicitly recognises experts in fire management, including First Nations Peoples, as having a vital role to play in the achievement of positive outcomes. The point being emphasised in the manifesto is that the established knowledge of experts needs to be accessed through relational learning processes that build trusted connections and promote the development of social-emotional skills, like communication and collaboration. This requires a learning environment where experts actively listen to students' views and perspectives, demonstrate empathy and understanding, and offer constructive and supportive feedback (Riddle & Hickey, 2025).

Over the last several years, the Harkaway students have engaged with a wide range of fire management experts from all over Australia, conversing with them in person, over videoconferencing, on the phone and via email. Many of those experts have engaged with successive cohorts of students and they have become well acquainted with the vision and intent of the manifesto, including its emphasis on relational learning. However, it must be acknowledged that relational learning represents a significant departure from the lecture-based 'sage on the stage' lessons that many experts are accustomed to delivering in schools. Therefore, it is helpful to clearly communicate the purpose of the engagement and provide supportive guidance and advice as required.

#4. We want to teach others. *A lot of people don't know very much about bushfire safety. Most kids don't know anything at all. We want to help educate everybody – our families, our teachers, and the other kids at our school. We also want to educate kids from other schools. Teaching others improves our own learning. Sometimes you don't know how much you know until you try to teach someone else, so it's a good way to find out if you actually know what you think you know. Teaching others can show us that we have really learned a lot and that makes us feel proud. We can also help teachers and fire agencies create better bushfire education programs for kids. We can tell them what is important to us because we are experts in our own lives and we can show them the best ways to teach us because we are experts in our own learning.*

At Harkaway, students use a wide range of platforms and mechanisms to share their knowledge and educate others. They create brochures, posters, websites, films, digital games, picture storybooks, dioramas and interactive workshops. The school also hosts learning exhibitions where the students showcase their creations to their school community and invited guests. As highlighted in the manifesto, these activities provide valuable opportunities for authentic assessment of student learning, but they also build school pride and cultivate a culture of genuine student participation in bushfire risk management within the school. Perhaps most importantly, educating others reinforces the ethic of care and empowerment that is instilled in the students during the learning process. Returning to the first point in the Manifesto, "Kids need to know what to do to be safe and they need to learn it from a young age". By encouraging and supporting students to educate others, Harkaway Primary School is helping to ensure that occurs.

The Manifesto has also served as a valuable educational tool in itself, providing policy-makers and practitioners with essential advice on the design and implementation of school-based bushfire education programs. It has informed the design of *Schools in Fire Country*, the Victorian Government's flagship bushfire education program for upper primary school-students. It has been highlighted as an example of good practice in the Australian Government's handbook on *Disaster Resilience Education for Young People* (Australian Institute of Disaster Resilience, 2021). It has also been presented at a wide range of academic and industry conferences, seminars and workshops where it has resonated strongly with educators and fire managers from around Australia.

4 Case Study #3 – Portugal: Fogo Frio – Cold Fire, a Mobile Exhibition-Game

Fogo Frio – Cold Fire[5] is a mobile exhibition resulting from a national participatory budget initiative in 2017 where citizens presented and then voted on their ideas. This pedagogical exhibition explored the use of fire in the cold season by farmers, shepherds and prescribed burning technicians as a wildfire prevention strategy. Since the topic was so specific, the National Agency of Ciencia Viva invited the University of Lisbon – School of Agriculture to develop the winning project. A key message from this particular program is that fire is not the enemy. Rather, it is a tool used by rural traditional communities for more than 2000 years, or by the foresters for almost 200 years. Furthermore, it is a valuable tool in land stewardship by helping to decrease extreme wildfire impacts – through the use of cultural burning or prescribed/technical fire. The opening of the mobile exhibition took place in October 2022. It was initially expected that the Fogo Frio – Cold Fire mobile exhibition would tour Portugal for only 1 year. Expectations were exceeded in terms of public interest and demand. Therefore, Fogo Frio – Cold Fire continued with the final exhibition occurring in July 2025. More than 10,000 visitors, mostly grade schools, have participated in the mobile exhibition. The exhibition has been travelling in the northern and central regions of Portugal with more than 30 municipalities as host communities. Informal evaluations about Fogo Frio – Cold Fire has shown that teachers have been using the mobile exhibition as a starting point to explore wildfires, ecology and climate change with their students. How to safely burn in rural areas is one of the key topics discussed by the public.

Fogo Frio – Cold Fire disrupts the status quo in Portugal because it is not common to talk so freely in public forums about fire. Usually, the message is that fire is something that our society needs to banish. However, in this mobile exhibition, we

---

[5] https://www.youtube.com/watch?app=desktop&v=SOnY9DP0L_o. Accessed 12 june 2026.

are shifting the narrative to saying that fire can be a good thing, and we can use it to prevent extreme wildfires. Simultaneously, we are also discussing the different causes of wildfires to demystify that most sources of uncontrolled fires are due to arson.

The mobile exhibition is visited by all ages from rural and urban areas. One of the main challenges is how to create clear and consistent messages that are understandable by everyone, as well as scientifically accurate. Fire education and awareness key messages must be both professional and engaging, while remaining relevant and accessible to members of both urban and rural communities, so they see themselves and their perspectives reflected in the reality of the Fogo Frio – Cold Fire exhibition. To overcome this challenge, an interdisciplinary team was created – bringing together educators with a fire/forest/education expert and museum and mediation experts. All the fire education and awareness communications content were peer reviewed by practitioners and internationally recognized scholars on forestry and fire. The interactive dimensions of the mobile exhibition with the public were in the form of "townhall"-style questions and answers plus simple games (for example, game of differences or a puzzle to create a safe and resilient landscape) that would summarize the key messages about fire.

This exhibition-game has four modules to be explored based on different topics:

First module: The urgency to manage the territory, finding solutions to change the landscape into a more resilient to wildfires; Do you know which are the main causes of wildfires in Portugal?; What is "Fogo Frio [Cold Fire]" and their objectives?; How does a fire start? – Fire Triangle.

Second module: Which are the drivers that influence fire behaviour?; Vegetation – What burns easily – Rules of safety; Meteorology – When can I burn – Rules, permits; Who can help us to burn safely?; Did you have the right answer about the main causes of wildfires in Portugal?

Third module: Technical fire – Prescribed Burn Technician; Prescribed burning – technical fire with a strong planning and very specific objectives; To burn we need to protect ourselves! – Personal protection equipment; Tools to perform a safe burn – Driptorch, Macleod tool and Fire beater.

Fourth module: Table game – Discover the 10 differences between a cold fire and a wildfire; The different impacts, on soil, air and plant and animals response; As cold fires do not reach very high temperatures and have a superficial and rapid progression, they have minimal negative impacts on plant roots, microorganisms in the soil, insects, mammals and birds; Wildfires will never cease to exist, so it is up to each one of us to manage the territory!

5 Case Study #4 – Edufire ToolKit – a European Training Course

Worldwide, countries who experienced large wildfires have developed several pedagogical materials related to wildfires (Pacheco et al., 2024). In Europe, the most active countries are the ones with Mediterranean climate like Portugal, Spain, Italy and Greece. However, climate change is creating the conditions for the occurrence of wildfires where these types of materials are not common.

With this in mind, in 2022, a consortium coordinated by Pau Costa Foundation (Spain) with two universities, University of Lisbon (Portugal) and Catalonia Open University (Spain), worked together with Leitrim County Council in Ireland to create a pedagogical toolkit on the topic of wildfires funded by the Erasmus + European program.

This European project entitled "EduFire Toolkit" aims to raise awareness among secondary school students (12–16 years old) about the link between wildfires and climate change, and how individual and collective actions can reduce the risk of wildfires. As wildfires and climate changes are a multidisciplinary and complex problem, the project developed a set of multidisciplinary pedagogical resources following a Project Based Learning (PrBL) methodology, together with activities designed to encourage community participation and engagement. Considering the connection to real-world and local challenges, a transversal approach incorporating concepts related to fire, wildfires, and climate change was adopted, contributing to the curricula of various STEAM (science, technology, engineering, arts, and mathematics) disciplines while aligning with European educational competencies.

In order to create materials that teachers and students could adapt and scale as their own, the team asked various audiences, unrelated to fire or forestry, about questions they had regarding these two topics. The content was then built on top of these questions resulting in the following nine pedagogical units:

(1) Understand your environment and connect it to the impacts of climate change.; (2) History of fire; (3) Fire and wildfires; (4) The impacts of wildfires and climate change on different ecosystems around the world; (5) Preparedness for wildfires; (6) Humans and their interactions with the landscape; (7) Response to wildfires; (8) Impact of wildfires; (9) Wildfires in the future.

Simultaneously, a set of 32 challenges using the PrBL approach were created for students. These challenges allowed them to engage hands-on, supported by resources the team created including videos, a teacher's guide, and a collaborative map.

In order to better understand how these pedagogical contents would be useful to the European target groups, an advisory board was created giving voice to worldwide experts not only on forest and fire, but also in education and psychology. The needs presented by the educators and the comments from scientific and pedagogical experts were taken into account, reshaping the format of the materials.

To make sure that the different pedagogical materials were easy to understand, useful for schools and achieve the objectives of the project, each partner invited a local school to implement a pilot test. Teachers and students would choose one or several challenges, putting in practice the suggested methodology. As a result, four pilot tests (two in Catalonia – Spain, one in Portugal and one in Ireland) were developed. Once more, new thoughts, ideas and adjustments were made on the toolkit.

A small guide was developed to encourage schools to leave the school grounds and engage with their local community. An open access platform with pedagogical resources aligned with the school curriculum is available online for the academic community to use (https://www.edufiretoolkit.eu). The platform is available in four languages: English, Portuguese, Spanish and Catalan.

Edufire toolkit was a project that faced many challenges. As the team had diverse backgrounds and expertise, there was a need to share knowledge and experiences, deconstruct preconceived ideas about fire, education, curriculum design and reach consensus. There was both progress and setbacks. Although challenging, it led to the creation of a pedagogical toolkit that is important for schools and the surrounding community to improve their resilience to fires.

6 Case Study #5 – FireExplorer – a Fire Education Platform

On the internet, numerous websites and platforms provide information on fire and wildfire. Even if the source is credible, most of the information that is available targets experts, researchers, and stakeholders, and fails to engage wider non-specialized audiences who seek simpler, practical and more intuitive ways to access and learn about the topic.

In this context, an online educational platform (https://fire-explorer.isa.ulisboa.pt/) is currently being developed in the framework of the European Project FIRE RES with the objectives to improve communication and make fire information easier to understand and accessible to everyone. The ultimate goal is that society is more educated about wildfires and becomes more prepared to face extreme wildfires by having more knowledge about how to create resilient landscapes and reconnect and recognize the fire's cultural heritage. The platform will be available in five languages.

6.1 A Platform Made with and for the Community

In order to design a platform that would bridge the gap between science and the needs, motivations and expectations of the target audiences, a large survey was disseminated in Europe. Once the needs were identified, the contents of the platform were created in reply.

Simultaneously, for the structure and format of the platform, a comparative analysis on fire communication best practices on the web and possibilities of multimedia technologies was performed. A performance analysis of the platform prototype was performed focusing on different aspects of user experience such as usability, user retention, and accessibility (UX Benchmarking) of selected solutions (Oliveira, 2024).

6.2 The FIreExplorer Platform

This fire education platform uses various approaches to reach diverse target audiences. Five main sections were designed to better reach the following key audiences:

1. To teach (educators), contains pedagogical materials to support teaching activities in the classroom/in the field. This includes videoclasses, training courses, webinars and community engagement examples;
2. To do (practitioners and professionals) – Repository of technical materials with detailed guidelines and documents for different actors of society. These actors include Land and Urban Planners, Landowners, Business, Public Administration and Inhabitants;
3. To learn (students and general public) – Easy fire knowledge for everyone, contains adjusted, general fire information, interactive exercises/games, podcasts and a special section for kids.
4. Fire Guide – Resulting from the survey main results, a 5-min fire guide was created with all the basic essential information regarding wildfires – what to do before, during, after and recovery.
5. To recognize: Fire Culture Heritage – Fire has always been a part of our daily life, through traditions, stories, events, practices, religion and several art forms. In this section, the visitor is invited to travel throughout the heritage of fire.

This educational tool aims to bring people closer to fire, exploring both its beneficial aspects (good fire) and its destructive potential in the case of out of control wildfires. To prevent extreme wildfires and mitigate their impacts, it is essential to adopt a diplomatic approach towards this natural force – understanding its demands and negotiating what can and cannot be burned. By focusing on various key audiences and age groups, the platform serves as an important educational and engagement tool for fostering a more fire-resilient future.

7 Final Remarks

Five case studies, across different cultures and demographics, were presented with various approaches to **fire education and awareness communications**. However, there are several common features that represent the shift on fire education towards a more fire resilient society. Notable lessons learned and promising practices across all five case studies are the following:

1. **Know your audience and be inclusive**. A 'general public' does not exist. Rather, there are various groups in society, all with different demands, values, priorities, needs, motivations and capacities. Therefore, in fire education and awareness communications, it is important to know your audiences to (re)frame your key messages;
2. **Involve the community**. The answers are often in communities. They know their communities and land much more than what one may think. They should be part of the solution – *nothing about us without us*;
3. **Be precise, clear and scientifically correct**. Peer review and working with a multidisciplinary collaborative team will facilitate enhanced learning and related experiences;
4. **Bottom-up approaches, or at least transversal** methodologies support projects to be more effective in creating spaces for community solutions; and
5. **Fire is not the enemy**. Understand and respect that wildfire is on a spectrum. It is a natural-force and one benefits from adopting a diplomatic approach to fire.

In closing, honouring community members' perspectives as recognized ways of knowing greatly aid in advancing fire education and awareness communications. Across the globe, societies benefit from the use of participatory approaches to fire education and awareness communications that are based on collaboration, partnerships and engagement by and for local communities (for example, Indigenous Peoples, rural community members). This ensures that culturally safe/safer strategies are used when developing and sharing fire education and awareness communications "in a good way" from start to finish. We invite readers to adapt, scale and customize the learnings from the five case studies as an opportunity to communicate and (re)connect to land, people, place and culture in relation to fire.

Acknowledgments MCC is funded by DOI 10.54499/2020.01072.CEECIND/CP1591/CT0004 and she would like to thank the projects DecisionES (Grant Agreement number: 101007950—H2020-MSCA-RISE-2020) and FIRE RES (grant agreement No. 101037418). MCC thanks to her colleagues, that worked on the presented European educational projects, namely Letícia Oliveira, Iryna Skulska, Renata Pacheco, Ana Catarina Sequeira (case studies #4 and #5), Miriam Arenas Conejo and Helena Ballart (case study #5), Mapa das Ideias (case study #3).. BT and LJ acknowledge the Boonwurrung and Wurundjeri peoples of the Kulin Nations as the Traditional Custodians of the land on which they live and work, and they pay their respects to elders past and present. They extend their gratitude to the students of Harkaway Primary school for their valuable contributions to research and practice in the field of school-based bushfire bushfire education. They also extend their gratitude to the following school staff, researchers, emergency management practitioners and community members who have supported the students in their learning and action: Sam McGough, Lisa Harrison, Chris Adams, Lea Gard, Rennay Wood, Dr. Raphael Blanchi, Justin Leonard, Dr.

Michelle Villeneuve, Dr. Rachel Westcott, Dr. Marta Yebra, Yvonne Cowling, Dr. Joshua Whittaker, Lisal O'Brien, Antonia Mackay, Matt Henry, Mitchell Emmet, Neil Munro, Nathan Maddock and Tom Lowe. The work at Harkaway Primary School was supported by funding from the Bushfire and Natural Hazards Cooperative Research Centre, the Victorian Country Fire Authority, Victorian Department of Environment, Land, Water and Planning, Safer Together and Natural Hazards Research Australia (case study #2). NC and ACC extend their heartfelt gratitude and appreciation to the following individuals and groups who were instrumental during the co-writing and peer review journey for the "Blazing the Trail: Celebrating Indigenous Fire Stewardship" book: Elder Nklawa, David Diabo, Katie Ellsworth, Brady Highway, Shalan Joudry, Lorne L'Hirondelle, Waylon Skead, Michelle Vandevord, Ray Ault, Walter Andreeff, Solomon Carrière, George Chalifoux, Paul Courtoreille, Winston Delorme, Jeff Eustache, Harold Horsefall, Brad McDonald, Rolland Malegana, Gerald Michell, Herman J. Michell, Gerry Rempel, Ellen Simmons, Amber Simpson, Madeline Walker, FireSmart Canada, and the Government of Alberta. Furthermore, creative arts greatly assisted in extending knowledge about Indigenous fire stewardship to a wider and diverse audience. Many thanks to the following artists who shared their skills in the design of the "Blazing the Trail" book: Tom Spetter, Trinda Marie Cote, Andrew (Enpauuk) Dexel, Karen Erickson, Daniel Harrington, Darren Matasawagon, and Stacey Orr (case study #1).

References

Australian Institute of Disaster Resilience. (2021). *Disaster resilience education for young people.* Commonwealth of Australia.

Bartlett, C., Marshall, M., & Marshall, A. (2012). Two-Eyed Seeing and other lessons learned within a co-learning journey of bringing together indigenous and mainstream knowledges and ways of knowing. *Journal of Environmental Studies and Sciences, 2*, 331–340. https://doi.org/10.1007/s13412-012-0086-8

Christianson, A. C., Johnston, L. M., Oliver, J. A., Watson, D., Young, D., MacDonald, H., Little, J., Macnab, B., & Gonzalez Bautista, N. (2024). Wildland fire evacuations in Canada from 1980 to 2021. *International Journal of Wildland Fire, 33*, WF23097. https://doi.org/10.1071/WF23097

de Oliveira, E., Colaço, M. C., Fernandes, P. M., & Sequeira, A. C. (2023). Remains of traditional fire use in Portugal: A historical analysis. *Trees, Forests and People, 14*, 100458.

de Oliveira, E., Sequeira, A. C., Fernandes, P. M., & Colaço, M. C. (2025). The use of vegetation fire in Portugal: Historical legislative and normative analysis. *Environment and History, 31*, 15–39. https://doi.org/10.3828/whp.eh.63835725711833

Donovan, G. H., & Brown, T. C. (2007). Be careful what you wish for: The legacy of Smokey Bear. *Frontiers in Ecology and the Environment, 5*, 73–79.

Greenler, S. M., Lake, F. K., Tripp, W., McCovey, K., Tripp, A., Hillman, L. G., Dunn, C. J., Prichard, S. J., Hessburg, P. F., Harling, W., & Bailey, J. D. (2024). Blending indigenous and western science: Quantifying cultural burning impacts in Karuk Aboriginal territory. *Ecological Applications, 34*, e2973. https://doi.org/10.1002/eap.2973

Harkaway Primary School Bushfire Safety Committee, Towers, B., & Johnson, L. (2020). *Learner agency is an essential element of effective bushfire education for upper primary school children.* Bushfire and Natural Hazards CRC.

Hoffman, K. M., Davis, E. L., Wickham, S. B., Schang, K., Johnson, A., Larking, T., Lauriault, P. N., Quynh Le, N., Swerdfager, E., & Trant, A. J. (2021). Conservation of Earth's biodiversity is embedded in Indigenous fire stewardship. *Proceedings of the National Academy of Sciences, 118*(32), e2105073118.

HPSBSC. (2021). *Bushfire safety for kids: A manifesto from Harkaway Primary School.* Harkaway Primary School/Leadrrr. https://www.youtube.com/watch?v=FP5R-gYZUPI

Johnson, V. A., Ronan, K. R., Johnston, D. M., & Peace, R. (2014). Evaluations of disaster education programs for children: A methodological review. *International Journal of Disaster Risk Reduction, 9*, 107–123.

Mariani, M., et al. (2022). Disruption of cultural burning promotes shrub encroachment and unprecedented wildfires. *Frontiers in Ecology and the Environment, 20*(5), 292–300. https://doi.org/10.1002/fee.2395

McGee, T. K., & Christianson, A. C. (2021). *First nations wildfire evacuations: A guide for communities and external agencies*. Purich Books.

Mitra, D. (2018). Student voice in secondary schools: The possibility for deeper change. *Journal of Educational Administration, 56*, 473–487.

Oliveira, L. (2024) *Plataforma de Informação para o Fogo—Como melhorar o Engagement através de tecnologias interativas e Design Centrado no Utilizador*. Mestrado em Multimédia da Universidade do Porto, Portugal.

Pacheco, R., Skulska, I., Sequeira, A. C., & Colaço, M. C. (2024). Wildfire education: A review across the globe. In J. Rodrigo-Comino & L. Salvati (Eds.), *Fire hazards: Socio-economic and regional issues*. Springer. https://doi.org/10.1007/978-3-031-50446-4_3

Ralls, D., Lahana, L., Towers, B., & Johnson, L. (2022). Reimagining education in a pandemic: Children and young people as powerful educators. In R. Turok-Squire (Ed.), *COVID-19 and education in the global north*. Palgrave Macmillan, Cham. https://doi.org/10.1007/978-3-031-02469-6_1

Reed, S., Mullen, C., & Boyles, E. (2021). *Problem-based learning in elementary school*. Springer.

Riddle, S., & Hickey, A. (2025). *Unlocking the potential of relational pedagogy: Reimagining teaching, learning and policy for contemporary schooling*. Routledge.

Sauvé, L. (2002). L'éducation relative à l'environnement: possibilités et contraintes, Connexion. Revista de Educação Científica. *Tecnológica e Ambiental da UNESCO, XXVII*(1–2), 1–4.

Towers, B. (2019). School-based bushfire education: Advancing teaching and learning for risk reduction and resilience. *Australian Journal of Emergency Management, Monograph, 5*, 71–74.

Towers, B., Perillo, S., & Ronan, K. (2018). *Evaluation of survive and thrive*. Country Victorian Country Fire Authority, East Burwood and the Bushfire and Natural Hazards CRC.

Turco, M., Jerez, S., Augusto, S., et al. (2019). Climate drivers of the 2017 devastating fires in Portugal. *Scientific Reports, 9*, 13886. https://doi.org/10.1038/s41598-019-50281-2

UNDRR. (2023). *Closing climate and disaster data gaps: New challenges, new thinking*. United Nations Office for Disaster Risk Reduction.

UNEP. (2022). *Spreading like wildfire—The rising threat of extraordinary landscape fires*. A United Nations Environment Programme Rapid Response Assessment.

Victorian Department of Education and Training. (2019). *Amplify: Empowering students through voice, agency and leadership*. Victorian Government.

Vinyeta, K. (2021). Under the guise of science: How the US Forest Service deployed settler colonial and racist logics to advance an unsubstantiated fire suppression agenda. *Environmental Sociology, 8*(2), 134–148. https://doi.org/10.1080/23251042.2021.1987608

Vinyeta, K., & Bacon, J. M. (2024). Backfire: The settler-colonial logic and legacy of Smokey Bear. *Environmental Politics, 1–26*. https://doi.org/10.1080/09644016.2024.2353544

Educommunication and Climate Crisis Literacy: Transmedia Artivism i5

Carolina Fernández-Castrillo

Abstract The research approaches citizen engagement for climate action from the perspective of educommunication and (post)digital innovation. A focal point of this chapter is the role of experiential learning, cyberactivism, and intermedial participatory strategies to promote an educational and sociocultural scenario sensitive to the need for profound changes in our lifestyles. Based on over two decades of experience, the introduction of the Transmedia Artivism i5 methodology is based on the following key concepts: investigation, intermediality, immersion, interactivity, and intercreativity. This specific educational procedure aims to stimulate the students' creative skills from their role as "produsers" and committed citizens by exploring new ways of cocreation both in connective and face-to-face contexts. In fact, by designing blended strategies based on the opportunities offered by the space in between online and offline environments, the chances to reach inclusive socialization and to develop more empathic tactics seem to substantially increase. The examination of case studies will elucidate the progression of this educational methodology in response to contemporary societal issues, illustrated by several critical topics related to the climate emergency, including climate obstructionism, environmental delinquency, and ecoanxiety. This research establishes an epistemological framework centered on cyberactivism, immersive art, and experiential communication, emphasizing the significance of media artivism within the educational context and its ability to drive behavioral and social change.

1 Introduction

The collective construction of knowledge and decision-making through social networks and platforms represents an unprecedented communicative challenge for the collective construction of a future based on ecological awareness. Over the past few decades, the failure of the scientific, academic, political, and journalist fields to

C. Fernández-Castrillo (✉)
Department of Communication and Media Studies, Universidad Carlos III de Madrid, Getafe, Spain
e-mail: carolfer@hum.uc3m.es

J. J. Conde et al. (eds.), *Practices in Education for Climate Emergency*,
https://doi.org/10.1007/978-981-92-2304-6_12

communicate the risks of Climate Change (CC) has led us to an emergency, jeopardizing the survival of our planet. This study suggests that the fifth power can play a leading position in creating new ecomedia literacy strategies through disruptive classroom actions that raise awareness of the need for social change. From an approach centered on the potential of media artivism, we will focus on the role of produsers to consume, produce, and viralize innovative content that generates the necessary impact to establish some of the fundamental data needed to understand the current state of the climate crisis. Likewise, the aim is to provide solutions that mitigate the feeling of exhaustion and eco-anxiety. Hybrid strategies, which combine actions in real life with the implementation of initiatives in virtual worlds, help to increase the reach and experience of a transmedia approach in the (post)digital era.

The proposal of an educommunicative methodology called "Transmedia Artivism i5" consists of the implementation of media artivism in formal education and sociocultural dissemination has the possibility to enhance the capabilities of citizens, students, and produsers within a meta-category that surpasses the distinctions between content receipt, production, and distribution. The presence of the User-Generated Content (UGC) entails rethinking the function of scientists, specialized journalists, or sociopolitical agents and reviewing the function of citizens in the public sphere.

Given the breadth of the subject matter, this proposal focuses on crowdsourcing dynamics in connective environments through appropriationism; that is, by utilizing pre-existing materials on the web and by activating experiential dynamics of a collaborative nature. The objective is to contemplate the climate emergency by creating new narratives that render essential knowledge accessible and articulate primary concerns to foster action in pursuit of potential solutions. Consequently, the objective will be to assess the appropriateness of utilizing the principle of intercreativity (Berners-Lee, 1997) to enhance civic consciousness for the purpose of facilitating direct social action.

2 Educommunication and Experiential Learning for Climate Emergency

CC and educommunication are interlinked. There is an increasing need to introduce educational strategies in the audiovisual and (post)digital context, with a particular focus on addressing the impact of the spread of fake news on CC through disinformation literacy programs. Issues of great social significance, such as ecological awareness, the current state of environmental rights, stopping greenwashing, or fighting against abusive environmental policies should also be transversal content in all educational levels, whether formal, non-formal or informal contexts.

Unfortunately, "only 7% of references to climate change in primary education in the UNFCCC country submissions focused on social an emotional learning." (UNESCO, 2024). The *Global Education Monitoring Report* urges focusing on action-oriented learning to motivate CC mitigation and to monitor curriculum content to better understand country efforts on establishing national educative frameworks and syllabi to promote CC literacy. An effectively structured curriculum, appropriate pedagogy, and relevant learning tools can play a greater role in combating the climate emergency (Feinstein & Mach, 2020). Addressing psychosocial factors that hinder action, such as feeling grief, anxiety, and impotence, is essential in instructional design (Hargis et al., 2021; Hickman et al., 2021). Therefore, action-based, and attention-seeking initiatives are required to mobilize the communities and change social norms from (post)digital initiatives to activate and reinforce citizen participation and civic engagement.

From the perspectives of social justice and youth empowerment, understanding the role of new generations is essential for research related to natural disasters, particularly in enhancing their resilience and addressing their feelings of powerlessness and despair amid the ecological emergency (Bessaha et al. 2022; Haynes & Tanner, 2015). On the other hand, youngers demand a greater presence of technologies at the instrumental, strategic, and emotional levels throughout their educational process and a more solid training for the proper management of the information available on the network, thus surpassing self-training (Ballesteros Guerra et al., 2020).

Based on previous studies on the socio-communicative interaction dynamics of digital natives —millennial, centennial, and Alpha generations (Martínez, 2018; Quintana, 2016) —, some of the main trends in social mobilization on the internet, as cyberactivism or hacktivism, may be applied for educational purposes (Fernández-Castrillo, 2021, 2023, 2024, 2025, 2026; George & Leidner, 2019; Mejía Dávila, 2023; Sandoval-Almazán & Gil-García, 2014). The resolution of the new issues arising in the (post)digital era requires the promotion and protection of creative thinking. In fact, it is one of the fundamental commitments undertaken by the United Nations and its member states to ensure the maximum freedom of expression today (UNESCO, 2018). Regarding the jobs of the future, the World Economic Forum (2025) supports this need by placing creativity in the fourth position among the top twenty-six core skills for the Fourth Industrial Revolution, compared to the tenth position it held on to the same list in 2015 (Gray, 2020) in the Age of Intelligence. For this reason, it is urgent for the educational system to promote the mastery of creative thinking to train professionals capable of anticipating the present and solving new challenges through multiple solutions. To train this type of intelligence, the subject we are addressing aims for students to feel in control of their own learning process, creating a strong sense of community using new information and knowledge technologies (World Economic Forum, 2020) to approach the ecological challenges of a global network.

To achieve this objective, interactive instructional methods are employed in which the teacher performs her role as an expert, but also as a mediator from a collaborative interaction model. In consequence, education is more than just a simple transmission of information by the educator, as it is an interactive communicative process (Amat, 2000).

The teaching methodology Transmedia Artivism i5 is a hybrid model based on experiential learning with some elements from service and cooperative learning, and by applying horizontality as a basic principle for green education in phygital environments. In the social sphere, horizontality occurs "(...) as dialogue and structural organization that develops or stimulates the equal participation of individuals."[1] (Cañizares Galarzaga et al., 2017, p. 128) to foster cooperative innovation and the establishment of new community relationships in decision-making based on access, dialogue, and participation as its key components (Beltrán, 1981). In education, the principle of horizontality is considered the basis of "(...) a shared relationship of attitudes, responsibilities, and commitments toward successful achievements and outcomes." (Adam, 1997).

In a hyperconnected context, a new structure of knowledge is required, one that is more accessible and participatory. Certain scholars predict that increasingly, "teachers will be regarded as facilitators and mediators of the learning experience, as well as curators of knowledge to be conveyed, which may originate from their own work or that of others, encompassing a variety of advanced products, languages, and technologies." (Solé Blanch, 2020, p. 105). In light of this future vision, it must be emphasized that teachers should not assume the role of coaches or facilitators in the classroom under any circumstances (Larrosa & Rechia, 2018). Their contribution as subject matter experts will always be essential as references and guides during the learning process regarding the updating, deepening, and contextualization of the content, especially on issues of such broad social significance as CC.

Among the educational methodological trends, service learning has begun to gain greater presence in the Anglo-Saxon context since the beginning of the twenty-first century to promote the participation of younger students in community service (Annette, 2005) and civic engagement (Sandia & Montilva, 2020). As Rodríguez Gallego expresses, "It is a form of experiential education in which students engage in community service activities while facilitating the learning of a subject and the development of professional competencies" (2014, p. 96). In this case, the interest in understanding the climate emergency justifies the need to generate practical exercises focused on designing awareness campaigns for citizen mobilization or studying the communicative strategies carried out through social media.

On the other hand, the cooperative learning method is also integrated, whose premise is that students accomplish their objectives if all group members have reached theirs (Pujolàs Maset, 2012; Delgado Peña et al., 2010). According to Aramendi Jauregui et al. (2014, p. 426), "a student, a citizen, a person who feels supported by their group (peers, friends, family...) at an academic, emotional, or professional level will likely find a safer, more inclusive, and motivating context to develop their own life project in a globalized society". From the conviction that collaborative work inside and outside the classroom enriches the analysis and reflection on contemporary social phenomena, the proposed methodology aims for students to be able to propose specific solutions to concrete cases through the development of projects and creative proposals. As the main goal, it aims to educate citizens with initiative and integrity to face the new challenges of social coexistence. Therefore, case analysis and cooperative learning can provide significant advantages in the development

[1] The author of this chapter has translated all the quotations from Spanish to English.

of competencies – knowledge, processes, and attitudes – both professionally and socially.

Based on Dewey's educational model (1938), the final phase of experiential learning would consist of the ability to transfer new knowledge to other situations. As Ausubel (1976, p. 259) indicates, only when learning is relevant, the deliberate intention to acquire knowledge arises. Therefore, experiential learning is particularly significant as it influences the student from the improvement of their cognitive structure to the modification of their values and perceptions. Gutiérrez Fernández et al. (2011) argue that "it is through active, meaningful, and experiential participation that students construct new and relevant knowledge that influences their development and leads to responsibility and commitment to their own learning".

Although a robust theoretical framework for experiential climate change learning or green educommunication in the (post)digital era is lacking, the Transmedia Artivism i5 encourages transcending a passive engagement with reality by generating a series of experiences that facilitate reflection and conceptualization, thereby activating novel communicative dynamics rooted in interaction and immersion within interconnected environments.

3 (Post)digital Popular Culture and Climate Crisis Literacy

The social role of the "user", designated Person of the Year by *Time* magazine in 2006, has emerged as the focal point of contemporary major social movements, facilitating collective actions via the network. The pivotal role of users and its development in the digital society originates from Toffler's (1980) theories on the prosumer and Bruns's (2008) research on the produser, highlighting the transition from passive consumption to active participation in content production and distribution via the Internet. Consequently, the produser epitomizes the pinnacle of read-write culture (Lessig, 2008), wherein the behaviors of remixing and the advocacy of the copyleft movement are contextualized. Produsage thus cultivates novel intercreative processes within the context of a participatory culture, predicated on the collaborative generation of knowledge.

In the dawn of the 4.0 age, the (post)digital condition exposes the fluid essence of the digital frontier, accompanied by an enhanced critical perspective on technodeterminism. The integration of the phygital realm in our daily lifes marks a progression in the post-media era by overcoming the barriers between the online and offline worlds within an intermedial continuum. The Transmedia Artivism i5 seeks to provide learning activities that empower students to become produsers and engaged citizens, rather than passive voters, to foster pro-environmental attitudes and behaviors (Fig. 1).

The suggested educommunicative methodology encompasses works derived from the UGC, transmedia tactics, and intercreative dynamics, prompting reflection on the existing copyright legislation. The proposal may be relevant from five main perspectives concerning the growing prominence of the UGC in (post)digital popular culture

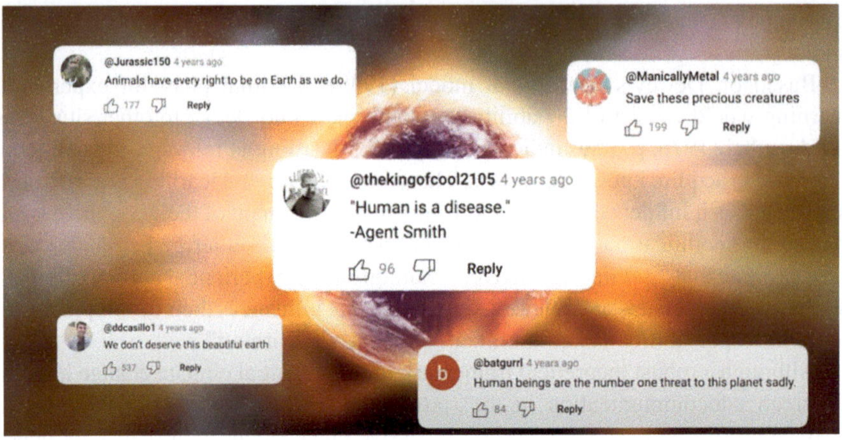

Fig. 1 Screenshot from an example of mash-uptivism. Intercreativity and Emerging Media, 2023–2024 (IE)

(Fernández-Castrillo, 2014) by enhancing civic engagement through the cultivation of collaborative dynamics in connective environments. This method's exploratory and multidisciplinary characteristics emphasize a creative approach to utilizing new technologies for social change, aiming to impact public perceptions and attitudes towards the climate catastrophe, while also being relevant to various civic issues.

4 Transmedia Artivism i5

The educommunicative method known as "Transmedia Artivism i5"—an updated version of Transmedia Artivism i4 (Fernández-Castrillo, 2025)—arises from over two decades of experience as a researcher, professor, and cultural manager in media artivist practices. It involves the implementation of five foundational concepts for socio-communicative innovation, influenced by the creative avant-gardes from the early twentieth century to the present: investigation, intermediality, immersion, interactivity, and intercreativity.

1. **Investigation:** constitutes the strategic intelligence layer of the project through a process of inquiry and data harvesting designed to transform information into an architecture of new meanings. It is based on the analysis of scientific publications, journalistic projects, governmental and civil society initiatives, sociological and psychological studies, and cultural strategies developed by experts.
2. **Intermediality:** novel expressive formulas that facilitate the exploration of subjects of general relevance through the principles of immediacy, simultaneity, and media hybridization.
3. **Immersion:** the construction or enhancement of an environment to simulate a new realm of reality, whereby multisensoriality, and role-playing dynamics.

4. **Interactivity:** the capacity to establish a connection between the user/audience and the production/content/environment through mutual influence.
5. **Intercreativity:** interconnective environments allow the access to information, and co-creative practices by interacting with other produsers, initiating, and/or participating in creative processes already existing.

This learning model reviews the notion of transmediality, asserting that UCG – in general – and appropriationism – in particular – is assuming an increasingly significant role in the intercreative dynamics prevalent on the web, based on the adaptation of prior concepts. The initial premise is derived from Jenkins' theoretical contributions regarding the impact of fandom dynamics on the formation of participatory culture, as well as his ongoing research at the Annenberg Innovation Lab at the University of Southern California. Jenkins founded *The Civic Imagination Project* within the study group Media, Activism, and Participatory Politics, supported by the MacArthur Foundation. His principal theoretical contributions include his research on media convergence, collective intelligence, participatory politics, and activism, which are vital to suggest provocative initiatives in education for climate emergency (Jenkins, 2006; Jenkins et al., 2013).

Jenkins introduced the term "transmedia activism" to refer to "The coordinated co-creation of narrative and cultural expression by various constituencies who distribute that narrative in various forms through multiple platforms, the result of which is to build an ecosystem of content and networks that engage in community-centered social" (2016). In 2011, Srivastava developed and provided a template for internet users to facilitate the organizing of transmedia activist campaigns, aiming to enhance civic consciousness through Creative Commons licenses (2013).

Transmedia Artivism i5 presents a civic awareness framework grounded in intercreative appropriation and experiential communication, in accordance with the effort initiated by Jenkins and Srivastava. This methodology aims to integrate the presence of UGC as an essential requirement for delineating contemporary transmedia practices, as I stated in 2022 alongside Scolari at the "International Conference on Transmedia Activism: Creativity & Expanded Information" (Fig. 2).

Provocation, guided improvisation, and the intervention of a produser-audience take on a special role in Transmedia Artivism i5. Its disruptive nature is the result of the influence of performance art and guerrilla activist actions. In its development, various methodological strategies are integrated:

- *Simulation/role-playing*: the development of critical thinking and empathy through staging, by dramatizing a situation while adopting various roles.
- *Case studies*: search, analysis, and resolution of case studies to subsequently reflect and debate about the conclusions drawn.
- *Problem-based work*: identification of current or future problems related to the contents of the subject matter under study. Search for documentation, reflection, analysis, and debate of possible options for resolution.
- *Project-based work*: development of interdisciplinary and transmedia projects to address novel content as a strategy for expanding and updating the concepts introduced in the lectures (Fig. 3).

Fig. 2 Poster of the "International Conference on Transmedia Activism: Creativity & Expanded Information", February 25th 2023 (UC3M)

Problem-based learning focuses on its resolution, whether simple or complex, emphasizing the importance of cooperation and the responsibility of the students themselves in the current social context. In the case of CC literacy, it is essential to approach the didactic activities from a glocal perspective to communicate, reflect, and find solutions from local-based experiences and site-specificity creative strategies.

Another key framework for the creation and evolution of the Transmedia Artivism i5 is the influence of media artivism,[2] a concept that I introduced at 59th Biennale Arte

[2] For a deeper understanding of the concept of Media Artivism, it is suggested to consult the special issue "Media Artivism: On the Archaeology and History of Digital Culture for Social Change" – published in 2024 in *Artnodes. Journal of Art, Science and Technology*—, and its introduction (Fernández-Castrillo & Mantoan, 2024).

Fig. 3 Transmedia Artivism i5. Guided role-playing performance. Communication and Participation of Civil Society in the Internet, 2023–2024 (UC3M)

Venezia to define the crossroads between activism, art, and media/technology. It is also important to link this novel method to the development of civic webs, living labs, meaningful play in ludified transmedia activities, and serious video games, among other media creative expressions, that can explore the nexus of civic participation and sustainable development (Fernández-Castrillo & López-Varela, 2024).

Examples of creative appropriationism and cultural sabotage arise from (post)digital popular culture, including *mash-uptivism*, a term that I coined in 2022 to express the natural evolution of remix practices through the reuse of pre-existing content in interconnected environments and the incorporation of original materials for the creation of original audiovisual montages with the aim of fostering civic awareness about social causes of interest in the resulting productions. This concept entails the reconfiguration of found footage and audiovisual collage by employing reappropriationist dynamics typical of social networks (Fernández-Castrillo et al., 2022). The principle of re-contextualization is a significant narrative and socio-cultural trend today, as seen in the development of fake news through the reuse and manipulation of previously published information. The modernization of remix practices in connective environments enables the re-signification of existing content in popular culture, while also incorporating original materials to augment civic awareness among produsers.

In the context of the climate emergency, the production of YouTube Poops or *mash-uptivist* clips[3] facilitates the resignification of any kind of audiovisual popular

[3] Regarding the concepts of *mash-uptivism* and transmedia ecoliteracy, it is suggested to access the interview in which the author explained both concepts in the program "La Vuelta a la España

productions, encompassing traditional media, films, television series, music videos, video art, video games, and social media material and discussions, among many others. Each academic year, the procedures for the CC are modified to correspond with current conditions, after a collaborative brainstorming session in which each working group identifies the most relevant ecological viewpoint. Prominent topics associated with climate change include: the influence of new generations to introduce transformations in consumption practices; the escalating loss of biodiversity; the impact of plastics in our ecosystem; detrimental effects of the fashion and food industries —including ecological food—; global natural disasters; disinformation and climate obstructionism; repercussions of technological progress on natural resources; the reorganization of urban environment to reduce air and water pollution; greenwashing and environmental evidence, etc. Depending on the number of students, a subcategorization within each team may be implemented to explore the issue using seven methodologies that enrich the storytelling: typographic play; visual symphony; media coverage; demonstrations; landscapes; protagonists; and fears/desires (Fig. 4).

This teaching innovation project involves a revision of design thinking emphasizing two unique attributes of modern communicative processes: vernacular creativity and the application of lateral thinking through the recombination of materials from various sources and authors on the World Wide Web. The Transmedia Artivism i5 integrates appropriationism as an educational strategy to enhance creative competency in the (post)digital era – adopting mash-up as the primary instrument – alongside experiential learning.

The objective is to experiment with new didactic formulas that facilitate the integration of the sociotechnological dynamics inherent in daily interactions, thereby creating transmedia campaigns that also involve experiential communicative strategies, both within and beyond the classroom to promote civic awareness regarding the climate emergency. The role of gamification and the creation of interactive and immersive experiences to investigate disruptive methods to break the fourth wall that divides the content creator from the audience. In this way, the audience becomes part of the campaign itself, activating co-creative strategies that, may appeal to an emotional engagement with the activist action, from discomfort to enjoyment.

Digital" [Episode 12, min. 17:13–23:09] in Radiotelevisión Española (RTVE), the Spain's national state-owned public television broadcaster (RTVE, 2025).

It is also available a selection of the *mash-uptivist* projects developed with the students:

- *Ecoliteracy & MOJO. New journalistic strategies for raising awareness about climate change* (2024–2025, Mobile and interactive narratives – UC3M. 22nd Call to support teaching innovation experiences, 2024–2025 – UC3M). https://youtu.be/XZ7Z-fZBx6g?si=b1ezA031mrtpI5dH
- *Plan B the change* (2024–2025, Cultural Creativity – UC3M). https://youtu.be/eYMIGaMm_Ic
- *Melting Point: The planet is ticking* (2023–2024, Cultural Creativity – UC3M. 21st Call to support teaching innovation experiences). https://youtu.be/Ur1zblOYYAA?si=zGYEtq2vlAfG_o2y

For further examples, it is suggested to access the YouTube channel *@CFCASTRILLO*, 2026. https://www.youtube.com/@CFCASTRILLO

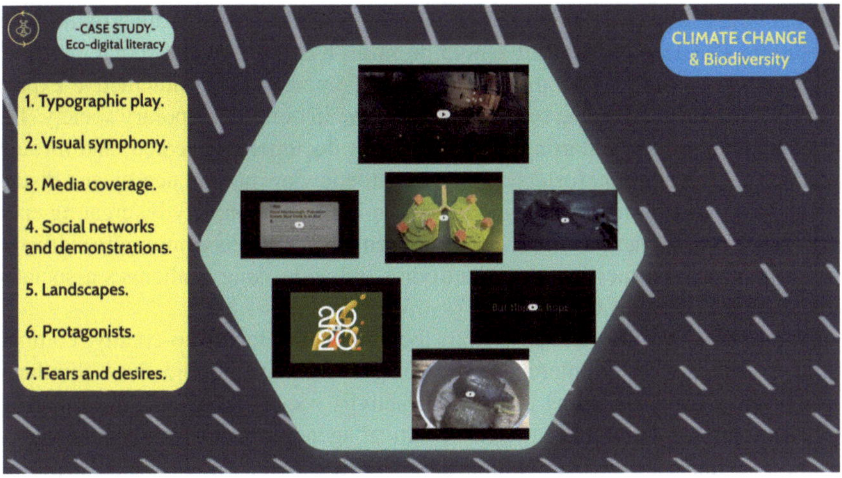

Fig. 4 Transmedia Artivism i5. Screenshot of a *mash-uptivist* clip selection. On Ecodigital literacy: Climate change & biodiversity. *Digital Media Culture*, 2022–2023 (IE)

The disruptive nature of this teaching methodology is integrated into guerrilla media artivist practices, fostering provocation – from discomfort to enjoyment – and critical thinking. The primary objective is to enhance awareness of the critical ecological crisis, while concurrently revising behavioral paradigms at both local and global levels by disseminating potential activities to alleviate or address concerns impacting our daily existence.

5 Conclusion

This study contends that CC literacy must evolve to fulfil its potential by surpassing mere dissemination of academic knowledge and anticipating significant effects on the social surroundings to which we are connected. As we have had the opportunity to observe in recent decades, the adoption of measures for a sustainable model within the capitalist production system has been inadequate. Likewise, the new challenges posed by obstructionism and climate (dis)misinformation require new strategies to identify those predominant narratives that lead us away from an environmentally respectful model of society.

Therefore, the learning paradigm must not depend exclusively on information transfer; it should also include social, emotional, and action-oriented learning.

Education has the potential to empower youth and adults as agents of change to mobilize their communities from blended contexts – online and offline – to change social norms and promote real commitment to our biggest challenge in planetary terms. The implementation of educommunicative strategies seeks to explore innovative media-based ways to improve a reductionist and stereotyped knowledge, raise awareness, and change the system of values to inseminate the main values of the green revolution. The Transmedia Artivism i5 proposes a creative use of social platforms and networks, together with the introduction of immersive and interactive experiences to focus on the leading role of the produsers in our contemporary society. The creation of total events, starting from a rigorous documentary search, and through creative experimentation, fosters a more empathetic view of both nearby and distant phenomena. This form of dynamics seeks to mitigate eco-anxiety by creating environments for the exchange of concerns. Engagement with one's community cultivates intergenerational partnerships for identifying and addressing challenges associated with the local effects of the CC.

This method may also attract scholars focused on developing co-creative communication strategies inside interconnected contexts for societal objectives. Similarly, the intercreative model aims at the cyberactivist sector, focusing on influencing and achieving concrete changes in the political agenda and the (post)digital public sphere. This may also involve communication professionals who specialize in project organization and the design of innovative campaigns intended to generate social impact, as well as those focused on developing new strategies to enhance intergenerational connections and restore trust in the media. Moreover, given the research funding initiatives of the European Commission Research Executive Agency, aimed at improving citizen engagement, it is likely that in the coming years, this area of study will attract the attention of a growing number of researchers and organizations pursuing these opportunities.

The creation of a collaborative communicative prototype for civic awareness in interconnected environments through information reappropriation techniques together with experiential actions have highlighted the need to improve the synergy among educational programs, artivism, communicative innovation, and sociocultural innovation for social transformation. This study's interdisciplinary nature significantly contributes by exploring innovative approaches to enhance social mobilization through creative methodologies and the dissemination of current knowledge regarding the global ecological catastrophe with considerable social impact.

Acknowledgments This research is supported by the projects *Innovación en formatos digitales para afrontar la ecoansiedad intergeneracional* (Carlos III University, 2025/00756/001) and *IBERIFIER2 – Iberian Digital Media Observatory* (European Commission Research Executive Agency, 101,158,511).

References

Adam, F. (1997). *Algunos enfoques sobre Andragogía.* Universidad Nacional.

Amat, O. (2000). *Aprender a enseñar. Una visión práctica de la formación de formadores.* Capellades.

Annette, J. (2005). Character, civic renewal and service learning for democratic citizenship in higer education. *British Journal of Educational Studies, 53*(3), 326–340. https://cutt.ly/9kugXQG

Aramendi Jauregui, P., Bujan Vidales, K., Garín Casares, S., & Vega Fuente, A. (2014). Estudio de caso y aprendizaje cooperativo en la universidad. *Profesorado. Revista de Currículum y Formación del Profesorado, 18*(1), 413–429. https://cutt.ly/Tkug8uR

Ausubel, D. P. (1976). *Psicología educativa. Un punto de vista cognoscitivo.* Trillas.

Ballesteros Guerra, J. C., Calderón Gómez, D., Kuric Kardelis, S., Megías Quirós, I., & Sanmartín Ortí, A. (2020). *Barómetro Jóvenes y expectativa tecnológica 2020.* Centro Reina Sofía sobre Adolescencia y Juventud. https://cutt.ly/PjqZj1Y

Beltrán, L. S. (1981). Adiós a Comunicación y Sociedad. *Comunicación y Sociedad, 6*, 5–35.

Berners-Lee, T. (1997, March 12). Realising the full potential of the Web. *W3C meeting.* London. https://cutt.ly/cxLREdS

Bessaha, M., Hayward, R. A., & Gtanas, K. (2022). A scoping review of youth and young adults' roles in natural disaster mitigation and response: Considerations for youth wellbeing during a global ecological crisis. *Child and Adolescent Mental Health, 27*(1), 14–21. https://doi.org/10.1111/camh.12517

Bruns, A. (2008). *Blogs, Wikipedia, Second Life, and beyond: From production to produsage.* Peter Lang.

Cañizares Galarzaga, F. P., Quevedo Arnaiz, N., & García Arias, N. (2017). Análisis sobre la horizontalidad en la educación universitaria. Apuntes de una valoración sobre su aplicación educativa en UNIANDES. *Revista Publicando, 4*(13), 123–147. https://cutt.ly/AkuhIrt

Delgado Peña, J., Romo Parra, C., López Romero, L., Vera Balanza, M. T., Meléndez Malavé, N., Sánchez Alarcón, M. I., & Gutiérrez Lozano, J. F. (2010). Innovación educativa a través del aprendizaje cooperativo y significativo: una visión multidisciplinar. In C. Ruiz Román (Coord.), *Innovación educativa y nuevos modelos de enseñanza-aprendizaje en el contexto universitario.* (pp. 149–200). Universidad de Málaga.

Feinstein, N. W., & Mach, K. J. (2020). Three roles for education in climate change adaptation. *Climate Policy, 20*(3), 317–322. https://doi.org/10.1080/14693062.2019.1701975

Fernández-Castrillo, C. (2014). Prácticas transmedia en la era del prosumidor: Hacia una definición del Contenido Generado por el Usuario (CGU). *CIC. Cuadernos de Información y Comunicación, 19*, 53–67. https://doi.org/10.5209/rev_CIYC.2014.v19.43903

Fernández-Castrillo, C. (2021). La condición transreal: información expandida y hacktivismo en el Media Art. *Artnodes, 28*, 1–10. https://doi.org/10.7238/a.v0i28.377879

Fernández-Castrillo, C. (2023). Hacktivism as an artistic practice: Visual evidence for social change. *Tecmerin. Journal of Audiovisual Essays, 11*(1) https://cutt.ly/PrenXumr

Fernández-Castrillo, C. (2024). Intermedialidad y activismo: Artistas, hackers y produsuarios. *Revista De Comunicação e Linguagens, 60–61*, 189–207. https://doi.org/10.34619/xdqp-bsmn

Fernández-Castrillo, C. (2025). Artivismo transmedia I4 en el ecosistema mediático: remediación, reapropiacionismo y viralización en la era del produso. *Itinera, 29*, 237–258. https://doi.org/10.54103/2039-9251/29690

Fernández-Castrillo, C. (2026). Ecoalfabetización mediática y movilización social en la era de la IA generativa. In C. Fernández-Castrillo & J. Flores (Eds.), *Horizontes de la inteligencia artificial para la acción climática. Perspectivas, desafíos y soluciones* (pp. 39–56). Fragua.

Fernández-Castrillo, C., & López-Varela, A. (2024). Intermedial participatory culture: Ludification and gamification for sustainable development. *Arte, Individuo y Sociedad, 36*(4), 779–788. https://doi.org/10.5209/aris.92679

Fernández-Castrillo, C., & Mantoan, D. (2024). An archaeology of media artivism: Attempting to draft the history of digital culture for social change. *Artnodes, 33*, 1–12. https://doi.org/10.7238/artnodes.v0i33.420928

Fernández-Castrillo, C., Rogel, C., & Romero Martínez, S. J. (2022). El *mash-up* como recurso educomunicativo para la concienciación social sobre feminismo. *ArtsEduca, 32*, 177–192. https://doi.org/10.6035/artseduca.6339

George, J. J., & Leidner, D. E. (2019). From clicktivism to hacktivism: Understanding digital activism. *Information and Organization, 29*(3). https://doi.org/10.1016/j.infoandorg.2019.04.001

Gray, A. (2020). *The 10 skills you need to thrive in the Fourth Industrial Revolution.* World Economic Forum. https://cutt.ly/WjqZl9p

Gutiérrez Fernández, M., Romero Cuadrado, M., & Solorzano García, M. (2011). El aprendizaje experiencial como metodología docente: aplicación del método Macbeth. *Argos, 28*(54), 127–158. https://cutt.ly/BkuUXNB

Hargis, K., McKenzie, M., & LeVert-Chiasson, I. (2021). A whole institution approach to climate change education: Preparing school systems to be climate proactive. In R. Iyengar & C. T. Kwauk (Eds.), *Curriculum and learning for climate action: Toward and SDG 4.7 roadmap for systems change* (pp. 43–66). Brill.

Haynes, K., & Tanner, T. M. (2015). Empowering young people and strengthening resilience: Youth-centred participatory video as a tool for climate change adaptation and disaster risk reduction. *Children's Geographies, 13*, 357–371. https://doi.org/10.1080/14733285.2013.848599

Hickman, C., Marks, E., Pihkala, P., Clayton, S., Lewandowski, R. E., Mayall, E. E., Wray, B., Mellor, C., & van Susteren, L. (2021). Climate anxiety in children and young people and their beliefs about government responses to climate change: A global survey. *The Lancet Planetary Health, 5*(12), e863–e873. https://doi.org/10.1016/s2542-5196(21)00278-3

Jenkins, H. (2006). *Convergence culture. Where old and new media collide.* New York University Press.

Jenkins, H. (2016, January 9). Telling stories: Lina Srivastava talks about transmedia activism (Part One). *Pop Junctions. Reflections on Entertainment, Pop Culture, Activism, Media Literacy, Fandom and More.* https://cutt.ly/7wLXDM0V

Jenkins, H., Ford, S., & Green, J. (2013). *Spreadable media. Creating value and meaning in a networked culture.* New York University Press.

Larrosa, J., & Rechia, K. (2018). *Elogio de la escuela.* Miño y Dávila.

Lessig, L. (2008). *Remix: Making art and commerce thrive in the hybrid economy.* London Bloomsbury Academic.

Martínez, N. (2018). *La generación zeta es autodidacta, sensible con el medio ambiente y adicta a la tecnología.* Universitat Oberta de Catalunya. https://cutt.ly/gjqCG0g

Mejía Dávila, M. R. (2023). Ciudadanía digital docente y ciberactivismo. *REDISED. Revista Diálogo Interdisciplinario sobre Educación, 5*(1), 169–176. https://cutt.ly/ErebjPeu

Pujolàs Maset, P. (2012). Aulas inclusivas y aprendizaje cooperativo. *Educatio Siglo XXI, 30*, 89–112. https://cutt.ly/TkuIZjJ

Quintana, Y. (2016). Generación Z: vuelve la preocupación por la transparencia online. *Revista de estudios de juventud, 114*, 127–142. https://cutt.ly/frebQV7t

Rodríguez Gallego, M. R. (2014). El Aprendizaje-Servicio como estrategia metodológica en la Universidad. *Revista Complutense de Educación, 25*(1), 95–113. https://doi.org/dt98

RTVE. (2025, April 13). *La Vuelta a la España Digital.* Pinacotech. C. de Madrid [Episode 12]. https://www.rtve.es/play/videos/la-vuelta-a-la-espana-digital/episodio-12-pinacotech-madrid/16539423/

Salehizadeh, M. R., & Behin-Aein, N. (2013). Effects of cooperative learning plus inquiry method on student learning and attitudes: A comparative study for engineering economic classrooms. *European Journal of Engineering Education, 38*(1), 23–35. https://doi.org/f4wz

Sandia, B. E., & Montilva, J. A. (2020). Tecnologías Digitales en el Aprendizaje-Servicio para la Formación Ciudadana del Nuevo Milenio. *RIED. Revista Iberoamericana de Educación a Distancia, 23*(1), 129–148. https://doi.org/10.5944/ried.23.1.24138

Sandoval-Almazán, R., & Gil-García, J. R. (2014). Towards cyberactivism 2.0? Understanding the use of social media and other information technologies for political activism and social movements. *Government Information Quarterly, 31*(3), 365–378. https://doi.org/10.1016/j.giq.2013.10.016

Solé Blanch, J. (2020). El cambio educativo ante la innovación tecnológica, la pedagogía de las competencias y el discurso de la educación emocional. Una mirada crítica. *Teoría de la Educación. Revista Interuniversitaria, 1*, 101–121. https://doi.org/f4w2

Srivastava, L. (2013). Narrative design for social impact: The project model canvas. *Transmedia Activism*. https://cutt.ly/ccyG8YE

Toffler, A. (1980). *The third wave*. Morrow.

UNESCO. (2018). *Re|pensar las políticas culturales. Creatividad para el Desarrollo*. United Nations Educational, Scientific and Cultural Organization. https://cutt.ly/njq0 Be7

UNESCO. (2024). *Global education monitoring report. Education and climate change. Learning to act for people and planet*. University of Saskatchewan. https://doi.org/10.54676/GVXA4765

World Economic Forum. (2020). *Schools of the future: Defining new models of education for the fourth industrial revolution*. World Economic Forum. https://cutt.ly/CjqL4i0

World Economic Forum. (2025). *Future of jobs report*. World Economic Forum. https://cutt.ly/5rens7sc

Co-Designing and Evaluating Two Practical-Oriented Climate Change Teacher Training Programs: Game-Based Learning for In-Service Teachers in Southern Spain

Enrique González-Muñoz and José Gutiérrez Pérez

Abstract This chapter presents the co-design and development of an in-service teacher training program on climate change education, emphasizing disruptive methodologies and Game-Based Learning. Developed with two Teacher Training Centers, the program targeted Primary and Secondary teachers, aiming to enhance their scientific understanding of climate change and provide engaging, emotionally resonant tools – such as escape rooms, board games, and role-playing games – for classroom use. Adopting an emergent research approach, the program used participatory action research and co-design to adapt its content and methods to local needs. Teachers were actively involved in shaping the program, testing educational games, and reflecting on their pedagogical impact. The following chapter details the program's phases, from needs assessment and theoretical grounding to iterative refinements based on feedback. Preliminary results show positive evaluations of the training and resources, with teachers valuing hands-on experimentation and collaborative reflection. Despite climate change education being included in Spanish education policy, the study identifies ongoing gaps in teacher training. These findings highlight the importance of sustained and specialized programs that support both scientific literacy and innovative teaching practices. The success of the program is attributed to the collaborative design process and strong partnerships with teachers and training centers, showcasing the effectiveness of Game-Based Learning in climate education.

E. González-Muñoz (✉) · J. Gutiérrez Pérez
Departamento de Métodos de Investigación y Diagnóstico en Educación, Universidad de Granada, Granada, Spain
e-mail: egmunoz@ugr.es; enriquegonzalezmunoz@gmail.com

J. J. Conde et al. (eds.), *Practices in Education for Climate Emergency*,
https://doi.org/10.1007/978-981-92-2304-6_13

1 Introduction

1.1 The Urgency of Climate Change Education in the Context of an Ecosocial Emergency

The notion that climate change is a remote and amorphous phenomenon has been discredited by the preponderance of direct and indirect ecosocial impacts that have been witnessed in recent years (Chu & Yang, 2019). Extreme heat waves have besieged Asia, catastrophic floods have devastated Brazil, Kenya, and some countries in Europe (as in the case of the DANA in Spain, Valencia), and voracious wildfires in the middle of winter have ravaged western Canada and the United States; even arid regions such as the United Arab Emirates have recorded record rainfall (OECD, 2025). This context of a climate emergency, characterized by global risks of great magnitude and the threat of tipping points that could accelerate and exacerbate the crisis, requires an urgent and transformative response at all levels of society. Training teachers in these issues should be a priority, considering their capacity to influence numerous future generations.

The dissemination of information and its visibility in the media, in conjunction with the pressure exerted by climate movements, have resulted in mounting social concern and a demand for increased government involvement. This demand calls for the implementation of policies that are equitable, sensitive, and committed to addressing citizens' demands regarding climate change. While the denialist discourse has evolved to address the argumentative gaps generated by visible data and phenomena (no longer denying the existence of the phenomenon but minimizing its severity or the need for systemic changes) a growing number of citizens have begun to disentangle themselves from these fallacies (Klein, 2014; Leiserowitz et al., 2018). This shift in the prevailing social narrative regarding climate change has led to a heightened societal awareness of its gravity, as demonstrated by the increasing media coverage and the emergence of prominent movements such as Fridays for Future, Extinction Rebellion, and Teachers for Future. These movements have effectively mobilized a wide range of societal sectors. However, despite this collective awakening, the awareness and participation in climate action remain fragile and subject to fluctuations, necessitating constant effort to sustain momentum towards real and sustainable change (Chen et al., 2023). In this context, formal education emerges as a pivotal element to fortify climate resilience competencies, and in-service teachers play a pivotal role in this regard.

1.2 Climate Change Education Within the Regulatory Framework: Challenges and Opportunities

In the international context, the Sixth Assessment Report of the IPCC underscores the pivotal function of education in cultivating a culture of sustainability, underscoring that education must be transformative and oriented to nurture the critical competencies and skills imperative for climate action (IPCC, 2024). The 2030 Agenda (United Nations, 2015) and the Sustainable Development Goals (SDGs) underscore these challenges. Specifically, SDGs 4 and 13 aim to integrate climate change into education, fostering the development of an informed and prepared citizenry to respond to climate emergencies. SDG 4 underscores the imperative for inclusive and quality education, with the capacity to equip future generations with the necessary skills to confront global challenges, including the climate crisis. Additionally, SDG 13, which is dedicated to take immediate measures to address climate change and its effects, further reinforces the notion that education plays a pivotal role in implementing global strategies to mitigate the repercussions of climate change.

At the European level, the European Sustainability Competence Framework (Joint Research Centre, 2022) delineates the essential competences to address climate change challenges, proposing an educational approach that enables students to make informed decisions and actively participate in the creation of sustainable societies. This framework establishes an educational roadmap to equip young people with the necessary tools to effectively address environmental and social challenges. At the Spanish level, similar steps have been taken to integrate climate change into its educational system. The current Spanish educational legislation or LOMLOE (Spain., 2020) establishes climate change as a transversal axis in compulsory education, also promoting educational innovation. This law encourages the use of active and participatory methodologies that favor the development of competencies needed to face climate challenges in a dynamic way and adapted to local realities. However, despite the opportunities and responsibilities arising from regulatory advances, significant challenges persist in the effective implementation of these frameworks in the educational sphere. A significant challenge is the paucity of specific teacher training on climate change. A systematic review by García-Vinuesa (2024) identified curricular limitations as a major obstacle to in-depth treatment of climate change in the classroom. These limitations include the lack of alignment of content with curricular standards and its confinement to specific subjects such as Geography or Earth Science, hindering a broader, multidisciplinary discussion of climate change. Furthermore, a significant number of teachers exhibit a lack of confidence and knowledge in addressing these subjects effectively, leading to an overreliance on textbooks and a paucity of contextualization appropriate to local challenges.

1.3 Disruptive Methodological Innovations for a Climate Change Education Paradigm Shift

The pressing necessity to transform education in the face of the climate crisis demands, to a considerable extent, a reconceptualization of teacher training methodologies. The implementation of innovative methodologies that incorporate disruptive, playful, and emotional approaches is imperative for teachers to effectively address climate change. Conventional educational approaches, which prioritize the transmission of knowledge through a cognitive and objective lens, have proven inadequate in this context. These approaches fail to adequately address the complex emotions, attitudes, and motivations that climate change evokes among students (Brosch, 2021; González-Muñoz et al., 2024; Jones & Davison, 2021).

The employment of disruptive approaches, such as Game-Based Learning, which integrates gaming into the educational process, enables students to experience the effects of climate change and reflect on their decisions in simulated scenarios. This methodology not only fosters cognitive understanding but also facilitates the management of emotions in the face of climate change's consequences, offering a more comprehensive and balanced education. Recent studies have demonstrated that Game-Based Learning fosters heightened intrinsic motivation, facilitates collaborative learning, and develops key competencies, such as decision-making, creativity, and empathy, which are fundamental for effectively dealing with climate challenges (Dahalan et al., 2024; González-Muñoz et al., 2025; Veldkamp et al., 2020). Furthermore, the integration of the emotional dimension in climate change teaching is presented as a pivotal instrument for galvanizing action. Emotions can function in a detrimental manner, leading to paralysis or pessimism, or in a beneficial manner, propelling motivation and commitment towards climate action. Educational approaches must thus be meticulously designed to assist students in adeptly managing their emotions in a healthy manner, achieving a balanced perspective between perceptions of the climate crisis and practical, optimistic solutions. This approach is imperative to circumventing apathy or misguided optimism that can emerge when confronted with the intricacies of this multifaceted issue (Brosch, 2021; González-Muñoz et al., 2024; Jones & Davison, 2021).

1.4 The Training of In-Service Teachers: Research Transfer and Co-Design

The continuous training of in-service teachers is a crucial aspect for ensuring the implementation of disruptive methodologies, such as Game-Based Learning, in climate change teaching. However, one of the main challenges faced by education systems is the lack of adequate training, both in climate change education and innovative methodologies. This limitation restricts teachers' ability to effectively incorporate these contents in the classroom.

Caride et al. (2025) emphasize that, despite the growing relevance of climate change, teacher training in this area remains inadequate. A significant proportion of teachers lack the requisite expertise to seamlessly integrate climate change across various subjects, impeding its application in a more comprehensive manner. This training disparity hinders the effective teaching of climate change, necessitating the development of effective interdisciplinary approaches. The dearth of training in climate change education is further compounded by a lack of preparation in innovative methodologies. As García-Gómez and Gil Flores (2022) demonstrate in their study on innovative teaching approaches in Spanish Secondary Education (equivalent to the K-12 system), innovative pedagogical methods remain in the minority, particularly at this educational level. Despite the evident advantages of employing active and participatory methodologies, most teachers in Secondary Education do not implement them in their classrooms, due to a lack of training and confidence to do so. The synthesis of these two studies underscores a dual imperative: firstly, there is a necessity for enhanced preparation in climate change education for teachers; and secondly, there is a pressing demand for comprehensive training in active methodologies, such as Game-Based Learning, which have been shown to foster the development of crucial competencies essential to addressing the climate crisis.

García-Gómez and Gil Flores (2022) identify several factors that favor the use of innovative approaches, such as teachers' perceived self-efficacy, their collaboration with other teachers, and their participation in professional development activities. These factors are related to an increased ability of teachers to address complex issues, such as climate change, in a cross-cutting manner and with active methodologies. This finding underscores the necessity for ongoing training that not only enhances teachers' understanding of climate change but also empowers them to integrate innovative and transformative approaches into their pedagogical practice, which is consistent with the conclusions of a review by García-Vinuesa (2024).

The development of training programs that integrate climate change education with innovative methodologies, crafted through a collaborative design process with the Teacher Training Centers, is identified as a pivotal strategy to address these deficiencies. The co-design process, involving active involvement of teachers in the development of training programs, ensures that these programs are tailored to their real needs and the unique characteristics of their respective educational contexts. This approach enhances the relevance and effectiveness of the training, leading to increased interest and motivation among teachers to utilize these innovative methodologies in their pedagogical practice. Furthermore, it fosters their confidence and autonomy in implementing these methods in their classrooms. This formative approach is particularly relevant in secondary education, as young people are at a critical stage in shaping their understanding of the world and developing critical skills. The integration of climate change education and disruptive methodologies in continuing education is essential for empowering teachers, enhancing their pedagogical practices, and equipping students with the necessary tools to effectively address the climate crisis.

In light of the prevailing global climate emergency, the importance of climate change education as a pivotal instrument in cultivating conscientious citizens capable of informed decision-making regarding climate change and its associated risks is paramount. The adoption of innovative methodologies, such as Game-Based Learning, presents a distinctive opportunity to transform the pedagogical approach to climate change teaching, fostering a dynamic, participatory, and emotionally engaged learning environment. However, for this transformation to be effective, it is necessary to overcome the structural challenges faced by educational systems, especially in terms of continuous teacher training and the development of innovative pedagogical competencies. Only through a comprehensive and transformative educational approach, supported by global and local policy frameworks, can we prepare future generations to meet the challenges of the climate emergency in an effective and sustainable manner.

2 Method

The aim of this chapter is to describe and compare two models of teacher professional development in relation to the climate emergency and disruptive methodologies, specifically in the use of Game-Based Learning. To this end, two training programs were designed and implemented in collaboration with two Spanish Teacher Training Centers, located in two Andalusian contexts, Granada and Almería. These programs were aimed at Primary and Secondary Education teachers. The training programs were designed to foster the development of competencies and the utilization of ludo-emotional resources in climate change education, employing innovative tools such as escape rooms, board games, and role-playing games. The primary objective was to provide teachers with a scientific update on climate change, complemented by practical, first-person training experiences that enabled the integration of these disruptive methodologies into their daily teaching practice.

The training program design adopted an emergent research approach, allowing constant adaptation to the changing needs and realities of teachers and the educational context. The course was based on a participatory action research and co-design approach, involving teachers in the process of creating and modifying the training program content and structure. This participation facilitated the contents and methodologies to be adjusted to local needs, and empowered teachers, allowing them to evaluate, reflect, and apply innovative methodologies in their classrooms with autonomy and confidence.

The present study has been reviewed and endorsed by the Ethics Committee of the University of Granada (nº 3252/CEIH/2023), in accordance with the ethical principles outlined in the Declaration of Helsinki. This ensures that all participants have been informed of the study's objectives, that their data will be kept confidential, and that the study adheres to international ethical standards for scientific research.

2.1 Research Objectives

The objectives of this study are as follows:

- To promote the co-design of situations involving disruptive methodologies that favor teacher professional development in the treatment of climate emergencies.
- To compare the training proposals developed, analyzing the strengths and weaknesses of each model in terms of content, structure, and methodology.
- To evaluate the adequacy and effectiveness of the selected pedagogical materials, considering their pedagogical value and their feasibility for implementation in primary and secondary education classrooms.

The participatory action research and co-design process developed follows a sequence of five phases, which are described below.

2.2 Description of the Sample

The final sample comprised 28 active teachers from Granada and Almería, who voluntarily enrolled in the study following an enrolment process. The participants were distributed among 14 teachers from Granada (from 12 schools) and 14 teachers from Almería (distributed among 5 schools, with one pilot school representing 64.3% of the sample). Demographic and professional characteristics of the participants are outlined below:

Granada: The mean age was 49.36 years (SD $\pm$ 6.27 years), with a range between 39 and 58 years. Of the teachers, 64.3% were female and 35.7% were male. Most of them taught in Compulsory Secondary Education (*Educación Secundaria Obligatoria*, ESO, official Spanish designation), with a notable presence of teachers from the Biology and Geology department.

Almería: The mean age was 45.71 $\pm$ 8.20 years, with a range from 31 to 56 years. The sample was balanced between men and women (50% each). Of the teachers, 78.6% worked in Primary Education, while the rest worked in ESO level. A high percentage of the primary school teachers were tutors, and in ESO level, all belonged to the Biology and Geology department.

2.3 Instruments and Data Analysis

A total of six data collection instruments were utilized, and a process of quantitative and qualitative data analysis was carried out to obtain a comprehensive and complex view of the co-design process, the effectiveness of the pedagogical resources, and the impact on the professional development of teachers in the management of disruptive methodologies and the motivation to address aspects related to climate change in their classrooms (Table 1).

3 Co-Design Process and Implementation of the Training Process in Climate Change Education for Climate Emergency and Game-Based Learning

Prior to the initiation of this section, a visual depiction of the design, dissemination, implementation, and evaluation process, with a focus on enhancing future training initiatives, is provided in Fig. 1. The subsequent sections will provide exhaustive explanations of each step involved in the figure.

3.1 Phase 1. Theoretical and Evidence-Based Foundation

The training process on Game-Based Learning applied to climate change education is grounded in a robust body of academic evidence and an empirical diagnosis, integrating both scientific literature and the results of an initial needs assessment of teachers and students. In alignment with the principles of the *Paradigm of Complexity* (Bonil et al., 2004; Morin, 1992; Salazar, 2004), which emphasizes the non-linearity, uncertainty and interdependence of real systems (especially in social and educational contexts), the process was designed to be flexible, adapting to the characteristics of Andalusian teachers and the inherent differences in the profiles of participants. This adaptability ensures the training program's relevance and enhances its potential effectiveness in practice.

The development of this process began with a comprehensive literature review that explored key areas: climate change education, teachers' and students' perceptions and needs regarding climate change, advancements in environmental education, and the application of Game-Based Learning to teach complex topics like climate change. This review identified both barriers and opportunities in climate change education, establishing a solid theoretical foundation for structuring the training process. Simultaneously, the current educational regulations were analyzed, including Spain's mandatory curriculum legislation and its regional implementation in Andalusia, to align the training content with official educational frameworks concerning climate topics.

Based on this analysis, the *Resclim@tiempo Questionnaire* was refined to explore profiles in relation to the climate emergency and its teaching. The revised instrument retained its original function but was expanded to investigate emotional dimensions, pro-environmental behaviours, and perceptions of climate change education more deeply. Additionally, the adapted version allows for the evaluation of transformations resulting from the training interventions. The questionnaire was structured into two

Table 1 Implemented tools for evaluating the training programme: from initial diagnosis to final assessment

Instruments	Technical description	Application and purpose
Resclim@tiempo Questionnaire	An *ad hoc* instrument designed to explore ecosocial profiles in the context of the climate emergency and its pedagogical implications. This enhanced version of a third generation questionnaire broadens the original framework by incorporating emotional aspects, activism, and climate change education, while also evaluating the transformative effects of pedagogical interventions	To diagnose needs at the start of the training program and assess changes at its conclusion
Resource Evaluation Scale	An *ad hoc* instrument consisting of 13 quantitative items that assess the quality, relevance, and motivational aspects of the resources utilized. Statistical analyses, including the KMO test (0.85) and Bartlett's test of sphericity ($p = 0.00$), indicate appropriate sample adequacy. The scale reveals two primary dimensions: (1) Pedagogical value ($\alpha = 0.91$, $\omega = 0.91$), and (2) Feasibility of implementation ($\alpha = 0.86$, $\omega = 0.87$). It also includes a qualitative SWOT (Strength, Weakness, Opportunities and Threats) analysis to deepen the reflection on the strengths, weaknesses, opportunities, and threats of the resources	To continuously evaluate resources throughout the training program, focusing on pedagogical quality and practical feasibility for classroom implementation
Focus group sessions for resource evaluation	Semi-structured group reflection sessions conducted after interactive activities, during which teachers assess the effectiveness and applicability of the resources used. These sessions provide valuable qualitative data on the potential for integrating these resources into classroom settings	To facilitate ongoing evaluation of resource effectiveness and applicability during the training program
Final portfolio	An *ad hoc* instrument enabling teachers to document the use of resources in their classrooms. The portfolio includes observation scales, rubrics, and other tools to evaluate various aspects of the educational implementation from the teachers' perspectives. The collected data serves as the foundation for critical reflection in the training program's final phase	To assess the implementation of resources in the classroom, providing a mixed evaluation of their educational impact

(continued)

Table 1 (continued)

Instruments	Technical description	Application and purpose
Final focus group on implementation and course evaluation	A semi-structured group session conducted at the end of the training program, where teachers discuss their experiences with resource implementation and evaluate the training program. This activity identifies strengths and areas for improvement in course design and development	To conduct a final evaluation of resource implementation and the overall effectiveness of the training program, offering recommendations for future improvements
Official evaluation survey administered by the Teacher Training Centers	An official survey administered by Teacher Training Centers, consisting of 9 quantitative items and 3 qualitative items gather feedback on areas for improvement, key learnings, and other comments	To evaluate the training program, including its design, development, and implementation, as well as the instructor's performance

main versions: one for in-service and pre-service teachers, and another for primary and secondary students, with linguistic adjustments tailored to each educational level. Both versions shared a core set of items focused on ecosocial profiles related to climate change, complemented by pedagogical adaptations addressing the teaching and learning of the climate emergency.

The findings from both documentary and field-based diagnostics revealed critical insights that informed the structuring of the training process. One of the most significant aspects identified was the emotional dimension of climate change. The reviewed studies (Brosch, 2021; González-Muñoz et al., 2024; Jones & Davison, 2021) consistently highlighted the crucial role of emotions in shaping climate motivation and action. Emotions can act as either barriers or catalysts for pro-environmental behaviors, yet this dimension remains underexplored in educational contexts. Training interventions should therefore integrate strategies that promote not only cognitive understanding of climate change but also emotional regulation, helping students navigate feelings of anxiety or helplessness. It is essential to strike a balance by avoiding catastrophic narratives devoid of practical solutions, while also steering clear of overly optimistic perspectives that may induce complacency. A balanced pedagogical approach that encourages collective action is critical to preventing emotional paralysis and fostering sustained, meaningful participation in climate solutions.

Regarding the impact of Game-Based Learning on climate change education, the reviewed studies (Dahalan et al., 2024; González-Muñoz et al., 2025; Veldkamp et al., 2020) provided strong evidence of the effectiveness of serious games in facilitating the assimilation of complex and systemic concepts like climate change. Serious games have been shown to enhance conceptual understanding, foster both intrinsic and extrinsic motivation, and promote positive emotional experiences that reinforce learning. Moreover, they cultivate transversal competencies essential for building

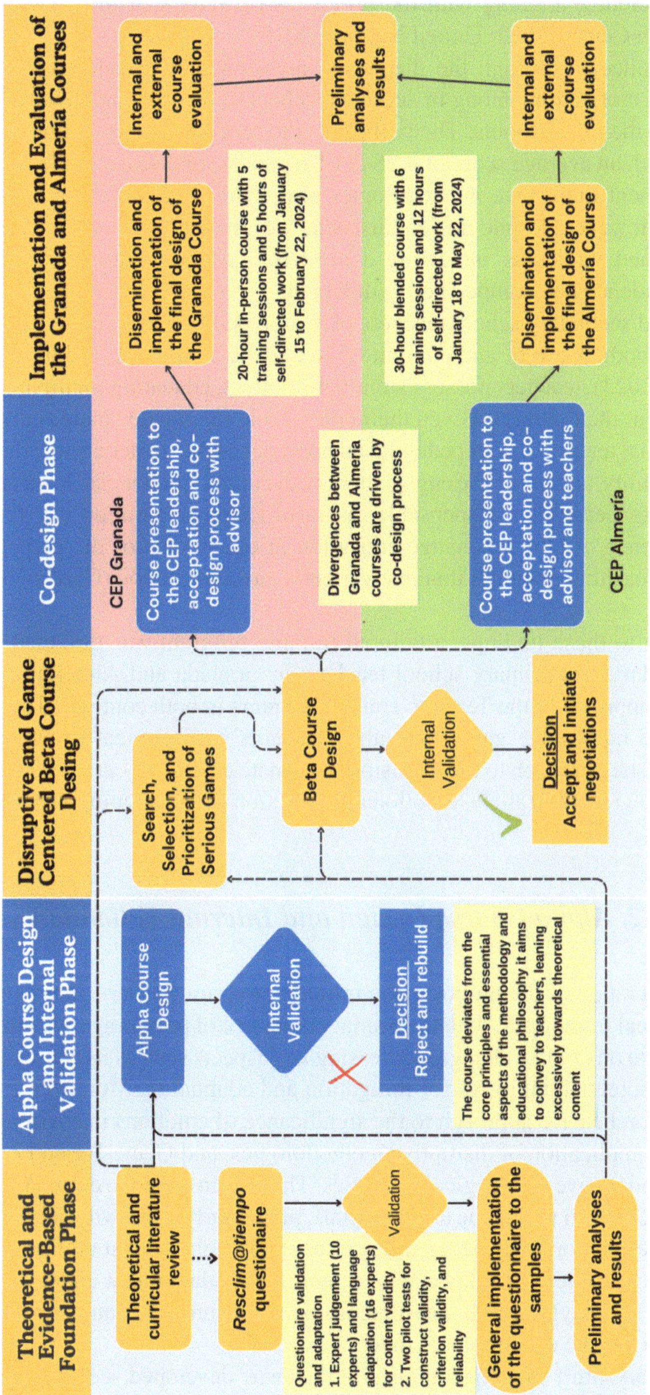

Fig. 1 Phases of the research and training process

sustainable societies, aligning with the principles of SDG 4 of the 2030 Agenda (Joint Research Centre, 2022; United Nations, 2015).

The data collected through the diagnostic questionnaire revealed a high level of interest and motivation among in-service teachers to incorporate climate change education into their classrooms. The willingness to receive training in this area was substantial, with an average score of 4.28 ± 0.76 on a 5-point scale, while the intention to implement innovative methodologies also scored high, with an average of 4.25 ± 0.80. However, despite this enthusiasm, self-assessed knowledge of climate change remained moderate, indicating a need for further training to strengthen teachers' confidence and competence in this field.

Primary and secondary school teachers play a pivotal role in shaping future citizens who are both aware of and committed to climate action. According to Aczel and Makuch (2023), teachers possess a multiplier effect, educating multiple generations throughout their careers. Given their central role in climate change education, it is essential that teachers adopt pedagogical approaches that foster critical thinking, civic responsibility, and proactive engagement with climate challenges. In this regard, teacher training becomes indispensable, as insufficient preparation can result in either the absence of climate change education in classrooms or the perpetuation of ineffective teaching practices that inadvertently sustain inaction (García-Vinuesa, 2024).

In response to these findings, a tailored training program was designed for in-service secondary and primary school teachers in Granada and Almería, and was submitted for approval to the Teacher Training Centers in both contexts. The course aims to bridge knowledge gaps, strengthen teachers' pedagogical strategies, and promote a holistic approach to the inclusion of climate emergency in the educational system that integrates cognitive, emotional, and action-oriented components.

3.2 Phase 2. Alpha Course Design and Internal Validation

In the initial iteration of the proposal, the training program was structured around a robust theoretical framework, with the primary objective of providing a scientific and technical update on climate change, addressing key aspects such as its causes, consequences, and potential pathways for mitigation and adaptation. Moreover, it incorporated a comprehensive approach to the significance of emotions in environmental education, the application of disruptive methodologies, and Game-Based Learning, among other innovative pedagogical strategies. The design aimed to eliminate potential biases of denialism within the teaching staff, particularly those who, albeit subtly, attempt to downplay or delay necessary actions in the fight against climate change. Similarly, the objective was to lay a solid scientific foundation that underscored the importance of the playful-emotional dimension as a potentially transformative tool for the climate change education of students.

A preliminary draft of the training program was developed, comprising eight modules that outlined the pedagogical objectives, thematic content, methodology to be employed, session timelines, activities to be undertaken, required material and human resources, evaluation criteria, and reference bibliography. However, following a critical review and multiple discussions among the research team, it was determined that the proposed methodological approach adhered to a more traditional teaching model, potentially causing a disconnection between the information conveyed regarding the benefits of playful-emotional education and how this information would be applied with teachers. Consequently, despite the progress achieved in the initial structuring of the training program, it became evident that a re-evaluation of the proposal was necessary. The original design did not align with the principles of methodological disruption or the transformative pedagogical values that were central to the programme's objectives. Consequently, a redesign was undertaken to ensure that the training experience embodied educational innovation, promoting active participation, meaningful learning, and the integration of play-based methodologies aligned with the emotional and experiential dimensions of teacher professional development.

3.3 Phase 3. Disruptive and Game-Centered Beta Training Program Design

As a precursor to developing the beta version of the training program, a thorough review of the available serious games with potential for climate change education was conducted. This selection was based on scores derived from the needs assessment, which evaluated both teacher and student interest and motivation toward various types of playful resources through one of the dimensions of the *Resclim@tiempo Questionnaire*. Among the top-rated options by both groups were escape rooms, board games, role-playing games, and video games. To ensure the pedagogical appropriateness and feasibility of the selected resources, specific screening criteria were established and applied (Table 2). Game exploration was carried out through systematic searches on platforms such as Itch.io, Steam, CENEAM, Genially, InteRed, among others.

As a result, a structured database was created, containing over 60 serious games related to climate change. From this selection, the research team applied the *Resource Evaluation Scale*, an *ad hoc* tool designed to assess each game's pedagogical and instructional potential based on criteria such as its ability to promote active learning, alignment with curricular content, and ease of classroom implementation (Table 1). Following this evaluation, games with the highest scores were prioritized for inclusion in the beta version of the training program. However, the final selection remained subject to review by experts in teacher training and environmental education. This subsequent validation phase would allow for fine-tuning and optimizing the final selection of resources based on real-world applicability in educational contexts and the specific needs to be addressed.

Building on this foundation, an initial draft of the training program, titled *Disruptive Methodologies for the Playful-Emotional Teaching of climate change*, was

Table 2 Screening criteria for game selection

Criteria	Type	Description
Serious games	Exclusion	The selection of resources is based exclusively on those that have been designed and/or adapted for pedagogical purposes
Central theme: climate change	Exclusion	Climate change issue is systemic and complex, as it is interconnected with other issues, including social, environmental, and economic. However, the central theme of the resource should be climate change itself, which can be approached from many perspectives (e.g., causes, consequences, risks, planetary boundaries, energy, climate justice, activism, etc.)
Interconnected climate change approach	Exclusion	Resources must address climate change causes (e.g., greenhouse gases, deforestation, energy production, food production and consumption), consequences (e.g., health impacts, biodiversity impacts), and mitigation and/or adaptation solutions (e.g., energy transition, climate activism) in an interconnected way (systemic understanding, not isolated)
Non-scientific and/or climate change denial content	Exclusion	Games that promote non-scientific or climate change denial content (including traditional and new forms of denial) are excluded
Potentially harmful climate change message	Exclusion	Games whose message was of a catastrophic and disabling nature (that is, those with an extremely high negative impact, without a mobilizing or empowering approach that allows the player to know what they can do) or of an unrealistic and hyper-optimistic type (that is, those with a high positive impact that discourages action and delegates the climate fight to vague hopes [e.g. techno-optimism]) were excluded
Language	Exclusion	The selection of games is limited to those available in Spanish and/or English
Duration	Exclusion	The games selected were those with a duration per game session of less than or equal to 45 minutes, which could be completed within the time frame of a single class session
Ease of implementation	Exclusion	The selection of games is based on the simplicity of their fundamental rules and mechanics, which are essential for their implementation. Additionally, interactive tutorials are a requisite for the selection of digital games
Non-functional media (digital only)	Exclusion	Digital games that were based on non-functional digital media (e.g., Java, Adobe, etc.) were discarded
Playable on browser (digital only)	Exclusion	The digital games should be playable on browser. These resources generally have lower hardware requirements and facilitate ease of implementation in the classroom, as they require no prior preparation

(continued)

Table 2 (continued)

Criteria	Type	Description
Age-appropriate difficulty	Exclusion	Only games whose level of difficulty (in terms of mechanics, concepts, content, among others.), has been explicitly designed to be suitable for primary and/or secondary education students are selected. This ensures cognitive and curricular alignment with the target age group
Realistic collective solutions	Priority	Preference is given to games that present feasible, systemic, and collective solutions for climate mitigation and/or adaptation
Low-cost or free access	Priority	Preference is given to free or low-cost resources (e.g., open access, print-and-play) that require minimal materials or installation
Games suitable for the entire class	Priority	Games that require multiple players are preferred, particularly those designed for whole-class engagement (e.g., 25–30 students)
Cooperational objectives	Priority	Preference is given to games that foster cooperation and shared goals both within individual groups (intra-group) and across different groups (inter-group), promoting a non-competitive, collaborative learning environment
Teacher guidance or supplemental materials	Priority	Games that provide teaching guides, curricular links, and/or complementary learning activities to support educators are given priority
Inclusive and intercultural perspective	Priority	Preference is given to games that reflect social, ethnic, gender, and geographic diversity, and address climate justice through an inclusive lens
Appealing visual design	Priority	Preference is given to games with appealing, well-developed visual design (e.g. tokens, cards, characters, environments, inventory, among others)

designed to serve as the starting point for the co-design phase with pedagogical advisors. The course was structured into 5–6 sessions, each lasting approximately three hours. In the first four sessions, teachers received theoretical updates on climate change, followed by engagement with and analysis of a serious game, culminating in a reflective space and final evaluation of the experience. The final two sessions were designed for participants to develop a teaching unit incorporating one of the evaluated resources, and then implement their plan in their classrooms, facilitating the practical transfer of learnings into their educational contexts.

3.4 Phases 4 And 5. Co-Design, Dissemination and Implementation of Granada and Almería Training Programs

Although the beta design was positively evaluated internally by the research team, the final design and development of the teacher training program on climate change education resulted from a collaborative co-design process with the Teacher Training Centers in Granada and Almería. Teacher Training Centers are specialized institutions that are responsible for the management, design, dissemination, and certification of official in-service teacher training at the non-university level in Spain. These centers, regulated in Andalusia by Decree 93/2013 (Junta de Andalucía, 2013), develop, based on a process of detection of training needs, face-to-face, blended or online programs aimed at the disciplinary, methodological and evaluative updating of teachers. This training is intensified during periods of pedagogical renovation, which are often precipitated by regulatory changes and the implementation of new legislative frameworks. The organization of Teacher Training Centers and the allocation of their training responsibilities may vary depending on the territorial and geographical characteristics of the provincial education districts within the same province. These centers are staffed by specialists who are tasked with the planning and execution of the training in a coordinated and progressive manner over time.

The primary goal of this co-design process was to ensure that the course's approach, content, methodology, and other components were aligned with the specific needs and contexts of the prospective teachers in each territorial area. Collaboration began with contact between the leadership of the respective Teacher Training Centers, as well as the submission of a formal document that included an overview of the project, a summary of the previous needs assessment phases, and a detailed description of the proposed beta version of the training program. This initial step laid the foundation for smooth and constructive cooperation among all parties involved.

The initial feedback received from both centers was positive, highlighting the growing demand for climate change training and its teaching methods, as well as emphasizing the interest in a training process that was not solely theoretical but also incorporated innovative and active approaches. This response aligned with prior observations made by the research team, showing consistency between the results from the initial phase of the study and the internal needs assessments carried out by the Teacher Training Centers. As a result, the leadership of the institutions approved the initiative and directed the contact to the specialized advisors in the subject area of each center to initiate the process of reviewing, negotiating, and co-designing the training programs. In Granada, the leadership decided that the training focus would align with the Secondary Education Advisory Service in the Scientific-Technological Field, while in Almería, it was assigned to the Primary Education Advisory Service. In the latter case, the advisor initially proposed direct collaboration with one of their reference schools, which had expressed interest in participating in the co-design process and subsequent training. From that point, numerous in-person meetings were held with both advisory services to co-design training modalities tailored to the needs of teachers in each region.

In Granada, one of the first decisions made during the co-design process was to change the training program title from *disruptive methodologies* to *active methodologies*. This change was driven by the perception that the term *disruptive* could create misunderstandings within the Spanish educational context, as the colloquial use of disruptive in education is often associated with students whose behavior is perceived as problematic. After discussions between the parties, it was agreed to implement a term in the title that highlighted innovation and active student participation, avoiding any potential confusion. In Almería, this concern did not arise from either the advisor or the reference teachers, so the term was retained.

Regarding the training modality and target audience, the co-design process led to some divergent nuances in each context. The Granada training program focused primarily on Secondary Education teachers, while the Almería training program, due to the collaboration with a Primary Education reference school and a Primary Education advisor, centered on this educational level, with an emphasis on fostering interaction between both educational stages. In Granada, due to the advisor's experience, the training program was fully in-person, while in Almería, a blended format was adopted, combining both in-person and virtual sessions. This format was chosen to facilitate participation and optimize logistics, as some of the playful sessions could be conducted without the physical presence of the research team, which helped with session organization. In-person sessions were held at the reference school or the Almería Teacher Training Centers, while virtual sessions were managed via the Virtual Teacher Training Platform.

In both contexts, co-design led to the decision to reverse the logic of the first four sessions, starting with playful and practical experiences, followed by an evaluation, and concluding with a brief theoretical learning module at the end of each session. This approach was based on the premise that when teachers are in training, they are not so different from any other student. Therefore, starting each session with a playful activity was expected to boost motivation and interest, a key element to ensure sustainability throughout the training program. Additionally, the theoretical component at the end of each session allowed the content to be connected and reinforced with the playful-emotional experience from the first part. In Almería, due to the blended format, two online game sessions and two in-person sessions were planned.

Regarding resources, in Granada, priority was given to Escape Rooms (both physical and virtual) and board games, which generated more interest among teachers in the earlier phases of the needs assessment. This decision was also supported by the advisory service's recommendation to focus on fewer game types. In Almería, the co-design process resulted in a broader variety of resources, including the introduction of role-playing games.

Once the co-design phase was completed, a detailed summary was prepared containing information about the training program title, modality, location, objectives, content, and other key aspects. This summary was published on the respective Teacher Training Centers' websites, where interested teachers within Spain could register. However, given that the courses were delivered in face-to-face or blended formats, registration was primarily constrained to local teachers from Granada or Almería. Additionally, information was disseminated via phone calls and email to numerous school leadership teams in public schools, prioritizing those schools that had participated in the needs assessment phase. The aim was to evaluate the impact of the training program on the participating teachers and students.

3.5 Implementation of the Training Program

During the implementation phase, the training program structure was consistent across both contexts, incorporating an innovative approach designed to capture the attention and motivation of the teachers from the beginning. The first four sessions in both training programs followed a similar format: initially, teachers engaged with one of the selected resources, followed by a brief evaluation of their experience. At the end of each session, a *bite-sized* theoretical training module was delivered, which focused on either the use of disruptive/innovative methodologies for climate change education or a scientific update on the causes, consequences, and/or solutions to climate change.

In both cities, the fifth session focused on how to implement one of the resources explored during the training program into a learning activity in the classroom. For this purpose, teachers participated in a guided session. At the conclusion of this session, teachers were provided with a portfolio to assess the various sessions of the activity they had designed and implemented. Between the fifth and sixth sessions, teachers took a break to carry out the implementation of the learning activity in their respective schools.

Finally, the sixth session in both training programs involved a collective reflection on the implementation of the learning activities. Teachers shared their experiences and reflections on how they integrated the course resources into their climate change lessons. Each teacher submitted their final portfolio, which included a report on their implementation experience. The Teacher Training Centers then conducted an official evaluation survey of the training process. This evaluation allowed for closing the training cycle with valuable feedback for future iterations, as well as providing a contextualized analysis that would aid in interpreting the results obtained.

4 Analysis and Discussion of the Results of the Evaluation Processes

4.1 Teachers' Evaluation and Selection of Educational Resources

The evaluation process of the educational resources used in the teacher training program enabled an analysis of their pedagogical potential and feasibility for classroom implementation. Through individual critical reflection, participants assessed each resource based on its effectiveness for climate change education and its applicability in real-world contexts. Figure 2 provides a summary of the results obtained, including key variables such as the target grade level, the training program in which the resources were implemented, the number of testers, pedagogical value, implementation feasibility, application intention, and the percentage of teachers who ultimately chose to incorporate the resource into the required educational intervention.

Note: The scale for the "Pedagogical value for climate education", "Implementation feasibility," and "Application intention in next 3 months" items ranges from 1 to 4. The "Testers' final decision" is a binary scale that ranges from Yes to No.

4.1.1 Pedagogical Value for Climate Change Education

The educational resources generally received positive evaluations regarding their capacity to address climate change education, particularly those that promoted critical analysis and decision-making. Overall, the evaluated resources scored above 3.63 out of 4, indicating their pedagogical effectiveness in addressing the topic (Fig. 2). Specifically, the role-playing game *Negociaciones de altura* [Negotiations at the

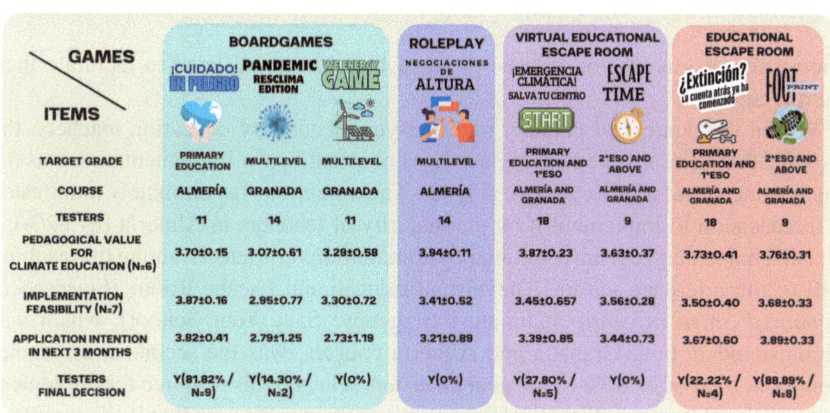

Fig. 2 Evaluation of serious games applied to climate change education

Top] received the highest ratings in terms of pedagogical value. This game allows participants to take on different roles in a negotiation context focused on reducing greenhouse gas emissions, fostering active, contextualized learning and debate to generate profound reflections on policies and practices surrounding the climate crisis.

On the other hand, the board games *We Energy Game* and *Pandemic: Resclima Edition,* used in the Granada training program, received lower evaluations concerning their pedagogical effectiveness. The former addresses climate emergency from the perspective of the energy transition, focusing on urban energy management. However, its narrow focus on climate change through this single dimension limits its capacity to offer a broader and more comprehensive view of the climate phenomenon. Although the latter explores the effects of climate change on health, particularly the spread of diseases, its emphasis on the health crisis and mitigation does not delve into the underlying causes of climate change or the solutions required, limiting its potential to foster a more holistic understanding of the issue.

4.1.2 Implementation Feasibility

Most of the resources received positive ratings, particularly for their ease of integration into the classroom without the need for extensive training (Fig. 2). The board game *¡Cuidado! En peligro* [Watch Out! In Danger] specifically designed for primary education, was the highest-rated resource in this category. The game's simple structure and clear rules made it easy to apply in various educational contexts, making it an attractive option for teachers. In contrast, the board game *Pandemic: Resclima Edition* received the lowest rating in terms of feasibility. The complex mechanics and rules posed a challenge for teachers, generating some reluctance to implement it in the classroom, especially among those with less experience using such resources.

4.1.3 Application Intention and Final Decision

The results of this section will be grouped by educational level to facilitate their interpretation.

Within the context of primary and first-year secondary education teachers, the resource that demonstrated the highest level of intention for implementation was the board game *¡Cuidado! En peligro* (Fig. 2). This intention was ultimately manifested in the decision to implement it by the majority of teachers in Almería (81.82%), a decision that emerged from a voluntary yet unanimous consensus among the teaching staff of the reference school. The Virtual Educational Escape Room *¡Emergencia climática! Salva tu centro* [Climate Emergency! Save Your School], which was administered in both Granada and Almería courses, was the second most implemented. A total of 27.80% of the general testers selected this resource for implementation in the classroom (Fig. 2). However, it is important to note that if the sample is restricted to teachers in Granada, the percentage increases to 71.43%. This indicates that the resource has a high acceptance rate, which can be attributed to its pedagogical merit and the ease with which it can be implemented in the classroom. As a result, it establishes itself as the preferred choice among Granada's participants. The third-place position is occupied by the Educational Escape Room *¿Extinción? La cuenta atrás ya ha comenzado...* [Extinction? The Countdown Is On...], which was selected by three teachers from Almería and one from Granada for its implementation. The board game *Pandemic: Resclima Edition* was chosen by only one teacher, who explained its selection based on a personal predilection for board games and prior experience in the classroom with similar games, despite its lower ratings in terms of pedagogical effectiveness and feasibility. Finally, the board game *We Energy Game* and the role-playing game *Negociaciones de altura* were not implemented by any primary or first year secondary education teachers.

The results of the evaluation and testing of educational resources by teachers of compulsory secondary education (with the exception of first-year teachers) indicate that the Educational Escape Room *Foot Print* emerged as the most popular option, exhibiting a general adoption rate of 88.89% and receiving high ratings across all evaluated dimensions (see Fig. 2). This percentage constitutes 85.71% of the testers enrolled in the Granada course and 100% of those participating in the Almería course. This resource was distinguished by its considerable pedagogical value and its feasibility for implementation in the classroom, thereby establishing itself as the evident preference of the group under consideration. In the second position is the board game *Pandemic: Resclima Edition*, selected exclusively by a single teacher. The justification for this selection was analogous to that of the primary educator, emphasizing personal preference and prior utilization of analogous games in the classroom. Regarding to the resources that were not selected in this group, *We Energy Game* was most likely rejected on account of its perceived limited approach to the issue of climate change. In contrast, Virtual Educational Escape Room *Escape Time* and the role-playing game *Negociaciones de altura* were most likely overlooked in favor of other resources, despite having received positive evaluations in several categories.

4.2 Evaluation of the Training Experience

Upon completion of the teacher training programs, the Official Evaluation Survey administered by the Teacher Training Centers was distributed to the participating teachers (Table 1).

The survey's findings, as depicted in Fig. 3, reveal an overwhelmingly favorable evaluation from the participating teachers. The mean score attained for the questionnaire was 9.45 in Granada and 9.47 in Almería, out of a maximum of 10 points, indicative of an exceptional level of satisfaction in both provinces. This high level of acceptance not only legitimizes the training proposal but also invites a more thorough exploration of the specific elements that have contributed to this evaluation.

A more thorough examination of the items deemed essential unveils a collective awareness among the teachers regarding the course's significant alignment with

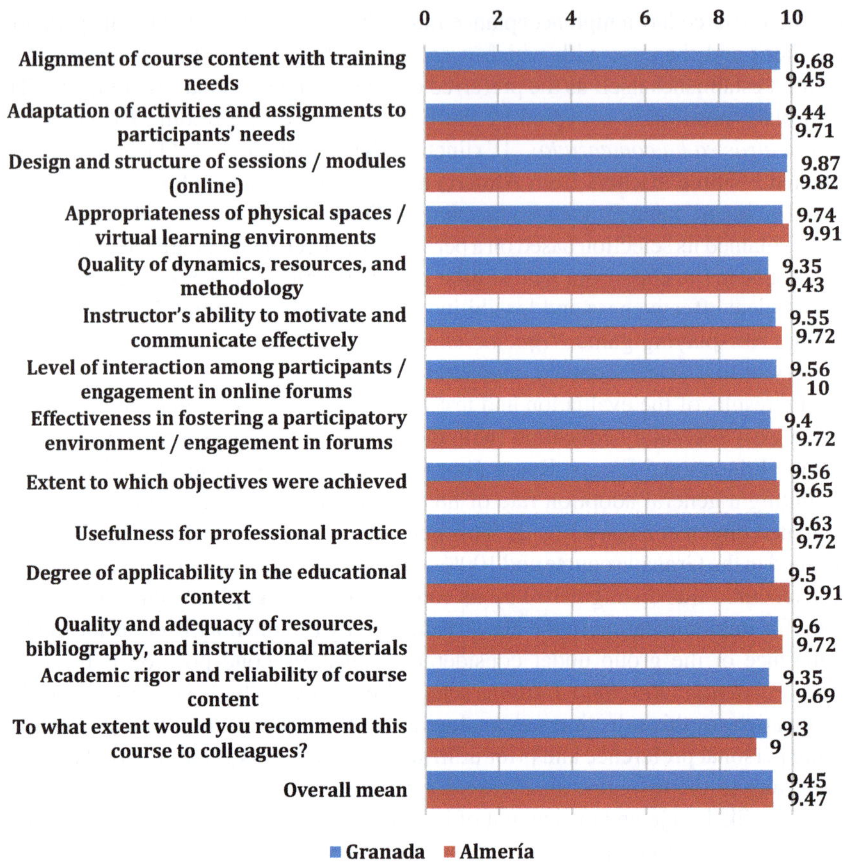

Fig. 3 Official evaluation survey administered by the Teacher Training Centers

their training requirements. The contents, activities, and resources utilized in the training program were also reported to be highly suitable, indicating a close alignment between the training offer and the professional expectations of the participants. This fit, which allows us to formulate a very positive preliminary judgment in terms of adequacy, may be explained mainly by the preceding process of needs detection, as well as by the collaborative approach in the design of the course. It is reasonable to also hypothesize that these results are associated with the explicit interest of the teachers themselves, who, when they voluntarily enrolled, already demonstrated a predisposition towards the training proposal.

Another aspect that was highly valued by the participants was the direct applicability of the course to their educational context and perceived usefulness for professional practice, indicating a meaningful connection between the course content and their professional responsibilities. These findings are particularly salient given that they serve to substantiate the pragmatic and situated nature of the training's design.

Moreover, the implications of these outcomes are amplified given that educators were tasked with implementing a pedagogical approach grounded in the resources that were presented during the course. This pedagogical approach provides a concrete illustration of the experiential nature of the training, offering empirical evidence of its efficacy in authentic educational settings.

The trainer's competence was also favorably appraised, particularly regarding their aptitude for motivation, effective communication, and the promotion of a participatory atmosphere. The participants' high degree of interaction and the construction of a collaborative environment were recurrently highlighted, which suggests a very supportive reception of the methodology used. The sessions, conceived from an active and practical perspective, appear to have furnished a dynamic and motivating learning environment, aligning with the expectations of the teaching staff.

Lastly, the expressed willingness of most of the teachers to recommend the training to their colleagues serves as a reliable indicator of the program's overall effectiveness. This intention, substantiated by elevated scores on all evaluative items, denotes a discernible indication of the positive assessment of the course and its prospective replicability in subsequent iterations.

5 Conclusions, Limitations, and Future Research Directions

The teacher training program on climate change received highly favorable evaluations, suggesting that the training model implemented holds great promise for future editions. This model, grounded in a solid theoretical framework and informed by a needs assessment, was significantly enhanced through co-design with the Teacher Training Centers and several teachers. This collaboration likely contributed to the training program's success by aligning it closely with the teachers' specific needs and the unique characteristics of the local educational context.

Current educational legislation in Spain recognizes climate change education as a cross-cutting theme across all educational levels. However, the results from the training program suggest substantial gaps in teacher training for addressing this issue effectively. Despite the growing interest among teachers in climate change education and innovative methodologies such as Game-Based Learning, the climate emergency remains insufficiently integrated into continuous professional development programs and initial teacher education. The needs identified by the teachers in the assessment, feedback from Teacher Training Centers leadership and advisory bodies, and studies such as those by García-Vinuesa (2024) and García-Gómez and Gil Flores (2022) highlight the urgent need for specialized and ongoing training in this area.

The disruptive approach, centered on experimenting with educational games, was a key factor in the training program's success. Teachers not only gained theoretical and methodological knowledge and skills but also had the opportunity to test the resources within their own contexts, reflect on their applicability, and exchange experiences. This approach, which fostered both individual and collective reflection, facilitated the integration of resources into teaching practice, underscoring the importance of providing opportunities for critical reflection in future editions.

Despite the positive evaluations of the training program and serious games applied in it, the participant sample presents a notable limitation. Although the training program was open to all teaching profiles, most of the secondary school participants were biology and geology teachers. This suggests that while climate change is a cross-cutting issue, it continues to be largely viewed as mainly relevant to the natural sciences. The limited participation from other departments may indicate a barrier to integrating climate change into other subject areas. It is likely that this perception arises from the belief that the topic is more suited to science disciplines or from a lack of clarity on how to approach it in other contexts. Addressing this challenge will be crucial for future interventions, where strategies should be developed to promote the inclusion of the climate emergency across all curricular domains.

Another challenge identified was the limited time allocated for reflection after the implementation of the games. The brief duration of the post-game phase constrained teachers' ability to fully adapt the resources to their needs. It would be beneficial to allocate between 20 and 30 minutes to this phase in future editions, allowing for a deeper evaluation and the development of more effective implementation strategies.

Looking ahead, this study outlines several avenues for strengthening teacher training in climate change education. Expanding the sample nationally and exploring the use of Game-Based Learning in other educational contexts could provide valuable insights into the approach's effectiveness and adaptability. Additionally, the development of tailored educational resources that address climate change from multiple perspectives (scientific, emotional, and creative) could contribute to a more comprehensive and accessible educational experience. Finally, it is essential to continue developing teacher training programs that not only address emerging conceptual needs but also foster disruptive educational skills as critical reflection, collaborative learning, and the creation of resources that are adapted to local contexts. Collaboration among Teacher Training Centers, specialized research teams, educational institutions, and environmental organizations will be vital in designing initiatives that effectively and transformatively prepare teachers to teach about the climate emergency.

Acknowledgements This research was supported by the University of Granada and the Spanish government under Grant FPU20/03615. We also thank the Teacher Training Centers of Granada and Almería, their management teams, and especially the advisors, whose dedication and collaboration were essential in co-designing and shaping the initiative. Finally, we are deeply grateful to the teachers who participated in the research and training processes, contributing their experience and insight to collectively build and enrich the project's outcomes.

References

Aczel, M., & Makuch, K. E. (2023). Climate change, young people, and the IPCC: The role of citizen science. *Elementa Science of the Anthropocene, 11*(1), 1–13. https://doi.org/10.1525/elementa.2022.00029

Bonil, J., Sanmartí, N., Tomás, C., & Pujol, R. M. (2004). Un nuevo marco para orientar respuestas a las dinámicas sociales: el paradigma de la complejidad [A new framework to guide responses to social dynamics: The complexity paradigm]. *Investigación en la escuela, 53*, 5–19.

Brosch, T. (2021). Affect and emotions as drivers of climate change perception and action: A review. *Behavioral Sciences, 42*, 15–21. https://doi.org/10.1016/j.cobeha.2021.02.001

Caride, J. A., Gutiérrez-Pérez, J., & Meira-Cartea, P. Á. (2025). Las narrativas climáticas en las agendas educativas: sobre el valor del tiempo para la investigación pedagógica y las políticas ambientales [Climate narratives in educational agendas: On the value of time for pedagogical research and environmental policies]. *Revista Española de Pedagogía, 83*(290), 83–103. https://doi.org/10.22550/2174-0909.4147

Chen, K., Molder, A. L., Duan, Z., Boulianne, S., Eckart, C., Mallari, P., & Yang, D. (2023). How climate movement actors and news media frame climate change and strike: Evidence from analyzing Twitter and news media discourse from 2018 to 2021. *The International Journal of Press/Politics, 28*(2), 384–413. https://doi.org/10.1177/19401612221106405

Chu, H., & Yang, J. Z. (2019). Emotion and the psychological distance of climate change. *Science Communication, 41*(6), 761–789. https://doi.org/10.1177/1075547019889637

Dahalan, F., Alias, N., & Shaharom, M. S. N. (2024). Gamification and game based learning for vocational education and training: A systematic literature review. *Education and Information Technologies, 29*(2), 1279–1317. https://doi.org/10.1007/s10639-022-11548-w

Junta de Andalucía. (2013). Decreto 93/2013, de 27 de agosto, por el que se regula la formación inicial y permanente del profesorado en la Comunidad Autónoma de Andalucía, así como el Sistema Andaluz de Formación Permanente del Profesorado [Decree 93/2013, of August 27, regulating initial and continuous teacher training in the Autonomous Community of Andalusia]. *Boletín Oficial de la Junta de Andalucía*, 169.

García-Gómez, S., & Gil Flores, J. (2022). El profesorado y los enfoques innovadores en centros de educación secundaria en España [Teachers and innovative approaches in Spanish secondary schools]. *Archivos Analíticos de Políticas Educativas, 30*(167). https://doi.org/10.14507/epaa.30.7419

García-Vinuesa, A. (2024). Empowering secondary education teachers for sustainable climate action. *Sustainability, 16*(18), 7941. https://doi.org/10.3390/SU16187941

González-Muñoz, E., Gallardo-Vigil, M. Á., & Gutiérrez-Pérez, J. (2025). "The Mysterious Disappearance": Assessment of a sustainability-themed virtual educational escape room in higher education. *Journal of Applied Research in Higher Education, 7*(5), 2204–2224. https://doi.org/10.1108/JARHE-09-2023-0430

González-Muñoz, E., Meira-Cartea, P., & Gutiérrez-Pérez, J. (2024). Looking for the emotional footprint of climate change in young people: Connections with education, information sources and climate action. *Environmental Education Research, 30*(12), 2315–2339. https://doi.org/10.1080/13504622.2024.2364799

Intergovernmental Panel on Climate Change. (2024). *AR6 Synthesis Report: Climate Change 2023*. IPCC. https://www.ipcc.ch/report/ar6/syr/

Joint Research Centre. (2022). *The European sustainability competence framework*. Publications Office of the European Union. https://publications.jrc.ec.europa.eu/repository/handle/JRC128040

Jones, C. A., & Davison, A. (2021). Disempowering emotions: The role of educational experiences in social responses to climate change. *Geoforum, 118*, 190–200. https://doi.org/10.1016/j.geoforum.2020.11.006

Klein, N. (2014). *This changes everything: Capitalism vs. the climate*. Simon & Schuster.

Leiserowitz, A., Maibach, E., Rosenthal, S., Kotcher, J., Ballew, M., Goldberg, M., & Gustafson, A. (2018). *Climate change in the American mind: December 2018*. Yale University and George Mason University.

Morin, E. (1992). From the concept of system to the paradigm of complexity. *Journal of Social and Evolutionary Systems, 15*(4), 371–385. https://doi.org/10.1016/1061-7361(92)90024-8

OECD (2025). *What works? Groundbreaking evaluation of the effectiveness of climate policies*. Accessed February 28, 2025 from https://www.oecd.org/en/blogs/2025/01/what-works-groundbreaking-evaluation-on-the-effectiveness-of-climate-policies.html?adestraproject=OEC%E2%80%A6

Salazar, I. C. (2004). El paradigma de la complejidad en la investigación social [The paradigm of complexity in social research]. *Educere, 8*(24), 22–25. http://www.redalyc.org/articulo.oa?id=35602404

Spain. (2020). Ley Orgánica 3/2020, de 29 de diciembre, por la que se modifica la Ley Orgánica 2/2006, de 3 de mayo, de Educación [Organic Law 3/2020, of 29 December, which amends Organic Law 2/2006, of 3 May, on Education]. *Boletín Oficial del Estado*, núm. 340, 30 de diciembre de 2020. https://www.boe.es/eli/es/lo/2020/12/29/3

United Nations. (2015). Transformar nuestro mundo: la Agenda 2030 para el Desarrollo Sostenible [Transforming our world: The 2030 Agenda for Sustainable Development]. https://unctad.org/system/files/official-document/ares70d1_es.pdf

Veldkamp, A., van de Grint, L., Knippels, M. C., & van Joolingen, W. R. (2020). Escape education: A systematic review on escape rooms in education. *Educationl Reseach Review, 31*, 100364. https://doi.org/10.1016/j.edurev.2020.100364

Educating for the Climate Through Film Literacy: *"Educinema Clima Tour Action"* Erasmus+ Project

Francisco Sóñora-Luna, Aitor Alonso-Méndez, and José Antonio Caride

Abstract This chapter is based on the Erasmus+ project "Collective Creativity and Community Education in Film Literacy for Climate Action Tourism" (Educinema Clima Tour Action), funded by the European Union in collaboration with the Spanish Service for the Internationalisation of Education (SEPIE), under the Spanish Ministry of Science, Innovation and Universities. Led by the research group in Social Pedagogy and Environmental Education (SEPA-interea) of the University of Santiago de Compostela, with the involvement of universities and schools from Spain, Italy, Poland and Portugal, the project ran between 2020 and 2024. Its core objective: to promote film literacy, collective creativity and community education by fostering good practices in environmental education that promote sustainable tourism as part of climate action (SDG4 and SDG13). Through an interdisciplinary social pedagogical approach, using STE(A)M methodology, the project outcomes are clearly novel in terms of both methodology (including the production of a film musical titled *Climantopia*) and the content dealt with. Following its evaluation, the project was recognised as a good practice by the European Commission.

1 Challenges Inspiring the Project

The project was designed in 2020 as part of the actions funded by the Erasmus+ Programme aimed at encouraging "cooperation for innovation and the exchange of good practices" in the field of innovation and creativity (KA227). It was the best-rated proposal of all those submitted in this category to the Spanish Service for the Internationalisation of Education (SEPIE) under the Spanish Ministry of Science, Innovation and Universities. Without ignoring the adversities caused by the Covid-19 pandemic, the first actions under the project started on 1 March 2021 and ended in March 2024. Three major challenges influenced their approach:

F. Sóñora-Luna · A. Alonso-Méndez · J. A. Caride (✉)
University of Santiago de Compostela, Santiago de Compostela, Spain
e-mail: joseantonio.caride@usc.gal

J. J. Conde et al. (eds.), *Practices in Education for Climate Emergency*,
https://doi.org/10.1007/978-981-92-2304-6_14

(a) The growing need to address environmental education and climate awareness in and with students aged 15–17 years old, in response to scientific evidence detailed in the IPCC 2007 report that the source of climate change is more than 90% likely to be anthropogenic. Mass protests by school communities around the world at the end of the second decade in the twenty-first century made climate change and climate emergencies a priority for the education system, mainly in secondary education and continuing education for young people and adults, through an approach based on social pedagogy and environmental education.
(b) The impact of the Covid-19 pandemic, highlighting the importance of promoting sustainable tourism and climate action to prepare societies for post-pandemic mass tourism in attractive yet low-stress landscapes or destinations, due to the pull effect of "travellers" in search of less crowded destinations. Environmental education and social pedagogy applied in schools create opportunities to foster educational processes that prevent the risks of mass spread of post-pandemic tourism in new destinations, sometimes characterised by considerable fragility or vulnerability in terms of ecology, heritage and society. Engaging tourists in climate action and the local communities they visit requires providing them with pedagogical opportunities that give them effective tools for preventing potential risks associated with replicating, in their own contexts, the socio-environmental imbalances caused by mass tourism throughout the twentieth century and the first two decades of the twenty-first century. Action from teenagers and young people in schools, trained to help raise awareness among those in their communities and incoming tourists after the pandemic, became a major challenge for this project.
(c) The pandemic was an obstacle to collaborative collective creativity processes in school communities due to lockdowns and social distancing measures in the classroom. Overcoming these barriers became an urgent necessity, with social pedagogy and environmental education arising as two areas requiring significant educational reflection and action. In this context, film literacy in the school environment emerged as an opportunity to encourage and strengthen collective creativity among students. This involved novel participatory methods that—beyond school education, in the wider community—gave rise to processes of environmental awareness and advocacy related to climate change and the importance of climate action, from both a local perspective (in their own context) and a global one (across the whole planet). A task that saw great potential in the use of the performing arts, video resources and, above all, films starring (where possible) young people from different countries, encouraging their collaboration and involvement in pedagogical practices, teaching and learning with a clear focus on innovation and transformation.

2 Approach and Objectives

The conception, design and implementation of initiatives responding to these challenges and their processes of collective creativity reinforced the development of key skills in the area of school education through the STE(A)M methodology, which combines school science with art, technology and social engagement. As is well known, the STE(A)M methodology, beyond a mere teaching model, identifies a way of constructing learning in a cross-cutting, interdisciplinary way, with a focus on critical thinking, creativity, problem solving, collaboration and applying knowledge to practice, linking educational content with the reality of everyday life (Trilling & Fadel, 2012). STEAM methodologies were introduced into the project through the development of STEAM learning ecosystems which connect schools, universities, research centres, and science outreach institutions. These ecosystems aim to transform learning processes and dynamics by incorporating the artistic dimension of collective creativity within STEAM environments for teaching and learning experimental sciences (Bonilla et al., 2024), while also emphasizing a socio-pedagogical and community-oriented perspective (Caride-Gómez, 2022).

Students were motivated and engaged through a participatory model cantered on the production of a film about the impacts of climate change. The objective was to generate emotional, behavioural, and cognitive impact, thereby improving students' perceptions of competency-based work in collective STEAM creativity (Maričić & Lavicza, 2024). The project began with a training for participating teachers in Aveiro (Portugal) and Bologna (Italy) in STEAM techniques linked to collective audiovisual creation. Teachers were trained in audiovisual production within a STEAM framework as a first step toward harnessing the transformative potential of STEAM pedagogy, grounded in a holistic and interdisciplinary approach to education. This process placed teachers within an innovative context aligned with the intended learning ecosystem (Deák & Kumar, 2024). Teacher training was considered essential to reconcile curricular overload with the necessary interdisciplinary engagement of students in key competencies for collective STEAM creation. This approach enabled the high level of personal commitment to climate change education required by the project. In this way, the difficulty of integrating climate change into non-scientific subjects was overcome (García-Vinuesa et al., 2026), incorporating not only new ways of teaching and learning but also new forms of integrating multiple disciplines: arts, sciences, languages, technology, and citizenship.

Inevitably, this STEAM educational innovation project on collective school creativity faced the challenge of balancing the "open schooling" needs demanded by its innovative proposals with the constraints inherent to school schedules and curricula. It was therefore necessary to balance an ecological model of school openness to the environment with management strategies aimed both at mitigating resistance to innovation and at fostering the institutional, familial, and community collaborations required at different stages of the project; for example, in international mobility initiatives, rehearsals, and the filming of scenes for the intended movie. To reconcile regular curricular activity with educational innovation, various organizational, pedagogical, and community dimensions of school openness were considered (Sarid et al., 2025). This required fostering a deep and meaningful dialogue about the advantages of opening schools, despite the disruptions this might cause to teaching activities, calendars, and school schedules.

Students participated in highly significant mobility experiences, particularly due to the high-impact tourist contexts visited, especially in the city of Lisbon and in the Canary Islands. During their implementation, these activities generated forms of resistance among teachers similar to those identified by Chambers (2020), such as concerns that students were spending excessive time outside the classroom. Such negative feedback was interpreted as potential resistance to innovation, since producing a film as the central thread of the project represents an ongoing process of innovation for schools engaging for the first time in film literacy and production activities (Chambers, 2020). The STEAM Open Schooling systems (Okada, 2023) encourages students to collaborate with various entities -such as organizations (e.g., shellfish gatherers' associations in this project)- or universities. In this regard, the approaches and methodologies of Climántica's campuses (Sóñora-Luna & García-Vinuesa, 2020) were considered, including the study of climate change impacts on cockle sustainability in Galicia or rice cultivation in Montemor (Portugal). Opening schools to society is essential to connect learning with real problems, fostering research skills, collective creativity, and cooperation with diverse external stakeholders: local administrations, third-sector organizations, media, companies, professional groups, artists and performers, and research centres.

In their respective contexts, the climate crisis represents the most complex social, economic, and environmental challenge facing humanity, requiring educational responses to climate change (García-Vinuesa & Meira-Cartea, 2019; Caride-Gómez & Meira-Cartea, 2020; González Gaudiano & Meira-Cartea, 2020; Sóñora-Luna & García-Vinuesa, 2020). From a pedagogical perspective oriented toward educational innovation, it is necessary to advance toward STEAM models based on the Open Schooling approach. This paradigm promotes the opening of educational institutions to the social, scientific, and productive environment, fostering connections between learning and real-world issues: wildfires and erosion in Galicia, urban overtourism in Lisbon, or mass tourism in the Canary Islands, where students identify key socio-environmental challenges linked to climate change in areas expecting high tourist pressure.

Science and language skills were integrated into screenplays, coming together in the production of a film created by schoolchildren in contexts or activities with a high socio-environmental impact, such as those linked to tourism and the risks inherent to climate change. In order to create the screenplays, a school musical theatre play was designed beforehand to build motivation, attitudes, skills and so on, seeking to provide students with educational "tools" to improve their knowledge and training in film literacy and interpretation. This would also prepare them for a new understanding of the world and its circumstances, between the symbolic and the material, the imaginary and the practical, both individually and collectively.

In this regard, and in order to guide the production of the musical theatre play and the film *by* and *for* schoolchildren, methodological guides were created for students from different countries involved in the project. These guides not only deal with the creation and editing of the film, but also other initiatives like the production of e-books and other resources on methodology and teaching in the field of school science, including the didactic exploitation of the film in the educational communities involved in the project. In any case, the aim was to ensure the project outcomes achieved the following general objectives:

- To develop a STE(A)M pedagogical model for secondary school students, focusing on climate change and collective creativity.
- To establish multi-sectoral consortia to promote environmental education and film literacy in order to boost green tourism in vulnerable areas affected by Covid-19 and climate change.
- To incorporate collective creativity, science and art in the promotion of health and the environment, with a focus on intergenerational and community collaboration.

These general objectives were based on the following specific objectives:

- To promote the inclusion of climate literacy in school curricula.
- To foster a culture of sustainability in secondary schools through environmental education in climate action.
- To promote entrepreneurship in the area of climate action ecotourism through short films.
- To produce environmental short films focusing on potential tourism hotspots near schools involved in the project in each country.
- To assess the relationships between global environmental impacts and health, particularly in the context of the Covid-19 pandemic.

3 Participants

The main actors involved in implementing the project were students aged 15–17 and their teachers, who were considered the primary target group. However, due to the interdisciplinary nature of the project, which combines elements of education, film and tourism, other secondary target groups (policy makers, professional groups, local actors, families, etc.) and society at large were also taken into account.

From the beginning of the project, specific activities were designed for 15–17 year old students and their teachers, highlighting their place in a certain educational community. This involved promoting international educational activities related to film and climate change, and the active involvement of students in the production of short films about climate change and sustainable tourism. The teaching staff played a key role in guiding and supervising the students in these activities, as well as in meeting the project's pedagogical and creative objectives.

The use of film as a resource, medium and communication system allowed the initiative's actions to reach beyond the school communities involved. The outputs and outcomes were shared online, and various multiplier events and film festivals were organised, including in Las Palmas in Gran Canaria, where the project was brought to a close. This made it possible to reach out to society at large, including families, other educational institutions and local communities interested in issues about climate change and sustainable tourism. The project's multidimensional approach contributed to its success, boosting its visibility and impact on public awareness of critical sustainability issues.

The University of Santiago de Compostela - USC (Spain) led the project and its overall coordination and was also responsible for the work on pedagogy and didactics through the research group in Social Pedagogy and Environmental Education (SEPA-interea). This included collaboration from the projects division of the Vice-Rectorate for Science Policy at the University of Santiago de Compostela. Also involved were the University of Aveiro - UA (Portugal), coordinator of STE(A)M processes, and the University of Bologna - UNIBO (Italy), coordinator of the methodological e-books on film and theatre. The three partner universities (USC, UA and UNIBO) played a key role in selecting participants considered to be leading institutions due to their commitment to educational excellence, promoting research and pedagogical innovation based on the arts and sciences of learning and the STE(A)M approach.

Finally, the consortium included three schools: Instituto de Educación Secundaria Villa de Firgas on the island of Gran Canaria (Spain), XXVI Liceum Ogólnokształcące in Lodź (Poland) and Instituto Español Giner de los Ríos in Lisbon (Portugal). These three schools (IES VDF, LICEUM XXVI and IEGRL, respectively) allowed direct access to their school communities, teachers and students, thus ensuring the active involvement of students and educators in the project.

Beyond these "formal" partners, a number of "non-formal" partners also played an important role in the project, including Instituto de Educación Secundaria Virxe do Mar de Noia (Spain), which made significant contributions by sharing school science research related to the promotion of green tourism. It also helped by promoting science outreach at cultural tourism events held in other European countries, and by organising a multiplier event in the municipality of Outes (Spain) alongside the local secondary school Instituto de Educación Secundaria Poeta Añón, which offered its facilities and the infrastructural resources required for shooting part of the film *Climantopia*. Other collaborators on the shooting of the film included Finisterrae Aquarium (A Coruña), the Lisbon Oceanarium, the Pabellón del Conocimiento de Ciencia Viva in Lisbon, the company Naturgy and the Elder Museum based in Las Palmas in Gran Canaria. Moreover, the Spanish Association for Earth Science Education (AEPECT) organised an official teacher training course about the project, with 24 hours of continuing education recognised by Spain's Ministry of Education and Vocational Training.

In Portugal, the Agrupamento de Escolas de Aveiro, a group of public schools with experience in European projects, actively collaborated on the project. It published the project outcomes in the context of school education in Portugal and invited accredited European schools to be part of an international campus for the project. At this event, the local public authority Câmara Municipal de Aveiro played a key role in providing facilities and transport for the activities on the campus. The president of the Câmara Municipal de Aveiro officially opened the campus where students from different countries were involved. It should be noted that the local government's commitment in Aveiro contributed significantly to the awareness and engagement of the local community in climate and sustainable tourism activities.

Under the coordination of the Universities of Aveiro (UA) and Santiago de Compostela (USC), the non-formal partners in Aveiro, particularly Fábrica Centro Ciência Viva de Aveiro, managed to bring students from the three partner countries together with those from schools in a significant number (10) of regions across Spain and several countries in three regions of the Americas. This student gathering in Aveiro was a success thanks to the Climántica platform, which played an active role in the partnerships established between the Universities of Santiago de Compostela and Aveiro over the course of the project.

Ongoing dialogue, collaboration, inclusion, communication and accessibility were fundamental principles in the execution of all the activities that enabled students from different countries to come together in Aveiro. Their involvement not only provided them with the chance to feel empowered and self-fulfilled but also saw them as key agents in "positive" environmental change through creative film and musical theatre projects, taking school science on climate change as a base. This contributed to a new model of interactive culture together with a culture of sustainability and community, allowing all people to unleash their human potential on equal terms, while fully respecting their identity in a scenario enriched by diverse cultures, languages, expectations, experiences, etc.

In addition to the campus set up in Aveiro, collaboration between the various organisations and entities involved in the project in a wide range of activities open to educational communities made it possible for the established objectives to boost (both quantitatively and qualitatively) the outreach and impact of the activities carried out, especially the following:

- A joint event at the University of Bologna for the training of 21 teachers and professionals in the field of education from Spain (10), Portugal (7) and Poland (4), aimed at producing methodological e-books on film and theatre designed by the teaching and research staff at the Italian university.
- Three short-stay mobility programmes, with the involvement of 53 secondary school students and 23 accompanying teachers, together with staff specialised in performing arts and film. Along with other activities, these programmes allowed for the film to be shot in different locations in Galicia, Portugal and the Canary Islands.
- Three group seminars with different collectives or interest groups linked directly or indirectly to the project, involving 91 participants from Spain, 40 from Portugal and 137 from Poland.
- A conference involving academic and institutional representatives of the partner organisations, where the film *Climantopia* was shown in the auditorium of the

Elder Museum in Las Palmas, Gran Canaria, officially bringing the project to a close.

All activities carried out involved a total of 2233 project beneficiaries or addressees, most of them through collectives, institutions, etc. as "non-formal" partners of the project, through training seminars, creative workshops, short film competitions and communication through newsletters/an ad hoc website.[1]

4 Methodology

The project, through its social pedagogical approach, opted for a STE(A)M methodology linking the areas of Science, Technology, Engineering, Arts and Mathematics. The project initiatives adopted approaches and strategies intended to encourage creation, collaboration and learning-by-doing in a cross-cutting, interdisciplinary way. The methodology used in the project consisted of a range of approaches to enhance its design, implementation, monitoring and evaluation. Essentially, these were as follows:

- The Logical Framework Approach (LFA) was adopted at the start of the project when designing the technical report: from the beginning, the project was based on the LFA, a management and planning tool that ensures a precise, coherent formulation of this type of initiative, making it possible to provide it with a solid structure and a clear picture of the objectives and activities.
- The STEAM Approach, which was part of the methodology in order to promote the inclusion of science and art disciplines, fostering creativity and collaboration between students and teachers in exploring topics related to climate change and film.
- The Participatory Approach (PA) with the explicit purpose of encouraging the active participation of students, teachers and local communities at all stages. Dialogue and collective decision-making were encouraged, leading to greater involvement and a sense of belonging.
- The Scrum Approach for the efficient management of the film production team, seeking to organise work in iterations known as 'sprints' and to foster ongoing collaboration between the production team, students and teachers. In line with the agile Scrum principles, which include transparency, inspection, adaptation, collaboration, value focus and self-organisation, the aim was to improve efficiency and flexibility in the making of the film.
- The hosting of an international online short film contest for students, including drafting its rules, an application form and evaluation criteria for the jury. This pilot contest attracted more than 1500 participants from partner schools and the project's pilot network of schools in different regions across Spain and in

[1] Available at: https://www.educinema-cta.org/

three regions of the Americas involved in the Climántica platform. This initiative encouraged students' active participation and creativity in making short films about climate change in the context of their respective countries.

5 Intellectual Outputs and Outcomes

All the intellectual outputs and outcomes of the project are free to access on the project's website[2], which also contains weblinks to information about the project, partner organisations, activities carried out, videos (including the film itself) and more. The project outputs included: a methodological e-book on the language of film; a methodological e-book for musical theatre and shows; a school film musical for climate action tourism; an e-learning unit for the didactic exploitation of the film; and a handbook on the design of creative citizen science practices with infographic models for school stands at tourist sites. Below, a brief description of some of the details relating to each output is provided (methodology, content, usage, etc.).

5.1 Methodological E-Book on the Language of Film

This is a methodological handbook for training secondary school teachers in film literacy, highlighting the climate emergency and the need to rethink post-Covid-19 recovery. Its contents underline the importance of fostering a culture of innovation that involves society as a whole for sustainable tourism management in new destinations that can be promoted after a health crisis such as the pandemic. The e-book focuses on film literacy to raise awareness of the challenges facing tourism through continuing education for young people and adults, rooting the process in film literacy in school communities. In this context, the STE(A)M (Science, Technology, Engineering, Arts and Mathematics) methodology plays a key role in helping students develop the skills and competences needed for innovation and action on critical issues such as climate change and sustainability.

This methodological e-book on the language of film is one possible response to the need to educate society from the stages of childhood and adolescence, and guide people towards climate action in the post-pandemic era in the face of a potential boom in mass tourism in their communities. It seeks to improve secondary education teacher training regarding the use of film as a pedagogical tool to teach new generations about the film industry and its ability to spread knowledge, share emotions and raise awareness, with a focus on the climate emergency and the Sustainable Development Goals (SDGs) that address it, especially SDG13 on Climate Action.

A novelty of the e-book is the focus on the language of film, from the perspective of school education, as a creative art form that is crucial to educating and raising awareness about climate change. Some particularly novel elements are: an interdisciplinary approach that integrates concepts of STE(A)M (Science, Technology, Engineering, Art and Mathematics) into environmental education; content that explores the role of audiovisual communication in climate change education and awareness-raising; practical examples and weblinks that foster a deeper understanding of film as a tool for educational and social change; and an interactive format that combines written text with video clips, interviews, behind-the-scenes footage and case studies. In addition, the e-book can be accessed for free (open knowledge) and is available in English, Spanish, Portuguese, Italian and Polish.

The e-book's table of contents includes the following chapters/sections: (1) The language of film and its relevance to environmental messages; (2) Storytelling and narrative in film; (3) Filming; (4) Editing and post-production; and (5) Organisation of work and roles in film.

It should be added that building and developing STE(A)M skills through film, as highlighted in this methodological handbook, makes it possible to have an explicit impact the following aspects:

(a) Linguistic competence: through the analysis of screenplay and narrative in film, students improve their skills in communication and comprehension.
(b) Competence in knowledge of and interaction with the physical world; film production involves technical and scientific aspects that foster interest in knowing about and interacting with the physical world.
(c) Information processing and digital competence; video editing, the use of software and digital tools are considered as key components of film production, developing digital skills.
(d) Competence in social, civic and citizenship matters; film represents a socio-cultural resource that activates critical reflection, social awareness, the values of community and empathy by exploring different contexts and perspectives.
(e) Competence in culture and art; recognising film as a form of artistic expression that enhances appreciation for cultures and different representations of art.

Beyond these aspects, this e-book aims to help teachers and students understand the complex processes involved in filmmaking, presenting them in a simple, accessible way. This is achieved through examples and weblinks that invite readers to go beyond the words on the page, encouraging dialogue elsewhere such as in film forums related to sustainable development, production and consumption, mass tourism, the role of social media, or the role of communities in protecting their natural resources, heritage, etc. Undoubtedly, this "output" promotes education in the European values of climate action and fosters a critical-reflexive attitude in students towards film and the information they get from the media and elsewhere. It therefore seeks to prepare them to face social and environmental challenges with confidence and awareness.

The e-book was presented as an educational resource to teachers at the partner organisations during a teacher training activity held at the University of Bologna, receiving a positive response and considerable interest in its use.[2]

5.2 *Methodological E-Book for Musical Theatre and Shows*

The EduCinema Clima Tour Action project based its concepts and practices on a fundamental principle of identity: school theatre and film musicals are pedagogical tools with huge educational potential in any teaching and learning process geared towards film literacy. Hence, this e-book represents an innovative methodological approach that uses theatre education and performing arts together with or as a main medium alongside music, play and science demonstrations to build skills in STEAM and socio-environmental values. Furthermore, from a gender-based perspective, it aims to strengthen female representation in technical and scientific fields, using theatre as a resource to boost the self-image and self-esteem of female students. The goal is to reap its potential in response to studies that have identified a gender gap in the perception of skills in STE(A)M subjects and relate it to greater social pressure on or towards women. To address this issue, the aim was to involve female students as leaders in the production of musical theatre and science shows, reinforcing their skills in science and mathematics.

In this e-book, musical theatre and science shows are presented as a novel pedagogical approach with multiple positive effects on students' personal and educational development. Using games as a teaching tool, its methodologies offer training opportunities that are particularly inspiring, motivating, creative, participatory and even fun. Students play a leading role, which reinforces their significance and involvement.

The e-book is structured in several sections covering different topics, following a prologue and introduction that seeks to motivate potential readers towards the world of musical theatre and science shows understood as educational tools. The following is a list of its various "Acts": (I) *The company*, detailing the roles involved in theatre production, from director to audience; (II) *"To play": how to start*, exploring the importance of play in theatre and how to develop confidence and listening skills; (III) *Theatre making*, aiming to guide readers through the process of creating a play, from conception to scriptwriting (IV) *Rehearsing: from the text to the stage*, looking at planning rehearsals and directing theatre scenes; and (V) *Finalizing your work*, dealing with technical aspects such as lighting and sound, costumes, dress rehearsals and how to prepare before the show. The handbook concludes with an epilogue highlighting the importance of theatre in education and the benefits of musical theatre.

[2] As mentioned, it is available for use in Portuguese, Italian, Spanish, Polish and English on the project's website via the following link: https://www.educinema-cta.org/ebook-metodologico-sobre-el-lenguaje-del-cine/

Designed as a methodological handbook on school theatre, it answers frequently asked questions and provides further clarification on theatre methodology and science shows, generating a significant impact on the education and development of schoolchildren. The resource has received positive feedback from teachers and education professionals who have had the chance to read and use it. The resource's outcomes include:

- Greater socialisation: through a methodology that encourages social interaction among schoolchildren, especially benefiting those with learning difficulties by creating an environment of support and collaboration.
- Reinforcement of self-image; by showing how musical theatre and science shows empower students, improving their self-image and self-esteem by enabling them to play an active role in the educational process.
- Better concentration and focus; thanks to a playful and participatory methodology that stimulates students' concentration and focus by actively involving them in the activities.
- Teaching values; theatre and science demonstrations teach about values such as teamwork, cooperation, creativity, the importance of science and gender equality.
- Gender promotion and empowerment; by involving female students in leadership roles, the approach promotes gender equality in technical and scientific fields, reducing the gender gap.
- Encouraging thought and reflection; the book's methodology stimulates critical thinking and reflection in students, strengthening their cognitive skills.
- Stimulation of collective and individual creativity and imagination; by offering initiatives, practices, dynamics, etc. in which students can express their ways of being, interests, motivations and expectations of their daily life.

This "output" is aimed at teachers in different levels of education (especially secondary education), education professionals (social educators, educationalists, educational therapists, etc.), extracurricular instructors and any professionals directly or indirectly involved in activities of an educational, artistic, recreational or socio-cultural nature. This e-book was also presented as an educational resource to teachers at the partner organisations during a teacher training activity held at the University of Bologna, receiving a positive response and a considerable interest in its use.[3]

5.3 *"Climantopia": School Film Musical for Climate Action Tourism*

This project output seeks to involve school children aged 15–17 in a film musical about climate action tourism in various countries, conceiving of film as a fundamental tool in raising awareness among tourists and people living in places susceptible to

[3] Available in five languages on the project's website via the following link: https://www.educinema-cta.org/ebook-metodologico-para-el-teatro-musical-y-shows-de-divulgacion-cientifica/

a potential boom in tourism, especially in critical contexts such as those caused by, for example, the Covid-19 pandemic or, more recently, climatic phenomena such as droughts and heat waves. The film's screenplay was designed to cover the key causes of climate change and its consequences on land and marine ecosystems, particularly in the urban context of cities where tourism is greatest, offering concrete proposals for climate change mitigation and adaptation. In this sense, the film was geared towards its use by teachers in secondary school curricula, with its subsequent transition towards citizen science contexts, with students trained in the classroom acting as agents of social pedagogy, in an attempt reconcile their role as citizens with the development of skills that stimulate their creativity and interest in the environment, on both a local and global level.

The film was made with an emphasis on raising awareness of the contents of the project, which comprises the short films of each partner organisation involved in the creation of this output, with students from Spanish, Portuguese and Polish schools and scenes shot in Galicia (Spain), Lisbon (Portugal) and the Canary Islands (Spain). This includes a supposed award-winning short film in a final contest of the project, made by the Polish students in their country. Each partner country also contextualised the performance of the musical theatre show on climate change, specifically in relation the part of the script that refers to their country.

Through the multiplier events and outreach activities carried out in each country, together with international mobility programmes and stays, the initiative involved visiting places that attract a lot of tourists, which are of special relevance due to the impact that climate change is having on them. This served to design and developed communicative scenes involving tourists as voluntary extras of environmental education. The scenes of the film were shot in museums that contribute most significantly to informing people about the circumstances of climate change, as well as in the main squares of the towns and cities involved. But above all, scenes were shot in coastal environments, due to how attractive these environments are for both visitors and the local population.

One of the main novelties of this "output" is the completely original idea of films being made collaboratively, created and developed through the active involvement of schoolchildren from different countries, focusing on this theme. It should be added that the film is being promoted by all the project partners, entities and organisations involved, reaching out to a wide audience. Moreover, the film inherently boosts motivation for schools, youth groups and associations that have an interest in involving students in local activities through their active involvement.

In designing and implementing the various tasks associated with this intellectual output—also a work of art, the following activities were carried out: (a) The creation of a teaching guide; (b) determining the kind of communication to be used between the project partners; (c) organisation of a short film contest and selection of participants; (d) design of the film (the creation of the plot, the screenplay and a script for the musical on climate change and green tourism); (e) identification of tourism practices in Galicia, Canary Islands, Poland and Portugal; (f) coordination and organisation of the scenes of the musical with those of the films in each country; (g) filming of the short films by the students in fiction format, including shots of nature (landscapes);

(h) putting together and editing scenes, including the translation and preparation of subtitles in different languages; (i) a strategy for the promotion and dissemination of the film to be shown during the mobility programmes that took place in Galicia, Lisbon and the Canary Islands; (j) presentation and premiere of the film at the Elder Museum in Las Palmas, Gran Canaria. During the event, students' opinions on the filmmaking process were gathered, encouraging them to explore the topic of climate change and their passion for film.[4]

5.4 E-Learning Unit for the Didactic Exploitation of the Film

The e-learning unit has proved to be a highly effective tool in the form of a 120-page e-book.[5] Offering a comprehensive teaching approach to the content of the film, the book represents a versatile tool that can be integrated into various subjects, addressing specific skills and interdisciplinary aspects, such as those proposed by the latest law reforms in education in Spain (LOMLOE). Through science practices and simulations, this output offers a comprehensive practical guide to help students acquire the knowledge and skills they need to successfully explain the main issues of climate change to citizens, especially in places in their countries and cities that may potentially see a boom in tourism.

Designed for teachers, educators and students, this unit is ready for use in secondary schools, seeking to achieve objectives such as the following: (1) Facilitating the understanding of methods of collective creativity and innovation; (2) promoting participatory processes that stimulate the collective creativity of local citizens and tourists for the management of socially shared ecotourism projects; (3) offering a comprehensive practical tool for implementing participatory activism in school and community settings; (4) fostering an interactive, creative and collaborative culture; (5) incorporating film literacy and the performing arts as tools to talk, act, feel and raise awareness about climate change.

In this respect, it should be noted that measures are being taken in Galicia to address the rise of forest fires due to climate change driven by increased eucalyptus farming, as well as the risks and effects of desertification. Lisbon is addressing the impact of mass tourism in major capitals, the risks of rising sea levels and the need to adapt to climate action. In the Canary Islands, they are considering the opportunities and demands of managing the expansion of renewables, the difficulties of unsustainable urban planning and responses to climate action. Finally, in Poland,

[4] The film, the trailer and the "making of" are available via the following link: https://www.educinema-cta.org/pelicula-educinema-clima-tour-action/

[5] This resource, titled *"Climantopia: The School Book"*, was published by the University of Santiago de Compostela's Publications and Scientific Exchange Service (ISBN 978–84–19,679-95-6). The publication is available to download for free in English, Spanish, Portuguese, Italian and Polish from the project's website via this link https://www.educinema-cta.org/unidad-didactica-sobre-la-pelicula/

reference is made to the country's difficulties in divesting from coal and recent progress in the green transition to renewables.

As an intellectual output for teaching and social pedagogical purposes in environmental education, the following features of the resource stand out:

(a) It has an interactive format as an e-book, which makes it easy to access and navigate. Interactive elements include weblinks to additional resources, practical activities and support materials.
(b) Use of visual resources such as infographics and photographs related to the film and musical theatre. These resources deepen understanding and make it easier to respond to the activities proposed.
(c) It addresses specific challenges of climate action in affected tourism hotspots, providing guidance on how to engage and act effectively. It general, it takes an in-depth look at the problems in coastal areas with excessive tourism, covering issues related to ecotourism, its economic impact on coastal communities, sustainability in activities such as fishing and shellfishing, and the challenges and opportunities in contexts like the health crisis caused by Covid-19.

The e-learning teaching unit is aimed at teachers, educators and students, especially in the area of lower and upper secondary education. Its versatile design and content mean it can used by a wide audience: education professionals, students and anyone interested in promoting creativity, collective innovation and awareness of climate change and sustainable ecotourism. Its potential impact can be seen in the possibilities it opens up for training teachers and students in methods of creativity and collaborative innovation, promoting climate action and sustainable ecotourism. Moreover, it has a high potential for international transferability, as it is contextualised in several countries and regions, and published in five languages. As such, it makes broad, diverse contributions to building a more sustainable world. It can therefore be argued that the e-learning unit for the didactic exploitation of the film is a highly effective tool that will inspire and guide teachers, educators and students in their quest towards climate action and sustainable ecotourism. With its comprehensive content and practical approach, it is a valuable resource offering great commitment and inspiring initiatives for education and community work.

5.5 *Handbook on the Design of Creative Citizen Science Practices with Infographic Models for School Stands at Tourist Sites*

This intellectual output takes the form of a handbook, with a marked practical approach, and seeks to promote open citizen science by encouraging interdisciplinary collaboration between scientists and students in school, academic, university and other related contexts.[6] This handbook was designed to cultivate core skills in students to attract the attention of tourists and explain the causes and consequences of climate change through science practices and simulations. Its content offers detailed strategies for organising climate change science fairs, together with showing the film at free screenings. With a focus on citizen science, this publication aims to expand scientific knowledge beyond the traditional boundaries of laboratories and fieldwork by involving students in the collective construction of knowledge. Consequently, it further aims to instil the spirit of collaboration and interdisciplinarity in the educational sphere, turning citizen science into a form of service-learning practice that, in a collaborative sense, involves the entire population through students and their role as agents of scientific outreach.

Its content addresses problems, contexts and situations that pose everyday challenges, inviting teachers and students to research and respond to questions of great socio-environmental importance: What solutions can the educational context offer in response to climate change? How can sustainable tourism be promoted in our local areas? How can we achieve accessible tourism in our communities? What is the greatest form of biodiversity in our areas? Why is the sea level rising in some areas despite melting ice elsewhere? What are the effects of ocean acidification on local sea life? Why are we experiencing more heat waves? Can we get hurricanes in our area? What's happening to coastal plants and how is it affecting local tourism?

The handbook provides a detailed step-by-step guide for designing science practices with infographic models to be displayed on school stands located in tourist destinations, with specific reference to the following:

- Conducting science practices, suggesting or encouraging the exploration of scientific questions related to climate change and sustainable tourism, with specific examples and activities.
- Designing and developing the methodology of these practices in relation to citizen action in tourism, encouraging the adoption of a strategic plan that describes how to implement these practices at stands located in different destinations. This involves the organisation of the material and the involvement of different collectives with an intergenerational approach.
- Highlighting good practices of citizen science projects and resource pooling, including examples of successful citizen science projects and promoting collaboration between scientists and citizens.

This project output stands out for its original approach to citizen science as a means to empower students as agents of change in their own communities and countries. It also offers an attractive visual layout, making it more understandable and accessible to educational communities. Finally, it stresses that citizen science projects (led by citizens without requiring the involvement of professional scientists) such as those

[6] Conceived as an 82-page e-book, titled *"School Citizen Science in Climantopia"* and published by the University of Santiago de Compostela's Publications and Scientific Exchange Service (ISBN 978–84–19,679-94-9), it is available to download for free from the project's website in English, Spanish, Portuguese, Italian and Polish via this link: https://www.educinema-cta.org/practicas-de-laboratorio/

detailed in this paper, foster skills such as initiative, teamwork and entrepreneurship, and can therefore be extended more broadly to the general population in relation to climate issues and sustainable tourism. This handbook has been successfully implemented and has had the following impacts: promoting citizen science in the school environment; encouraging climate action and promoting sustainable tourism in the school environment; fostering the capacity and skill needed to carry out projects independently; inspiring future research projects, generating new ideas and knowledge in society; contributing to science outreach among the school and local population; boosting the co-creation of transparent research; creating opportunities for social transformation towards sustainability; and being applicable/adaptable to a diverse range of contexts.

6 In Conclusion: Recognition of the Project as a Good Practice by the European Commission

After the project finished in 2024, the Spanish Service for the Internationalisation of Education (SEPIE) carried out a quality assessment on the project. After the justification, monitoring and control of its final report, linked to the grant agreement (ref. 2020–1-ES01-KA227-SCH-096314), the project outcomes were shared with to the senior officials of the project, in particular the Vice-Chancellor of the University of Santiago de Compostela. The report issued by SEPIE explicitly stated that the project was carried out in accordance with proper prior planning and internal organisation that allowed satisfactory outcomes to be achieved. The quality and completeness of the information provided in the document was assessed positively. As the project achieved at least 90 points out of a maximum score of 100 in the final evaluation, the decision was made to submit it for evaluation as a potential model of good practice by the European Commission, which was approved in resolutions adopted for this purpose.[7]

This project on collective creativity in film literacy has also generated tangible changes within the school, especially among participating students. Years later, now as university students, some participants return to the school to train students at Virxe do Mar High Shcool in film and digital literacy, transferring the competencies acquired during the project. Educational experiences based on film literacy foster the development of critical, creative, and cultural competencies, while promoting active student participation in the production and understanding of audiovisual language (Reia-Baptista et al., 2014), in line with media education, which aims to develop critical understanding and active participation in media culture practices (Buckingham, 2003). Among other outcomes, such initiatives contribute to

[7] The project was recognised as a model of good practice, as indicated by the text "Good Practice" under the left-hand side of the project title on the European Commission's Erasmus+ website, available at: https://erasmus-plus.ec.europa.eu/projects/search/details/2020-1-ES01-KA227-SCH-096314

connecting climate change and sustainability across subjects; encouraging students to engage society through environmental projects; and enhancing collaboration and involvement among families and the wider community (Sóñora-Luna et al., 2019).

References

Bonilla, A. E. G., Cadena, K. Á. Á., Martínez, D. P. G., & Carrasco, J. D. S. (2024). Tendencias del aprendizaje cooperativo en la enseñanza de ciencias experimentales: un enfoque bibliométrico. *Estudios y Perspectivas Revista Científica y Académica, 4*(4), 1940–1959. https://doi.org/10.61384/r.c.a..v4i4.771

Buckingham, D. (2003). *Media education: literacy, learning and contemporary culture.* Polity Press.

Caride-Gómez, J. A. (2022). La pedagogía social en la construcción de la creatividad colectiva. In J. R. Coca & J. A. Roche (Eds.), *Bases biológicas, psicológicas y sociales de la creatividad* (pp. 212–221). Libros de La Catarata.

Caride-Gómez, J. A., & Meira-Cartea, P. A. (2020). La Educación Ambiental en los límites, o la necesidad cívica y pedagógica de respuestas a una civilización que colapsa. *Pedagogía Social. Revista Interuniversitaria, 36*, 21–34. https://doi.org/10.7179/PSRI_2020.36.01

Chambers, J. (2020). Consolidating an experimental pedagogy: exploring ecologies of film education within France's cinéma cent Ans De Jeunesse and Scotland's understanding cinema project (s) between 2013 and 2019. *Film Education Journal, 3*(2), 138–159. https://doi.org/10.14324/fej.03.2.03

Deák, C., & Kumar, B. (2024). A systematic review of STEAM education's role in nurturing digital competencies for sustainable innovations. *Education Sciences, 14*(3), 226. https://doi.org/10.3390/educsci14030226

García-Vinuesa, A., & Meira-Cartea, P. A. (2019). Caracterización de la investigación educativa sobre el cambio climático y los estudiantes de educación secundaria. *Revista Mexicana de Investigación Educativa, 24*(81), 507–535.

García-Vinuesa, A., Meira-Cartea, P. A., Caride, J. A., & Bachiorri, A. (2022). El cambio climático en la educación secundaria: conocimientos, creencias y percepciones. *Enseñanza de las Ciencias, 40*(2), 25–48. https://doi.org/10.5565/rev/ensciencias.3526

García-Vinuesa, A., Palmieri, M., & Sóñora-Luna, F. (2026). Exploring barriers and needs: secondary teachers' perspectives on climate change education within an international environmental campus. *Sustainability, 18*, 2519. https://doi.org/10.3390/su18052519

González Gaudiano, E., & Meira-Cartea, P. A. (2020). Educación para el cambio climático ¿Educar para el clima o para el cambio? *Perfiles Educativos, XLII*(168), 157–174. https://doi.org/10.22201/iisue.24486167e.2020.168.59464

Maričić, M., & Lavicza, Z. (2024). Enhancing student engagement through emerging technology integration in STEAM learning environments. *Education and Information Technologies, 29*(17), 23361–23389. https://doi.org/10.1007/s10639-024-12710-2

Okada, A. (2023). *Inclusive open schooling with engaging and future oriented science: evidence-based practices, principles, and tools.* The Open University UK Rumpus Press.

Reia-Baptista, V., Burn, A., Reid, M., & Cannon, M. (2014). Literacía cinematográfica: reflexión sobre los modelos de educación cinematográfica en Europa. *Revista Latina de Comunicación Social, 69*, 354–367. https://doi.org/10.4185/RLCS-2014-1015

Sarid, A., Boeve-de Pauw, J., Christodoulou, A., Doms, M., Gericke, N., Goldman, D., et al. (2025). Reconceptualizing open schooling: towards a multidimensional model of school openness. *Journal of Curriculum Studies, 57*(2), 227–245. https://doi.org/10.1080/00220272.2024.2392592

Sóñora-Luna, F., Alonso, A., & García-Vinuesa, A. (2019). Respuestas juveniles al cambio climático desde los campus de Climántica. *Innovación Educativa, 29*, 5–26. https://doi.org/10.15304/ie.29.6341

Sóñora-Luna, F., & García-Vinuesa, A. (2020). Climántica: un proyecto pedagógico-social y de educación ambiental en la lucha contra el cambio climático. *Pedagogía Social. Revista Interuniversitaria, 36*, 63–69. https://doi.org/10.7179/PSRI_2020.36.04

Trilling, B., & Fadel, C. (2012). *21st century skills: learning for life in our times*. Jossey-Bass.

The Power of Networks: Citizenship and Climate Governance

CLIMACTIVA Project: Citizen Adaptation and Resilience to Climate Risks

Mónica Arto Blanco, Tania Merelas Iglesias, and Miguel Pardellas Santiago

Abstract Climate disasters are intensifying globally, with Spain facing increasing risks from heatwaves, droughts, floods and forest fires. These events disproportionately affect vulnerable populations, especially in urban areas where social inequalities, gender and age deepen climate impacts. Addressing this challenge requires enhancing both adaptive capacity and resilience, not just to extreme events but as a part of daily life. Adaptation involves reducing exposure and vulnerability, while resilience reflects the ability to cope and recover. These efforts must go beyond economic measures and incorporate lifestyle changes, such as community care, leisure and activism. Citizen participation is essential in building effective and inclusive climate governance. CLIMACTIVA project explores eleven self-organised citizen initiatives across different Spanish regions responding to key climate risks. These groups often employ holistic approaches, such as creating green spaces that mitigate both heatwaves and flooding. Their locally grounded solutions offer valuable insights into sustainable adaptation practices. By highlighting these initiatives, the study emphasizes the importance of community led responses in fostering resilience, complementing public policy and promoting adaptive, climate-conscious lifestyles.

1 Presentation

Climate disasters are increasingly present throughout the world (IPCC, 2022). Communities are not only dealing with gradual risks such as rising sea levels or increasing temperatures, but also with extreme and immediate events such as floods,

M. Arto Blanco
Universidade da Coruña, A Coruña, Galicia, Spain

T. Merelas Iglesias
Arabías S. Coop. Galega, Santiago de Compostela, Galicia, Spain

M. Pardellas Santiago (✉)
Feitoría Verde S. Coop. Galega, Santiago de Compostela, Galicia, Spain
e-mail: miguelpardellas@feitoriaverde.com

J. J. Conde et al. (eds.), *Practices in Education for Climate Emergency*,
https://doi.org/10.1007/978-981-92-2304-6_15

heatwaves or forest fires. In the case of Spain, the climate outlook suggests a future with more events of this type, and with greater intensity.

The well-being of people living in affected areas is compromised due to the degree of exposure and vulnerability to these events. This is where social, economic and cultural factors come into play and determine the extent of harm (Habtezion, 2016). For example, social inequalities exacerbate the impacts of climate crises, leaving certain groups even more vulnerable (Vera & Sordi, 2021). This is more pronounced in cities, where the concentration of population, especially fragile groups, multiplies the hotspots of vulnerability to heatwaves, droughts, floods or fires (Sanz & Galán, 2021).

Among the factors influencing this vulnerability, gender inequality is one of the most decisive, since women tend to be more exposed to and affected by climate impacts. Age also plays an important role, as both children and the elderly display certain risks in terms of their health, development and education.

In order to address this situation, it is essential to strengthen people's capacity to adapt and their resilience in the face of extreme events, but also in everyday life. Adaptation seeks to reduce exposure and vulnerability to climate risks, while resilience involves the ability to cope and move forward (Talbot-Wright & Vogt-Schilb, 2023). The question is how to make adaptation part of our lifestyles, or, if we turn the question around, how to adapt our lifestyles to the climate emergency. Lifestyle goes beyond economic activities, in terms of spending and consumption, and includes caring for people, leisure activities, volunteering and activism (UNEP, 2022). This is where community participation is essential for effective climate governance, as citizens have a key role to play in promoting social and adaptive solutions (IPCC, 2022).

In Spain, research on citizen initiatives addressing adaptation to climate change is still emerging, with the exception of the RESCITIES project (Satorras et al., 2020). Studying how different groups organise themselves in relation to the climate risks in their surroundings means that practical solutions can be identified, often based on local knowledge. Moreover, understanding these initiatives helps to raise awareness of good practices, evaluate their impact and replicate them elsewhere, thus fostering greater resilience and promoting citizen participation in environmental decisions. These actions can, and should, complement public policies, contributing innovative and sustainable approaches directly from communities.

In this context, our study, "Integrating climate risk adaptation and resilience into lifestyles: exemplary citizen initiatives. CLIMACTIVA", analyses the strategies of eleven groups from different regions of Spain in response to the four main climate risks: forest fires, floods, heatwaves and droughts.

The selection of cases was based on the eminently self-organised and participatory nature of the initiatives, that is to say, their reaction to the threat was independent to that of public administrations. Although each experience is related to a specific risk, many of them take a more holistic approach, working on several of these problems simultaneously. For example, creating green areas not only helps to combat heatwaves, it also improves the soil's ability to retain water, thereby reducing the impact of potential flooding.

We consider the experience accumulated by these initiatives to be a valuable source of knowledge. From a local perspective, their proposals are more effective and adapted to reality; they strengthen communities to face the challenges posed by climate change and promote solutions that make a difference due to their efficiency.

2 Citizen Initiatives

Citizen initiatives play a crucial role in adapting to and mitigating the impacts of climate change, especially at the local and regional level. These include diverse collectives—associations, informal groups, irrigation communities, etc.—that implement proactive responses from inside the communities themselves. These collectives possess in-depth knowledge of the territory, its resources and particular needs. By identifying and analysing these experiences, we can raise awareness of successful practices, identify common patterns and highlight effective strategies that can be replicated or adapted in other contexts.

At CLIMACTIVA, we set ourselves the objective of conducting a systematic mapping of citizen initiatives aimed at adapting to climate change throughout the whole of Spain. To this end, we defined a work methodology based on the selection and study of specific cases. These cases are citizen initiatives that respond to four climate risks identified in the National Plan for Adaptation to Climate Change (Ministerio para la Transición Ecológica y el Reto Demográfico [MITECO], 2020): floods, forest fires, heatwaves and droughts. Our methodology takes a community and gender approach, seeking the greatest possible territorial diversity.

Out of the 27 initiatives identified initially, and after conducting previous informative work on each of them using the resources available (e-mail, internet, news, etc.), we finally selected 11. These initiatives fulfilled the criteria set out in the research and were willing to participate in the study.

Once the sample was defined, we proceeded to collect and systematise contextual information on each of the cases and conduct personal interviews with key informants. Two key informants were identified per initiative (with the exception of one case), taking into account their experience and involvement, as well as the gender criterion, aiming for equal participation. The final interview sample included a total of 21 people, 11 of whom were women and 10 were men. The profile of informants is predominately middle aged, with higher education, belonging to the middle class, residing in the area of action and actively working in various different professions (seven of them related to the environment).

The data collected through the interviews were subjected to rigorous analysis involving the initial transcription and coding of all material, identification of significant themes and patterns using a qualitative data analysis programme, and also data triangulation to validate findings (Table 1).

Table 1 Research sample summary

Initiative	Climate risk	Place
Barrios Verdes [green neighbourhoods]	Floods	Telde (Canary Islands)
Berdesia Association	Floods	Tafalla (Navarre)
Flood watch group	Floods	Alcalá de Ebro (Aragon)
Community of Montes de Baroña	Fires	Baroña (Galicia)
Pego Viu	Fires	Pego (Region of Valencia)
Barrios por el clima [neighbourhoods for climate]	Heatwaves	Cordoba (Andalusia)
Escuelas de calor [hot schools] platform	Heatwaves	Seville (Andalusia)
Barrio Zofío Neighbourhood Association	Heatwaves	Usera (Madrid)
Aragon Seed Network	Drought	Zaragoza (Aragon)
Ocho Caños irrigation community	Drought	Aldeanueva de la Vera (Extremadura)
O Galiñeiro water community	Drought	Gondomar (Galicia)

3 Core Ideas

The compendium of citizen initiatives analysed aims to contribute to a better understanding of existing adaptation practices by facilitating the exchange of learning among key actors and promoting the implementation of more inclusive and participatory strategies. The recognition and dissemination of these initiatives is essential to enhance collaboration and move towards a more resilient and sustainable development model to address the challenges of climate change.

In this section, we present the core ideas that have emerged from the analysis of the case studies. These ideas, grouped into six subject areas, provide an overview of the most relevant issues, both in terms of the attention they received in our research questions and in terms of their importance in the accounts gathered. Moreover, recommendations are presented, addressed to two main actors: other similar initiatives and public administrations. Since the ultimate goal of this research is to disseminate and replicate these good practices in other contexts and territories, we consider it appropriate to address the main actors who will have to manage the climate risks.

3.1 Risk Perception and Adaptive Capacity

The National Plan for Adaptation to Climate Change (MITECO, 2020) states that the aim of adaptation is not to eliminate risks, but rather to reduce them to acceptable levels. The National Plan states that what is considered acceptable is conditioned by interests, values and expectations, and will guide the responses chosen. It is, therefore, of particular interest to hear the perceptions of the people involved in climate risk initiatives, since this will largely determine the response actions. It is also necessary to take into account the adaptive capacity of the community. This is defined as the potential or ability of an individual or group to successfully take action when faced with climate risks, using the necessary resources (Kuhlicke et al., 2011). Nevertheless, using resources indeed implies that they exist and have been made available to the community (Ballester, 2017), an aspect that has been included among the elements to be considered in the analysis.

There is no widespread awareness of risk that motivates people to take preventive measures on a regular basis. For the people interviewed, there is a prevailing perception that their neighbours are aware of the existence of climate change, given that the changes are becoming more and more evident. However, they also state that when the threat is more indirect or discontinuous, less attention is paid to it.

In rural areas, changes are perceived to be more evident due to the type of activities carried out and the more obvious threat of fires and floods. In this context, a loss of territorial connection and devaluation is also identified, related to the decline of the primary sector, which leads to turning one's back on climate risks.

In any case, initiatives can play an important role in perception. Their interventions in the community have led to both broader and more complex reflections on local issues and concrete actions in relation to risk.

> Of course, concern has increased, but solutions have not. Therefore, concern is not the issue. The issue is that we don't have a conflict with heat. We have a problem that causes us suffering, but we don't see it as a conflict that we have to resolve like others. (Barrio Zofío)

The damage caused by different climate catastrophes is present in the collective memory and helps foster action. Several initiatives highlight the solidarity response of the population to extreme events through mobilisations, community work for the recovery of spaces, or accommodation and support for evacuees.

The physical and/or emotional impact of fires, floods or heatwaves is identified as a catalyst for many of the initiatives that, given their sense of vulnerability or lack of organisation, decide to play an important role in collective risk prevention and protection.

Furthermore, the people interviewed express concern over various climate hazards, not only those directly addressed in their initiatives, with the risk of drought being particularly prominent. Thus, seven initiatives refer to drought, four to heatwaves and three to forest fires. With the exception of concern for forest fires, which is associated with more rural areas, the other two risks are of similar concern to both rural and urban initiatives.

Directly related to this concern is the fact that social awareness of water is increasing in parallel with the rise of an unconscious pattern of water consumption. Water is one of the issues identified as fundamental today, both due to the direct impact caused by its scarcity and the social confrontation generated by its management.

In this respect, there is a dissociation between the general diagnosis of water scarcity and the uses or demands of the population, based on the idea of water as an unlimited resource. As a result, sectors such as agriculture, which are more vulnerable due to their dependence on the resource and, at the same time, aware of its reduction, are increasingly demanding more water instead of choosing to adapt.

> There are people who still think that water is unlimited, since it comes out of the taps every day and there are no restrictions, except in very specific places. But there has not yet been a major, highly visible water crisis. (Aragon Seed Network)

Spatial planning, urban planning and land use are also related to the increased impact of climate risks. The people interviewed note that a series of factors or trends increase vulnerability in the territory where they live. These include the progressive abandonment of agroforestry practices as a factor that aggravates the risk of fires; the loss of traditional land uses in favour of industrial agriculture, which is more demanding of water; and the change in land use from agricultural to urban, as well as the lack of urban planning.

Heatwaves are particularly linked to a dense or even overcrowded urban planning model, the artificialisation of spaces and a lack of green areas, in addition to the low quality of buildings.

All of these factors condition a divided perception: half of the people interviewed feel prepared to face climate risks, while the other half admit they do not. Those who feel unprepared highlight the gaps in their training, both individually and collectively, as well as the negative emotions generated by the possibility of an extreme event. Drought is the climate risk for which preparedness is perceived to be lowest. On the other hand, when faced with floods and fires, there is a greater perception of knowing how to act in some of the initiatives.

3.2 Community Action and Network Building

Collective action is a characteristic trait of the citizen initiatives studied. Thus, one fundamental aspect of this project is the discovery and analysis of the details of this action.

Social research shows that adaptation to climate change is an opportunity to foster civic engagement, learning and capacity for change (Fazey et al., 2018). Similarly, the sense of community and the web of social relationships are essential to prevent responses to climate risks from being individualised and poorly articulated (Perdomo et al., 2019).

In this respect, it is worth noting firstly that adaptation initiatives are launched as a result of an organised social base that functions as a school of citizenship. Most of the people interviewed have previous experience of social activism or participation in different types of organisations. This active citizenship profile leads to either integrating adaptation actions into an existing organisation in which one already participates, or starting a new organisation with like-minded people that specifically focuses on addressing climate risks. In rural areas, the focus is placed on the neighbourhood relations that continue in this type of context, which favour the inclusion of adaptation actions.

Due to their size and objectives, the initiatives mainly seek the participation of citizens and various organisations at the local level. Vicinity is fundamental for participation in citizen initiatives, as they are shaped by and addressed to the neighbours of villages, places or towns in rural areas, as well as those of neighbourhoods in cities.

This proximity is closely linked to the creation of meeting spaces (for deliberation, but also for recreational and shared leisure purposes) and to the emotional ties that arise from the initiatives.

> I think there is a change in the way people relate to one another and this affects everything, how we look at the neighbourhood, the empathy with which we understand problems and how we deal with life, and in the end the community itself demonstrates this because when something comes up, they are there. (Barrios Verdes)

The initiatives analysed are genuine social experiments regarding response to the impacts of climate change. As such, they are exposed to constant conflict between the feasible and the essential. Thus, the willingness to work with other initiatives or collectives is present on a regular basis, but rarely translates into lasting ties. And yet, educational actions and initiatives to recover local heritage and memory are designed and implemented despite the huge effort involved.

They aim to create networks and broaden the social base, thereby emphasising the need for the community to feel that the adaptation actions are their own and to restore links with their surroundings.

> We want the Berdesia project to not only belong to the association, but for the people of Tafalla to feel that it belongs to them. That's what we have to try to do, because when you feel like something is yours, you take care of it, you defend it and you work on it. (Berdesia)

In any case, all initiatives are implemented under the premise of collectivising responses to climate risks. Several of the people interviewed stated that individual solutions are conditioned by socio-economic status, technicality and may even harm the community. Moreover, they explain that, in their experience, collective solutions require a certain level of personal commitment, which is not always easy to fit into their lifestyle.

3.3 Activities Undertaken

The actions undertaken by the citizen initiatives under study provide essential information for understanding how they work and what their objectives are. Understanding the scale of the activities, the resources implemented and the social levers mobilised constituted a focal point of the analysis. Nevertheless, any collective project will face significant challenges such as the existence of psychosocial problems typical of volunteer groups (interpersonal conflicts, lack of dialogue, etc.), leadership management (difficulty in finding a leader or in relinquishing leadership when the time comes), or limitations in promoting significant changes in the community (Azkarraga et al., 2012).

The initiatives analysed show a heterogeneity of actions ranging from manual interventions on the territory to education and awareness-raising activities, or the monitoring and warning of extreme events, in addition to protest actions. We can, however, identify a common link in collectivity and in the reconnection with natural processes, in many cases through the valorisation of traditional practices. Collectivity sometimes has a very specific component focused on protest and self-protection actions. In other initiatives, collectivity is consciously strengthened through recreational activities, celebration and meeting places.

With regard to nature-based solutions, reforestation actions with native species and urban rewilding projects are of particular note. Other initiatives decided to update traditional uses such as conservation and seed exchange, water management based on community infrastructure and the implementation of silvopastoral practices.

The diversity of the actions compiled reveals a profoundly contextual nature, that is to say, they are initiatives adapted to the socio-environmental setting where they take place. There is also a predominant trend towards long-term, process-oriented interventions. This reflects the awareness regarding the need to implement gradual changes to address climate impacts with greater assurance in the medium and long term. Nonetheless, in cases where risks are of a more urgent nature, more immediate self-protection and monitoring actions are also activated, as is the case with almost all initiatives linked to floods and fires.

Directly related to the tensions and opportunities that arise, the main value of the initiatives is associated with the voluntary and self-organised work in the community, in addition to the mobilisation capacity. This does not preclude a number of recurring challenges. Among the most prominent are the scarcity of resources on the social level (fluctuation in participation, lack of involvement), and on the economic level (difficulty in accessing economic resources, scarcity of own funds). A problem is also identified regarding the work overload of participants, especially those who take on more responsibilities, in terms of how to integrate it into their day-to-day life.

Another issue of concern for those interviewed is the low social outreach of the initiatives or the loss of convening power. Similarly, a shared perception is evident, in several initiatives linked to older rural contexts, regarding the problem of generational handover and the majority male profile of the members. The lack of young people (between 18 and 30 years old) in the initiatives is a unanimous concern of all interviewees. This collective is perceived as turning its back on the countryside and the forest, and conceives it as a space for recreational activities, not for agroforestry use. Moreover, there is a complete lack of information regarding the codes to entice them to participate and how the activities would fit into the day-to-day of this age group.

3.4 Vulnerable Groups

The impact of climate change on communities depends not only on the existence of hazards (such as heatwaves, fires, droughts or floods), or exposure to them, but also on the degree of social vulnerability. In turn, this vulnerability is determined, among other factors, by socio-economic status, age, gender, health status, existence of social services or participation in networks (Oficina Española de Cambio Climático [OECC], 2024). Therefore, we believe it is important to examine how citizen initiatives address issues of social vulnerability to climate risks in their operation and activities.

The people in charge of the citizen initiatives or participating in the activities all have a low social vulnerability profile. Citizen initiatives tend to be led mostly by middle-aged men, with medium or higher education, in active employment, from the middle class and with origins in the community where their actions are directed. Those who participate in the activities of the initiatives respond to a socio-economic status and social network integration profile that reduces their vulnerability. In urban initiatives, nevertheless, there is a greater awareness of social vulnerability in terms of the economic status and migrant status of the people involved. On the other hand, in rural areas, members are more affected by ageing.

Those interviewed agree that the groups most vulnerable to impacts are the elderly and those who earn their living from agriculture. In addition to these two groups, people with low economic resources, children or those living in areas close to watercourses or river mouths are also identified in some cases.

Paradoxically, the outreach of citizen initiatives among the most vulnerable groups is still low, especially among people engaged in professional agriculture, and there are even some cases of blatant disagreement. Regarding the elderly, several of the activities promoted have focused on encouraging participation from this group through day centres or retirement associations.

As far as children are concerned, the risk of heatwaves does emerge as a growing threat to the right to education, especially in autonomous regions such as Andalusia. High temperatures make schools unsafe in terms of thermal comfort due to the persistence of the heat, the large number of students exposed to it and the lack of adapted infrastructures. The lack of air conditioning in most of these public buildings aggravates the situation, with access to cool spaces limited to canteens, teachers' lounges and administrative offices. Extreme heat has a negative impact on classroom dynamics and student performance, leading to learning difficulties and general discomfort. Moreover, it highlights the inequality between public and private schools in terms of the availability of air conditioning systems, and also the disparity between public schools where some families have been able to afford air conditioning or awnings.

While the issue of heatwaves receives special attention in the case of the south of the peninsula, as mentioned above, schools and children are at the centre of the initiatives' activities. The initiatives actively collaborate with schools, implementing activities related to environmental education and also to natural and cultural heritage. There are even initiatives that have specific groups that coordinate, design and develop both the activities they offer to schools and those requested by teachers and parents' associations. The people interviewed explained that the aim is to ensure that schools gain autonomy, and only require support from the initiatives for certain actions, and to promote an active role among students.

> Children play an inspiring role because, when tidying up the places, families were inspired to create a meeting place. They also have decision-making power. We ask them every time we have to make decisions about, for example, the organisation of the park, what things to put on each side, what they want and what they don't want. (Barrios Verdes)

3.5 Gender Mainstreaming

Social and civic organisation arises from the creation of local interpersonal networks that have a place in a particular community (Raso, 2017). These participatory spaces are modulated by socio-environmental contexts, where it is possible to identify differentiated roles for women and men, in addition to gender gaps. Gender inequality persists globally as a systemic feature, leading to increased vulnerability to climate risks for women (UNFCCC, 2022). This is compounded by the lack of data and indicators in this area.

In citizen initiatives of a mixed nature, such as those analysed, it was considered important to increase differentiated knowledge regarding the role and experience of women and men in the functioning of the organisations and the actions implemented.

Among the conclusions is the fact that women have a majority participation in the initiatives analysed, however stereotypes and gender roles continue to mark certain dynamics. Traditional gender beliefs are present in everyday life and are more evident among older age groups and in rural contexts. Overall, women tend to be involved in administrative and, above all, organisational activities that take care of the community link or are oriented towards awareness-raising and prevention. They are the majority in activities linked to the educational community, cleaning up degraded areas, or organising community meals. Men, on the other hand, are more likely to take on actions that fit the male stereotype of strength or public notoriety, such as operating machinery, brush clearing or cleaning sprinklers, to cite a few examples.

Similarly, female participation tends to be lower in the governing bodies of the different experiences analysed. In representative and decision-making positions, there is not always a situation of parity, with a greater number of men assuming these types of positions of responsibility and visibility in the community. This inertia is also evident when addressing technical issues, with inequality observed in the recognition of women as experts in subjects related to STEM disciplines (Science, Technology, Engineering, and Mathematics). This trend goes hand in hand with the idea that women are more sensitive to issues related to climate change or to the risks analysed, thereby highlighting the emotional component that drives their actions. However, changes are beginning to take place, and some initiatives are moving towards greater equality, with prominent and visible female leadership.

> In general, I see that men and women get involved in different ways. I see that women's actions are not publicised so much, even if they do very powerful things, whereas men have a need to draw attention to absolutely everything they do, and even to give their opinion, regardless of whether they did it or not. (Barrios Verdes)

In any case, there is no widespread reflection as a collective or association regarding the differences that climate risks have on the lives of women and men. There is a general tendency to interpret the effects of climate change as affecting all people equally, regardless of gender. Nevertheless, in some cases, there is an awareness that social class or socio-economic status are much more relevant impact differentiating factors.

Given that participation in collectives is usually carried out in people's free time, the different responsibilities related to care in the home and family sphere, which are still mostly and unequally assumed by women, mean that they have less time available to be able to commit themselves to certain initiatives. This situation is aggravated at times of life that coincide with a greater demand for care, such as raising children or caring for dependent persons. Paradoxically, sometimes these tasks, essentially raising children, serve as a vehicle for reorienting priorities and actually lead to the participation of many women so as to leave a better future for their descendants.

3.6 Contribution to Climate Governance

Climate governance not only faces the challenge of integrating adaptation into public policies, but also of promoting comprehensive strategies or actions that consider the complexity of the contexts of intervention. In this scenario, citizen participation takes on a prominent role, not as occasional interactions or a mere result, but rather as a fundamental part of the adaptation process itself, through innovative ways of interacting. This, in turn, puts the very model of institutional functioning to the test (Wamsler, 2017).

Since we focused on citizen initiatives that emerged independently of public administrations, the public-citizen interface was one of the central elements in the search for information.

Many of the citizen initiatives express their disagreement with the risk management undertaken by the different levels of public administration. While it is acknowledged that some actions have been taken, many initiatives emphasize the lack of long-term institutional planning that would allow for major transformations. Even where planning projects exist, they often ignore problems such as heat islands and social vulnerability. Instead, they opt for generic actions that do not address the needs of the neighbourhoods and territories, and which, moreover, do not involve the participation of the population.

Citizen initiatives, whether local or regional in scope, implement actions that transcend the occasional, showing a firm commitment to medium- and long-term interventions. In several cases, they do not have clear support from public administrations and it is they themselves who establish dialogue with other social actors to coordinate broader actions. In others, although it is common for activities to be coordinated with local administrations, the link is merely on a financial level, thus establishing a relationship of dependence with regard to the financing of certain actions which limits the development of more ambitious projects.

The tension regarding planning with the administrations can be extended to other issues, such as the instrumentalisation to which they sometimes feel they are subjected, communication difficulties or the political polarisation around issues such as water management. Furthermore, several initiatives warn of the apparent competition for resource management in the case of some administrations. Several of the initiatives linked to the management of specific resources (water or territory) repeatedly mention the different view they have from the regional administration in terms of planning or the scale and impact of certain actions.

4 Conclusions

As in almost any research project, the motivation of CLIMACTIVA was twofold: on the one hand, the objective justification which focuses on the few existing research projects in Spain regarding citizen initiatives and climate adaptation processes. On the other hand, the more personal and subjective justification, which includes unanswered questions, or ones with vague answers, from other studies in which we have participated. It is the conjunction of both motivations that ultimately gives rise to the main research questions that guide the whole process, and for which we have obtained some guidance:

- Citizen initiatives addressing climate risks arise because either existing organisations collectively manage a certain resource/space (water, forest, etc.), which is directly affected by the effects of climate change, and particularly by a specific risk, or one or several people aspire to participate in risk management collectively, thus creating a formal or informal organisation, and initiate actions in relation to a certain resource or space.

  Both require a social base of people who integrate activism into their daily lives, along with the perception of an environment where resilience has been lost, who have become more vulnerable to climate change and/or who have experienced extreme events in recent years.
- The people who participate in the initiatives consider themselves better prepared to face climate risks when they perceive coherent coordination of actions at the individual-community level with emergency services and public administrations (local, regional, etc.). If they notice a fracture in the different dimensions, or if the role of the various actors in risk management is unclear, they feel less prepared.
- The experiences gathered show that citizen initiatives channel individual need for action and direct it towards collective solutions. In turn, they encourage other people or organisations to refocus their priorities. In many of the citizen initiatives, it is also explained that this citizen drive comes up against the barrier of public administrations. Either plans and protocols are not put in place or, if they are, they do not always stem from constructive dialogue with communities. Only in exceptional cases are spaces created that allow citizens to be part of the design and implementation of adaptation processes. Therefore, the citizen-administration

interface is one of the areas of action identified by those interviewed as having the greatest potential for improvement.

- Citizen initiative actions are conceived by their key players as an open and improvable process in terms of adaptation or rapid response to an extreme event. They may have a greater or lesser territorial scope or complexity of actions; they may have succeeded in getting people to take part in the initiatives and integrate adaptation into their lifestyles; or they may simply have involved people in the actions they promote. Nevertheless, in all of them, the idea of sustaining or building community, in addition to creating capacity to act or develop responses to climate change, appear as key achievements and as recommendations for other similar initiatives.
- The participation of women is prominent in the different initiatives at grassroots level, and although there are stereotypes and gender roles present in the day-to-day work of the organisations, it is equally true that a transition towards greater equality in decision-making spaces is taking place. One key challenge is to dismantle the belief that undervalues women as experts in STEM-related fields of knowledge.

Acknowledgments This text reproduces information from the dissemination report produced in the framework of the project "Integrating climate risk adaptation and resilience into lifestyles: exemplary citizen initiatives. CLIMACTIVA", funded by the Biodiversity Foundation of the Ministry for Ecological Transition and the Demographic Challenge (MITECO), through the Call for subsidies to carry out projects that contribute to the implementation of the National Plan for Adaptation to Climate Change (2021–2030).

In addition to the dissemination report, CLIMACTIVA project includes two further products: an educational proposal in line with the results of the analysis carried out, with indications for the development of pedagogically sound educational interventions focused on adapting to the impacts of climate change; and a guide with recommendations for gender mainstreaming in citizen initiatives.

References

Azkarraga, J., Belloy, P., & Loyola, A. (2012). Eco-localismos y resiliencia comunitaria frente a la crisis civilizatoria: las iniciativas de transición. *Polis. Revista de la Universidad Bolivariana, 11*(33), 15–40.

Ballester, A. (2017). *Participación pública para una gestión eficaz del riesgo por inundación: construcción de capacidades sociales en la Ribera Alta del Ebro (España).* (Doctoral thesis). Universitat Autònoma de Barcelona. https://ddd.uab.cat/record/187354

Fazey, I., Carmen, E., Chapin, F. S., III, Ross, H., Rao-Williams, J., Lyon, C., et al. (2018). Community resilience for a 1.5 °C world. *Current Opinion in Environmental Sustainability, 31*, 30–40. https://doi.org/10.1016/j.cosust.2017.12.006

Habtezion, S. (2016). *Gender, climate change adaptation and disaster risk reduction.* United Nations Development Programme.

IPCC. (2022). In P. R. Shukla, J. Skea, R. Slade, A. Al Khourdajie, R. van Diemen, D. McCollum, et al. (Eds.), *Climate change 2022: mitigation of climate change. Contribution of working group III to the sixth assessment report of the intergovernmental panel on climate change.* Cambridge University Press. https://www.ipcc.ch/report/sixth-assessment-report-working-group-3/

Kuhlicke, C., Steinführer, A., Begg, C., Bianchizza, C., Bründl, M., Buchecker, M., et al. (2011). Perspectives on social capacity building for natural hazards: outlining an emerging field of research and practice in Europe. *Environmental Science & Policy, 14*(7), 804–814.

Ministerio para la Transición Ecológica y el Reto Demográfico. (2020). Plan nacional de adaptación al cambio climático

Oficina Española de Cambio Climático. (2024). Documentos de trabajo plan nacional de adaptación al cambio climático. 1- integración de factores sociales en la adaptación al cambio climático.

Perdomo, M., Reyes, C., & Sánchez, F. (2019). Análisis del sentido de comunidad como promotor de la resiliencia comunitaria en habitantes de ciudades en riesgo de desastres naturales. *Indagare, 7*, 14–18.

Raso, R. (2017). *Justiça social, feminismo e movimentos sociais: a secretaría das mulleres do sindicato labrego galego.* (Doctoral thesis) Universidade de Santiago de Compostela. Minerva-Repositorio institucional da USC. http://hdl.handle.net/10347/15480

Sanz, M. J., & Galán, E. (2021). *Impactos y riesgos derivados del cambio climático en España.* Ministerio para la Transición Ecológica y el Reto Demográfico. https://www.miteco.gob.es/es/ceneam/recursos/pag-web/impactos-cambio-climatico-espana.html

Satorras, M., Lara, Á., del Moral, L., Oteros-Rozas, E., Berraquero-Díaz, L., et al. (2020). *Compilación de iniciativas comunitarias de resiliencia urbana en Sevilla y Barcelona: la sociedad civil frente a los efectos del cambio climático.* Universitat Oberta de Catalunya.

Talbot-Wright, H., & Vogt-Schilb, A. (2023). *Con el agua y el calor al cuello: nueve caminos hacia un desarrollo resiliente al cambio climático.* Banco Iberoamericano de Desarrollo. https://doi.org/10.18235/0005214

UNEP. (2022). Enabling sustainable lifestyles in a climate emergency.. https://wedocs.unep.org/20.500.11822/39972

UNFCCC. (2022). United nations climate change annual report 2022.. https://unfccc.int/about-us/annual-report/annual-report-2021

Vera, F., & Sordi, J. (2021). *Diseño ecológico: estrategias para la ciudad vulnerable.* Banco Iberoamericano de Desarrollo.

Wamsler, C. (2017). Stakeholder involvement in strategic adaptation planning: transdisciplinarity and co-production at stake? *Environmental Science & Policy, 75*, 148–157.

Intergenerational Commitment in the Climate Emergency: Deepening Child and Youth Agency

Ferran Crespo i Torres and Marta Beatriz Esteban Tortajada

Abstract This chapter explores the climate emergency as a deeply systemic and intergenerational challenge that goes beyond biophysical dimensions to permeate social, political and economic life. The authors advocate a critical, complexity-driven approach, calling for transformative environmental education that empowers real action beyond mere knowledge. By foregrounding the concept of co-agency, the chapter argues that tackling the climate crisis requires genuine collaboration between children, young people and adults, and not only recognises the agency of younger generations, but also incorporates intergenerational partnerships into decision-making and collective action. Drawing on ecofeminist perspectives, the analysis critiques anthropocentric and adult-centric systems, emphasising how both children and Nature are marginalised under current capitalist and patriarchal norms. The chapter also reviews the legal framework provided by the UN Convention on the Rights of the Child, emphasising the importance of participatory pedagogical approaches that encourage shared responsibility. Ultimately, the authors contend that new, hopeful forms of climate action must be grounded in mutual recognition and co-agency, where children and adults collaborate to imagine and build a more just and sustainable future together.

1 Climate Emergency as a Shared and Intergenerational Challenge

The climate emergency is one of the most significant challenges we face as a society and human beings. Behind what is deliberately presented as a biophysical problem, addressed with technocratic efficiency, lies an unprecedented systemic crisis. This

F. Crespo i Torres
Department of Language, Science and Mathematics Education, Universitat de Barcelona, Barcelona, Spain

M. B. Esteban Tortajada (✉)
Department of Pedagogy, Faculty of Educational Sciences and Psychology, Universitat Rovira i Virgili, Tarragona, Spain
e-mail: martabeatriz.esteban@urv.cat

J. J. Conde et al. (eds.), *Practices in Education for Climate Emergency*,
https://doi.org/10.1007/978-981-92-2304-6_16

crisis needs to be approached and confronted from a complexity perspective since it is well known that approaching climate change solely from a biophysical point of view would lead to an inaccurate and naive exercise. This means understanding that the crisis is systemic and permeates not only the biophysical realm but also affects relationships between people and the environment, the generation of knowledge, politics, economics and an extensive list of other interconnected spheres and interact according to systemic and complexity principles.

At the same time, this vision urges us to respond from a critical perspective and across different fields of action. For some time now, efforts have been made to advocate and act towards the production of knowledge that transcends mere climate literacy to become critical, action-oriented literacy.

We must be prepared to change our attitudes and transform our perspectives on the planet. Environmental education is well known to be one of the fields taking up these challenges in the face of the climate emergency. Evidence of this can be seen in recent trends in Environmental Education for Sustainability and its theoretical frameworks, which advocate the need to integrate climate change in a cross-cutting way to promote pro-environmental attitudes rooted in a culture of sustainability, to incorporate complexity and systemic knowledge, to develop the capacity to envision future scenarios, and to adopt new practices and discourses (Caride et al., 2025).

However, uncertainty is now an embedded part of our lives beyond the educational sphere. As such, we should develop effective political, social and environmental strategies. Promoting new paradigms of relationships with the environment and between people must acknowledge the uncertainty involved in the search for new alternative futures. Therefore, initiatives that confront the climate emergency also promote collective well-being by enabling new counter-hegemonic imaginaries, which must necessarily be intercultural and situated.

In the face of the complex amalgamation of eco-social challenges, we can state that the human right to a clean, healthy and sustainable environment is being violated. The struggle against the climate emergency is also a struggle for human rights, and it must be a struggle that recognises that not all regions or lifestyles have contributed equally to the current critical situation. Nor are the impacts felt equally in all parts of the planet, communities, or groups and identities. The axes of discrimination that shape our system also condition and exacerbate this violation of rights.

Fortunately, in response to the realities that expose this systemic crisis, people are coming together, uniting, and organising for the common good and in defence of our rights—children among them.

2 Anthropocentrism and Adult-Centrism: Questioning Eco-Dependence and Interdependence

Given the exemplarity of the disruptive initiatives led or co-led by children, we will delve into the asymmetrical position of childhood in our societies to examine the common elements shared by the environment and children under the current systemic order. The capitalist, patriarchal and anthropocentric framework that dominates most contemporary societies assigns value primarily on the basis of productive capacity and economic profitability. Under this logic, hegemonic forms of childhood (Morales & Magistris, 2020) are seen as 'incomplete' and therefore excluded from public life. Similarly, any aspect of Nature that cannot be monetised loses its social value.

At this crossroads, it is beneficial to explore alternative perspectives that offer a more nuanced view of our relationship with Nature, particularly concerning the devaluation of what is deemed 'unproductive'. Ecofeminist approaches could be of particular interest in this regard, as they emphasise the interconnectedness of all living beings and propose an ethic of care that challenges the prevailing capitalist, patriarchal, and anthropocentric worldview.

Ecofeminism confronts the foundations of the anthropocentric and androcentric order (Mies & Shiva, 1997, 2014). Furthermore, it is a pluralist paradigm that values and emphasises the interconnectedness of human well-being and the well-being of Nature, a concept often referred to as interdependence, which challenges the dominant anthropocentric discourses that assert human supremacy and recognise the value of Nature solely on the basis of its usefulness for human needs (Limón-Domínguez, 2019). Rather than accepting the absolute dominance of humans over other living beings, ecofeminism posits that environmental exploitation and social oppression are deeply intertwined.

One might consider the ethic of care to be an illustrative ecofeminist principle, with its emphasis on empathy, reciprocity and the mutual flourishing of humans and non-humans alike. However, this ethic is not always aligned with extractivist logics that reduce Nature to a resource for profit. Similarly, patriarchal structures may undervalue care work and marginalise those considered less 'productive', such as women and children –mirroring how Nature itself is dismissed when not deemed economically beneficial.

The denial of interdependence is equally evident in human relationships, manifesting in hierarchies and oppression based on gender, class, ethnicity, physical appearance, age, and more. Adult-centrism—the sociopolitical and economic structure wherein adults control governance while children are relegated to subordination and oppression (Morales & Magistris, 2020)—is a clear example of this negation. Like Nature is often overlooked unless it aligns with human interests, children's voices and competencies are often disregarded unless they fit with adult-defined notions of productivity and contribution.

Furthermore, capitalist societies not only uphold a specific conception of domination and extractivism toward the environment but also shape the dominant adult concept of the child. According to (Magistris & Morales, 2021), 'childhood' should be understood not merely as a way of being but as a hegemonic discourse—an intentionally constructed narrative infused with various axes of discrimination. This discourse ultimately denies many alternative ways of experiencing childhood. Similarly, any expression of children's agency that does not conform to capitalist, patriarchal, and anthropocentric norms tends to be overlooked or deemed 'incomplete,' much like the natural environments that fall outside the scope of economic profitability.

In societies that do not consider investing in life to be useful (Magistris & Morales, 2021) or a priority, it is urgent to reclaim interdependence and ecodependence as the foundations for a broader sense of belonging and stronger community bonds. Limón-Domínguez and Alcántara-Rubio (2019) illustrate the fertile intersection between environmental ethics and the ethics of care. The former seeks to establish values and paradigms that contribute to reconfiguring harmonious relationships between humans and Nature, while the latter is closely linked to democracy, its practice, and the functioning of global society.

3 The Eco-Citizen Child

It is clear to us that we need new relational paradigms around the diverse ways of being a child. We need a paradigm based on deep, systemic confrontation in defence of children, the environment and all other vulnerable identities and realities. In this process of social construction towards more democratic societies, we will find communities that are more caring, respectful and in harmony with the earth –and vice versa. Indeed, this is what children and young people are teaching us through their youth movements in response to the climate emergency: they have been and continue to be one of the most influential groups in the fight against the climate crisis. They have found a space in the climate struggle to assert themselves as citizens and agents capable of making decisions, promoting change and strategically self-organising to achieve their goals.

Children exercise and signify their lived citizenship differently from adults. Although it is most common for them to develop local initiatives in response to the climate emergency, their actions contribute to solving global problems (Heggen et al., 2019). Evidence of this can be found in the United Nations General Comment No. 26 (Committee on the Rights of the Child, 2023), which acknowledges that "Children's efforts to stand up for their environmental rights motivated the Committee to create this General Comment" (p. 1). It then goes on to say:

> Child human rights defenders, as agents of change, have made historic contributions to human rights and environmental protection. Their status should be recognized, and their demands for urgent and decisive measures to tackle global environmental harm should be realized. (p. 1)

However, despite their prominent role in responding to the climate emergency, this is only one of the many genuine arenas of participation where children and young people engage in initiatives that promote individual and collective well-being. According to Reis (2021), inclusion is achieved through structural change rather than simply by integrating children and young people into pre-existing dynamics in these initiatives. This phenomenon shows that children and young people are not only citizens in the present but have also succeeded in alerting society to critical environmental issues, thus contributing to the learning and practice of eco-citizenship across societies (Heggen et al., 2019). At the same time, they have succeeded in exposing the asymmetries concerning children present in our societies.

The collective challenge of imagining and promoting alternative futures, different from dominant socio-economic patterns, requires a complex civic response. All people and realities are equally needed within civil society –including children. Just as we defend the need to include all identities and groups in this endeavour, we must also be able to embrace all forms of political and social participation. This includes those who work within existing institutions and administrations and those who rely on alternative forms of self-organisation and civic mobilisation. If there is one lesson from the fight against climate emergency, it is that disruptive initiatives are both necessary and essential to promote change and influence communities and public policies.

Therefore, in this chapter, we argue that in the response to the climate emergency –whether led or co-led by children through various initiatives– they are crucial for the necessary eco-social progress. In the same vein, we observe a high degree of children's agency within these initiatives. Along with O'Brien et al. (2018), we note that:

> Despite diverse expressions, all forms of political dissent suggest a belief or presumption of agency, that is, the ability of individuals to imagine a different future and a sense of purposeful expression of opinions or actions that are at variance with dominant or commonly held beliefs. By drawing attention to multiple ways for youth to express their agency both within and outside of traditional political processes, we highlight the ways that they are challenging the interests and power relationships that are perpetuating an unsustainable future. (p. 8)

It is precisely this premise of inherent agency in these initiatives that leads us to believe that, while disruptive initiatives foster a sense of agency among children, it is in defending and nurturing this very sense of agency that new disruptive initiatives by children will emerge as they become aware of their possibilities and civic means. This also applies to initiatives related to the climate emergency.

4 Children's Agency as Part of the Environmental Commitment

In this regard, various studies and essays have examined child and youth activism in relation to environmental issues, and in some of them, we have identified certain relationships and arguments concerning child agency.

While Heggen et al. (2019) argue for the local nature of children's eco-citizenship practices, it is important to note that there are many global experiences in which children have reclaimed children and young people as political and transformative agents through climate activism. O'Brien et al. (2018) explore different forms of political dissent in child and youth climate activism, arguing that through activism, children and young people have found interrelated ways to express their agency, act and influence both their present and future.

Blanchet-Cohen (2008), in her study of children's agency in environmental action, concludes that "children's capacity to affect change comes from a combination of being optimistic about the planet's future, their own confidence and an awareness of their limitations" (p. 270). Optimism, self-confidence and self-awareness are, therefore, three key elements in the development of their agency. At the same time, understanding curiosity as a precursor to inquiry –and seeing inquiry and curiosity as crucial to imagining more sustainable futures– underscores this development.

Thus, climate activism appears to embody several characteristics that are essential for fostering children's sense of agency. It also represents an important aspect of eco-citizenship, as it enables them to actively participate in problem-solving rather than being confined to roles that rely solely on expert opinion (Reis, 2021). In line with this perspective, Heggen et al. (2019) highlight that: "Recognising, supporting and helping children's curiosity is about the child's right to participate as a child, being an eco-citizen and developing themselves as becoming eco-citizens" (p. 392).

If the goal is to 'change the system, not the climate', then, as O'Brien et al. (2018) suggest, we might consider the importance of nurturing and encouraging child and youth activism –and eco-citizenship exercise in the present. These authors propose that this process begins by recognising the many facets, forms, spaces and expressions of child and youth dissent. They suggest that these direct experiences of participation and mobilisation offer many important eco-social lessons that many education systems have yet to address adequately.

Building on these insights, it is suggested that further exploration be given to what this chapter considers to be the foundational concept that enables and sustains such activism –namely, children's sense of agency. It is hoped that by understanding agency, we can develop a framework to see how children not only participate but also transform their environments and social structures.

5 On the Concept of Children's Agency and its Relevance

Children's agency is a dynamic and multifaceted concept that extends beyond traditional notions of autonomy and participation. It has been widely recognised as a defining principle in childhood studies, emphasising children as active constructors of their social realities rather than passive recipients of adult decisions. Robson et al. (2007) define agency as "an individual's own capacities, competencies, and activities through which they navigate the contexts and positions of their lifeworlds" (p. 135), negotiating both individual and collective choices for their present and future lives.

Importantly, agency is not a fixed attribute that children possess but rather a relational process that is exercised and made visible through practice.

This perspective challenges historical liberal views, which often regard children as incomplete beings lacking rational deliberative faculties. Instead, contemporary understandings argue that children exhibit agency through their strategic actions, competence, and awareness, reinforcing their entitlement to greater participation and the recognition of their rights (Valentine, 2011).

However, agency is not exercised in a vacuum; it is relational, embedded in social structures and mediated by cultural, political, and material conditions. This is particularly relevant when considering children's capacity to act as transformative social agents in contemporary crises such as the climate emergency. As Wyness (2015) argues, children's agency must not be reduced to individual autonomy but rather understood as a socially embedded and relational phenomenon deeply connected to collective structures and intergenerational collaboration.

5.1 *From Moral Agency to Political Agency: Children as Transformative Actors*

Moral agency is fundamental to human beings' capacity to engage in social and political life. Moral agency involves the ability to make ethical judgments, to understand justice and reciprocity, and to act on moral values. At the very least, a moral agent is someone who perceives herself as persisting through time, reflecting on her present preferences in the light of future interests, personal values and norms of moral decency (Valentine, 2011). This ability to transcend mere desire and act according to prudence and ethical values is the basis of moral autonomy.

Traditionally, liberal philosophical traditions have positioned moral agency as a capacity attained in adulthood, requiring rationality, self-awareness, and a sense of futurity. Within these frameworks, children have historically been excluded from moral agency due to their perceived lack of deliberative faculties (Valentine, 2011). However, contemporary research in developmental psychology challenges these adult-centered assumptions, demonstrating that children, even from an early age, exhibit moral reasoning based on justice, reciprocity, and empathy.

On the other hand, children's political agency is expressed in multiple ways, including individual decision-making, everyday resistance, and collective mobilisation. Children exercise agency not only as social actors in their everyday environments but also as activists and members of participatory groups where they influence collective discussions and political agendas. This perspective aligns with the argument that children's agency should not be understood solely in terms of expressing views for adults to act upon but rather as self-directed engagement in decision-making and political action (Larkins, 2023).

A key component of children's political agency is their ability to act within and beyond formal institutions. As Rodgers (2020) points out, much of the research on children's participation in social movements has focused on either their representation in adult-led activism or their levels of awareness and citizenship. However, a fuller understanding of children as socio-political agents requires acknowledging their actual participation in social movements beyond institutionalised forms of civic engagement. Children engage in protest, advocacy, and alternative forms of political expression that challenge dominant structures and demand systemic change (Häkli & Kallio, 2018). The climate justice movement, for example, has demonstrated how children not only assert their voices but also mobilise intergenerational networks, influencing both their peers and the adults around them.

Despite this, institutional and social structures often constrain children's political agency. As Borgne and Tisdall (2017) highlight, children's access to political spaces is frequently controlled by adults who act as competence gatekeepers, requiring them to demonstrate their legitimacy as political actors. While NGOs and youth organisations sometimes promote children's participation, these intermediaries may inadvertently reinforce hierarchical structures and limit children's ability to act autonomously. Overcoming these barriers requires recognising that children's political engagement is not an exception but a fundamental component of their lived citizenship. A powerful illustration comes from Naomi Wadler, an 11-year-old activist against gun violence. Speaking at the March for Our Lives rally in Washington, DC, on 18 March 2018, Naomi captured the essence of both moral and political agency when she declared:

> People have said that I am too young to have these thoughts on my own. People have said that I am a tool of some nameless adult. It's not true. My friends and I might still be 11 and we might still be in elementary school, but we know. We know life isn't equal for everyone, and we know what is right and wrong. (Wadler, 2018, 01:58)

Wadler's testimony challenges the notion that children lack the capacity for moral reasoning and political action. Instead, they affirm that children are aware of injustice, capable of ethical reflection, and willing to act on their moral convictions. Her words remind us that children's voices are both valid and essential in shaping a more just and inclusive society.

5.2 *The Sense of Agency: Self-Perception as Influential and Transformative Agents*

For agency to crystallise, children need to develop a sense of agency –a critical element that enables them to act as transformative social actors.

> Sense of agency refers to an awareness of being the author of one's actions and their effects, combined with an understanding of the wider consequences of those actions. It also involves self-awareness of one's own capacities, values, preferences and resources and the ability to engage in metacognition and critical reflection. Ultimately, a sense of agency allows children and young people, either individually or collectively, to recognise their transformative potential and their ability to influence their environment (Esteban Tortajada, 2023, p. 157).

The development of a sense of agency is promoted through genuine and meaningful participation, which could be expressed in terms of civic and political citizenship identity and action. In this regard, the United Nations General Comment No. 20 (Committee on the Rights of the Child, 2016) states that participation is essential for fostering active citizenship.

> The Committee emphasises that States should increase opportunities for political participation to enhance adolescents' sense of agency, enabling them to make informed decisions and advocate for their rights. Policies that support political engagement, such as civic education and digital activism, are critical to ensuring that children and young people develop a strong sense of agency as active citizens. (General principle D, Art. 24)

However, for a sense of agency to be fully developed, it must be recognised as legitimate by significant –and non-significant others– in the generational order. This means that adults must not only recognise children's potential for agency but also actively see them as allies in social and political causes rather than merely as agents in the making or as simple agents with limited influence. Such shift in perception is crucial to breaking down adult barriers that limit children's agency and to promoting co-agency, where children and adults work together as equal participants in shaping their shared realities.

This perspective leads directly to the following discussion on co-agency, highlighting the need for intergenerational cooperation in decision-making and social action.

5.3 *Co-Agency and the Role of Adults as Allies*

Co-agency refers to the processes by which children and adults engage in shared decision-making, co-constructing actions that recognise and respect the agency of both parties. Tomasello (2022) argues that human agency is inherently shared and socially constructed, arising from the need to build shared agreements and relational understandings. In contrast to classical conceptions that associate agency with individual rationality, Tomasello emphasises that what defines human agency is our social and moral condition, which depends on mutual cooperation.

This understanding of agency as a collaborative process is reflected in contemporary participatory models that emphasise intergenerational alliances. For example, the OECD's Sun Model of Co-Agency (Gottschalk & Borhan, 2023) –developed by the OECD Student Focus Group, composed of students from ten countries who helped shape the Learning Compass 2030– builds on Hart's (1992) ladder of participation. However, rather than retaining a linear 'ladder', the students chose a circular 'sun' and placed the term 'co-agency' at its centre to better capture its cyclical and relational nature. By shifting the focus from mere 'participation' to genuine 'co-agency', this model challenges hierarchical notions of participation and proposes a process in which young people and adults make decisions together, rather than adults allowing children to 'have their say'.

Co-agency is, therefore, not only a mechanism for enhancing children's participation but also a necessary condition for the meaningful exercise of their rights. The Convention on the Rights of the Child (CRC) reinforces this idea in Article 5, which calls on adults –whether parents, guardians or members of the community– to provide children with appropriate guidance and support in the exercise of their rights, in accordance with their evolving capacities (a principle that has been the subject of debate, although this is not the place to address it). This recognition emphasises the responsibility of adults not as custodians but as facilitators of children's agency, ensuring that participation is substantive and not just symbolic.

Furthermore, the notion of co-agency or collective agency goes hand in hand with a conceptualisation of responsibility that is fundamental to sustainability and to addressing the challenges of the climate emergency. As ecosystems are increasingly recognised as 'common goods,' their care requires negotiated and responsible collective action (Colmenarejo, 2016). This perspective reinforces the idea that co-agency is not only about empowering intergenerational and collective action but also about promoting intergenerational and collective responsibility for environmental care and sustainability.

By framing children's participation within intergenerational relationships of support and recognition, co-agency emerges as a critical groundwork for the realisation of children's rights. In what follows, we explore how international legal frameworks, particularly the CRC, establish participation rights as essential to children's civic and political engagement.

6 Children's Rights to Participation and Their Political Implications

The Convention on the Rights of the Child (CRC) provides a basic legal framework that recognises children as having subjective rights. It establishes four key principles: non-discrimination, the best interests of the child, the right to survival and development, and the right to be heard (Article 12). This last point is particularly important and was a novelty at the time, as it requires that children's views be taken into account –in accordance with their age and maturity, which, as mentioned above, is a controversial aspect– in all matters affecting them.

However, the CRC has been criticised for its Eurocentric and universalist assumptions about childhood. Scholars argue that it imposes a Western developmentalist model that does not take into account cultural differences in childhood experiences. Consequently, the African Charter on the Rights and Welfare of the Child (ACRWC) was created in response to the under-representation of African perspectives in drafting the CRC. Unlike the CRC, the ACRWC integrates community responsibilities and emphasises the role of children within the family and society, offering a more collectivist perspective on children's rights.

These critiques highlight the importance of contextualising children's rights frameworks within different cultural, social and political settings. The existence of alternative charters, such as the ACRWC, challenges the CRC's premise of a universalised childhood and demonstrates the need to balance global standards with localised understandings of childhood and agency. However, this chapter will focus on the CRC because it is the one that is closest to our grasp and in which we have specialised over time, without detracting from the criticisms it has received.

Gaitán (2018) states that the CRC is "the first human rights treaty to explicitly recognise children's civil rights', although she qualifies this by saying 'but not political rights" (p. 40). Nevertheless, despite the fact that the participation rights of children are not considered to be political in the formal sense, they can be viewed as a form of legitimisation of children's political mobilisation, and it can be argued that it provides children with legal tools for collective action and advocacy in the public sphere.

Whereas Article 12 of the Convention on the Rights of the Child is central to recognising children's right to express their views and have them taken into account, participation rights go beyond this article. The CRC provides children with a range of democratic participation rights that enhance their agency and ability to engage in public and political life: Article 12: Right to express opinions and have them taken into account, Article 13: Right to freedom of expression, Article 14: Right to freedom of thought and belief, Article 15: Right to freedom of assembly and association, Article 17: Right to access to information, Article 29: Right to education that promotes responsible citizenship, Article 37b: Right not to be deprived of one's liberty unlawfully or arbitrarily.

Each of these rights provides children with the legal and institutional support to engage in the public sphere and creates the necessary conditions for children's active participation in civic life, whether through school councils, youth parliaments or grassroots activism. However, the extent to which these rights translate into real political agency depends on the mechanisms and support available to exercise them. Structural barriers –such as intergenerational and structural power relations, adult-led decision-making processes, tokenistic participation frameworks and legal restrictions– often limit children's ability to realise their participation rights fully. For example, while Article 12 requires that children's views be taken into account in matters that affect them, their actual influence in political and institutional decision-making remains highly dependent on the willingness of adults to recognise and support their agency to do so.

This awareness reinforces the need to ensure that participation mechanisms are not only accessible but also truly effective, enabling children to influence policies and practices that have a direct impact on their lives. The effectiveness of these rights depends on intergenerational cooperation, in line with the principles of co-agency where children are not only consulted but actively involved as partners and citizens in our societies.

7 Accompany. A Stand Rooted in Our Commitment to Children

From our perspective, the promotion and consolidation of children's disruptive initiatives seem closely related to the support that we, adults, from a variety of socio-educational contexts and backgrounds, provide to the numerous dissident initiatives in which children are involved. It would be beneficial for us to collectively reflect on whether we adults adequately support children in developing their sense of agency and exercising their participation rights and citizenship.

For all these considerations, disruptive initiatives led or co-led by children must be grounded in value-based socio-educational relationships, especially if we aim to promote such initiatives within our pedagogical intervention spaces. We refer to a value-based socio-educational relationship as defined by Úcar (2020): one that is situated, effective, sustainable and satisfying. In the face of the climate emergency, it makes sense to propose an accompaniment that is:

- **Situated:** The impact of climate change on each territory and community is different. Also, different factors and life histories determine each group's participation culture. We must engage with children by recognising and valuing their particularities.
- **Effective:** The goal is to empower communities by fostering critical thinking and transformative action. To achieve this, efforts must be combined to bring these elements to fruition and bring about change. As evidenced by the many scientific reports on the impacts and implications of climate change, it is of the utmost importance to advance our response to the climate emergency.
- **Sustainable:** Through these initiatives, we seek to promote a culture of sustainability in its broadest sense –a culture based on the principles of inclusion, care, and solidarity, both diachronically and synchronically. These elements must also be present in our accompaniment for it to be sustainable.
- **Satisfactory:** Socio-educational relationships should recognise and value both individual and collective knowledge in order to build and imagine alternatives together.

For this reason, we advocate a style of accompaniment that creates scenarios and experiences based on the specificities of each group and territory. These scenarios and experiences help resolve problematic situations by equipping children with the necessary tools and resources.

In this context, the authors advocate an approach to accompaniment that creates scenarios and experiences tailored to the specificities of each group and territory. This approach aims to address problematic situations by equipping children with the necessary tools and resources (e.g., competencies, knowledge, confidence, and freedom to act). The term accompaniment is understood as multidirectional, or non-hierarchical, in the sense that both children and adults walk alongside one another through shared processes and journeys. Rather than imposing an adult perspective, it involves understanding children's viewpoints, identifying what adults can contribute to creating favourable conditions for their development, and remaining open to children's own initiatives while explicitly rejecting paternalistic approaches. Consequently, accompaniment becomes a multifaceted practice entailing numerous processes that enrich the various dimensions of the socio-educational relationship with individuals and groups, carried out from specific ethical and methodological paradigms. Of particular significance is the realisation that children, too, have the capacity to accompany adults, thereby fostering an environment characterised by mutual learning and shared transformation.

Among the methodological elements that we as adults can and should implement if we want to accompany or promote disruptive initiatives with children in the face of the climate emergency, Reis' (2021) proposal is particularly compelling. The author argues that for these initiatives to flourish, adults must not actively work against them. Instead, he argues that experimentation, contact and the development of different participation models –as well as the competencies, skills and knowledge associated with each experience– are necessary. For disruptive initiatives to produce alternative and adequate solutions to socially relevant problems, activism requires responsibility and a shared commitment between different institutions, actors and communities (Reis, 2021). All this is done by acknowledging the elements that Blanchet-Cohen (2008) identified as essential for the development of a sense of agency in environmental activism: optimism, self-confidence, and self-awareness.

From these perspectives, accompaniment means recognising the other as a significant equal and capable of creating political alternatives and contesting power, just as we do (Montenegro & Shabel, 2021). It means adopting a perspective of 'childism' (Wall, 2024) that recognises children as capable agents in the present and commits to deconstructing the historically entrenched prejudices against them and rebuilding more just structures.

If there is one lesson the climate emergency teaches us, it is that time is of the essence. Given this certainty, it is a matter of social justice to work to reverse all social asymmetries. Consequently, our responsibility as adults and educators is to create spaces where this is possible –especially if we recognise and value the knowledge and competencies of children to confront the climate emergency. In a system in crisis that calls us to action, disruptive initiatives led or co-led by children have occupied the spaces that adult governance has failed to address.

8 Final Reflections on Accompanying Children's Disruptive Initiatives in the Face of the Climate Emergency

In order to foster the development of children and young people's sense of agency, it is crucial that socio-educational practices create 'enabling contexts'. These are spaces specifically designed to nurture and enhance the competencies required for transformative action. According to Esteban Tortajada (2023), such contexts should, at a minimum, include the following: (1) a safe and respectful environment, (2) the recognition of children and young people as transformative social actors, (3) real and meaningful participation, and (4) spaces and moments for reflection and meta-participation. In the same way that environmental education provides settings for acquiring skills in sustainability, scientific literacy, and proactive engagement, our commitment must extend to cultivating environments in which children can further develop their sense of agency through the exercise of agency. This is fundamental for nurturing new and future disruptive initiatives –whether through collective re-signification, innovative approaches with youth, or creative environmental practices driven by their tireless commitment.

Moreover, this sense of agency is deeply intertwined with both moral and political dimensions. In today's climate emergency, children are not only gaining the capacity to act but are also embracing roles that challenge injustice, assert their rights, and demand systemic change. When children's agency is rooted in a commitment to equity and sustainability, it acquires a moral significance; when it manifests in calls for transformative policies, it takes on a political dimension. Nurturing these facets is crucial because societal transformation depends on both the internal development of ethical self-awareness and the external expression of collective political action. Consequently, empowering children to act with moral clarity and political determination lays the groundwork for eco-citizenship and contributes to building a more just, sustainable, and inclusive future.

The human capacity for agency in children and young people is expressed through several key qualities. In her research, Esteban Tortajada (2023, p. 174) identifies and describes the following intrinsically human agency-related qualities –qualities for which the climate emergency becomes an enabling context:

- **Voluntariness and intentionality:** Children and young people act out of their own will, not by imposition or coercion. Therefore, they act and participate according to their interests and motivations.
- **Transparent and accessible information:** Children and young people have access to sufficient and intelligible information, which is communicated clearly and directly to inspire trust and enable them to make informed decisions.
- **Practical skills (know-how):** Children and young people have a good understanding of themselves; they are aware of their interests, preferences and personal values; they have sufficient and accurate knowledge about the physical and social world; and they develop competencies that enable them to act effectively.
- **Decision-making:** Children and young people can make decisions, which gradually become binding, based on their intentions and will. They also take responsibility for the effects and consequences of their decisions.
- **Individual and group self-organisation:** Children and young people can plan their course of action using the means and resources available, analyse and review the process to identify the adjustments and refinements needed to achieve the proposed goals.
- **Achieving goals:** Children and young people perceive themselves as self-efficacious, for they achieve goals they set for themselves either individually or in collaboration with others.

- **Influence or the perception of 'making a difference':** Children and young people perceive their transformative power and their ability to use their actions to influence their environment or community. They understand that by participating, they can change and improve things.

These qualities underlie the many disruptive initiatives that children launch to confront the climate crisis with deliberation, strategy, and purpose. Such initiatives reveal not only the struggle to defend a healthy environment but also the essential reclaiming of children's rights, affirming their status as full citizens capable of thriving in society.

Historical and contemporary examples provide vivid illustrations of the transformative potential of youth-led activism. Rodgers (2020) provides examples of Barbara Johns, who, at the age of 16 in 1951, led a student strike at Moton High School in Virginia to protest against the dire conditions in a segregated education system. Johns emphasised that seeking adult help before taking action would have stifled her initiative. Rodgers (2020) further observes that such acts of youth activism have often been overlooked in mainstream narratives, despite their profound impact on social change. Johns' courageous stand not only contributed to the dismantling of segregated schooling but also highlighted the transformative power of independent youth action in challenging systemic injustices.

In the current climate emergency, these same agentic qualities are powerfully expressed through contemporary movements. For instance, *Fridays for Future*, initiated by Greta Thunberg, has mobilized students worldwide to demand urgent reforms in climate –exemplifying voluntariness and intentionality. Equally, youth-led groups leverage social media to disseminate reliable climate data, ensuring that complex scientific information is accessible to broad audiences. Young activists convert theory into practice by organising local sustainability projects –such as urban community gardens and recycling initiatives– that yield tangible environmental benefits. Decision-making and self-organisation are further demonstrated by Vancouver's Youth Climate Advisory Council and by the adaptive tactics of Extinction Rebellion Youth in the United Kingdom. Moreover, persistent youth protests in Germany contributed to the government declaring a climate emergency, and the influential advocacy of the Sunrise Movement in the United States illustrates how children and young people can shape national debates on environmental policy.

As Rodgers (2020) explains, children's participation in social movements can take various forms: they may serve as strategic participants, where adults deliberately plan their involvement; become participants by default, accompanying parents or caregivers to movement events; or act as active participants, autonomously taking part in strikes, marches, and protests, and even assuming leadership roles. These different forms of participation collectively underscore that nurturing a comprehensive sense of agency –encompassing both moral conviction and political engagement– is vital for empowering children and young people as effective eco-citizens capable of driving transformative change.

Disruptive initiatives force us to self-organise in multiple and diverse ways, enabling incremental victories in both confronting the climate crisis and defending children's participation rights. The perception of one's own impact, in turn, fosters the replication and promotion of new disruptive initiatives. Nevertheless, we must acknowledge that child youth activism remains challenging, as it often contends with techno-optimist and capitalist narratives that encourage disengagement, frustration, and demobilisation.

Building on these agency-related qualities and the approach to accompaniment discussed above, it is vital to consider the role of the facilitator –a socio-educational agent who can create enabling contexts, provide resources and information, critically reflect on their own practice, share responsibility in decision-making, and encourage and multiply children's achievements and goals (see Fig. 1).

Finally, the need to critically analyse, review and re-signify relational approaches towards childhood and Nature is emphasised. The objective is to accompany children so they can learn about and engage with these problematic situations from a transdisciplinary and critical perspective in order to recognise the urgency and complexity of the issues at hand. In addition, we seek to comprehend the intricate network of interrelated factors that give rise to these realities, with the aim of promoting changes in the prevailing discourse that upholds the status quo and the structures of injustice and oppression. This discourse positions children as marginalised individuals and views Nature solely in terms of its productive and extractive potential. While acknowledging the considerable challenges associated with achieving change

Fig. 1 Socio-educational framework for adult accompaniment of child and youth alliances in disruptive initiatives facing climate emergency

within this context, we recognise the difficulty of assisting children in transforming frustration into organised, coordinated, and effective action.

Despite the global nature of the collective challenges outlined in this chapter, there are opportunities within educational practice to create spaces that encourage care for both children and the environment. These spaces should emerge from personal and collective questioning of interpersonal relationships and our relationship with Nature. It is imperative to recognise that social representations influence our perception of the world and that these representations can be deconstructed and reconstructed within ourselves and, ideally, collectively constructed. Consequently, in order to accompany children in their efforts to address the climate emergency, it is essential to re-signify our representations of childhood, adopting empowering perspectives that recognise children's agency and environmental commitment. From these novel approaches, we must engage with and position ourselves across all spheres of society.

References

Blanchet-Cohen, N. (2008). Taking a stance: child agency across the dimensions of early adolescents' environmental involvement. *Environmental Education Research, 14*(3), 257–272. https://doi.org/10.1080/13504620802156496

Borgne, C. L., & Tisdall, E. K. M. (2017). Children's participation: questioning competence and competencies? *Social Inclusion, 5*(3), 122–130. https://doi.org/10.17645/si.v5i3.986

Caride, J. A., Gutiérrez-Pérez, J., & Meira-Cartea, P. Á. (2025). Climate narratives in educational agendas: on the value of time for pedagogical research and environmental policies. *Revista Española de Pedagogía, 83*(290), 83–103. https://doi.org/10.22550/2174-0909.4149

Colmenarejo, R. (2016). Enfoque de capacidades y sostenibilidad. Aportaciones de Amartya Sen y Martha Nussbaum [Capability approach and sustainability. Contributions of Amartya Sen and Martha Nussbaum]. *Ideas y Valores, 65*(160), 121–149. https://doi.org/10.15446/ideasyvalores.v65n160.43084

Committee on the Rights of the Child. (2016). *General comment No. 20 on the implementation of the rights of the child during adolescence.* United Nations. https://www.ohchr.org/en/documents/general-comments-and-recommendations/general-comment-no-20-2016-implementation-rights

Committee on the Rights of the Child. (2023). *General comment No. 26 on children's rights and the environment, with a special focus on climate change.* United Nations. https://www.ohchr.org/en/documents/general-comments-and-recommendations/crccgc26-general-comment-no-26-2023-childrens-rights

Esteban Tortajada, M.B. (2023). *Sentido de agencia humana y participación democrático-ciudadana de la infancia y la adolescencia. Claves para el acompañamiento socioeducativo [Sense of human agency and democratic-citizen participation of childhood and adolescence. Keys to socio-educational support]* (Doctoral Thesis), University of Barcelona. http://hdl.handle.net/10803/692201

Gaitán, L. (2018). Representaciones sociales de infancia y ciudadanía: entre el antagonismo y el cambio [Social representations of childhood and citizenship: between antagonism and change]. In A. L. Berñe & A. M. N. Cámara (Eds.), *Impulsar la participación infantil. Los consejos de infancia y adolescencia* (pp. 33–50). Graó.

Gottschalk, F., & Borhan, H. (2023). *Child participation in decision making: implications for education and beyond*. Organisation for Economic Co-operation and Development.. https://doi.org/10.1787/a37eba6c-en

Häkli, J., & Kallio, K. P. (2018). Theorizing children's political agency. In T. Skelton & S. Aitken (Eds.), *Establishing geographies of children and young people. Geographies of children and young people* (pp. 1–23). Springer. https://doi.org/10.1007/978-981-4585-88-0_1-1

Hart, R. (1992). *Children's participation: from tokenism to citizenship.* UNICEF International Child Development Centre.

Heggen, M., Sageidet, B., Goga, N., Torunn, L., Bergan, V., Wallem, I., et al. (2019). Children as eco-citizens? *Nordina, 15*(4), 387–402. https://doi.org/10.5617/nordina.6186

Larkins, C. (2023). Deeping the roots of children's participation. *Sociedad e Infancias, 7*(1), 147–163. https://doi.org/10.5209/soci.88645

Limón-Domínguez, D. (2019). *Ecociudadanía. Retos de la educación ambiental ante los objetivos de desarrollo sostenible [Ecocitizenship: challenges of environmental education in the face of the sustainable development goals].* Octaedro.

Limón-Domínguez, D., & Alcántara-Rubio, L. (2019). Ética ambiental y ética del cuidado: una base democrática para los objetivos de desarrollo sostenible [Environmental ethics and ethics of care: a democratic basis for the sustainable development goals]. In D. Limón-Domínguez (Ed.), *Ecociudadania: retos de la educación ambiental ante los objetivos de desarrollo sostenible* (pp. 31–52). Octaedro.

Magistris, G., & Morales, S. (2021). Mañana será tarde [Tomorrow will be too late]. In G. Magistris & S. Morales (Eds.), *Educar hasta la ternura siempre. Del adultocentrismo al protagonismo de las niñeces* (pp. 3–12). Chirimbote & Ternura Rebelde.

Mies, M., & Shiva, V. (1997). *Ecofeminismo: teoría, crítica y perspectives [Ecofeminism: theory, critique, and perspectives].* Icaria Editorial.

Mies, M., & Shiva, V. (2014). *Ecofeminismo [Ecofeminism].* Icaria Editorial.

Montenegro, H., & Shabel, P. (2021). Aprender a contar el mundo. El acompañamiento escolar como práctica de transformación y participación de niñxs [Learning to narrate the world. School support as a practice for transformation and child participation]. In G. Magistris & S. Morales (Eds.), *Educar hasta la ternura siempre. Del adultocentrismo al protagonismo de las niñeces* (pp. 179–198). Chirimbote & Ternura Rebelde.

Morales, S., & Magistris, G. (2020). Hacia un paradigma otro: niñxs como sujetos políticxs co-protagonistas de la transformación social [Towards a different paradigm: children as political subjects and co-protagonists of social transformation]. In S. Morales & G. Magistris (Eds.), *Niñez en movimiento: del adultocentrismo a la emancipación* (pp. 23–47). Chirimbote & Ternura Rebelde.

O'Brien, K., Selboe, E., & Hayward, B. M. (2018). Exploring youth activism on climate change: dutiful, disruptive, and dangerous dissent. *Ecology and Society, 23*(3), 230342. https://doi.org/10.5751/ES-10287-230342

Reis, P. (2021). Cidadania ambiental e ativismo juvenil [Environmental citizenship and youth activism]. *Ensino de Ciências e Tecnologia em Revista–ENCITEC, 11*(2), 05–24. https://doi.org/10.31512/encitec.v11i2.433

Robson, E., Bell, S., & Klocker, N. (2007). Conceptualizing agency in the lives and actions of rural young people. In R. Panelli, S. Punch, & E. Robson (Eds.), *Global perspectives on rural childhood and youth. Young rural lives* (pp. 135–148). Routledge.

Rodgers, D. M. (2020). *Children in social movements: rethinking agency, mobilization and rights.* Routledge.

Tomasello, M. (2022). *The evolution of agency. behavioral organizations from lizards to humans.* The MIT Press. https://doi.org/10.7551/mitpress/14238.001.0001

Úcar, X. (2020). Principios sociopedagógicos y relaciones socioafectivas [Socio-pedagogical principles and socio-affective relationships]. In F. J. del Pozo Serrano (Ed.), *Intervención educativa en contextos sociales: fundamentos e investigación, estrategias e innovación en el ciclo vital* (pp. 3–22). Uninorte Editorial.

Valentine, K. (2011). Accounting for agency. *Children & Society, 25*(5), 347–358. https://doi.org/10.1111/j.1099-0860.2009.00279.x

Wadler, N. (2018, March 24). *11-year-old naomi wadler's speech at the march for our lives (Full)|NBC News*. YouTube. https://www.youtube.com/watch?v=C5ZUDImTIQ8

Wall, J. W. (2024). Infancismo y las políticas de empoderamiento social [Childism and social empowerment policies]. *Sociedad e Infancias, 8*(2), 205–225. https://doi.org/10.5209/soci.97639

Wyness, M. (2015). *Childhood*. Polity Press.

Community Climate Education for the Transition from the Mundane to the Ecosocial Utopia

Miguel Pardellas Santiago, Júlio J. Conde, and Pablo Ángel Meira Cartea

Abstract This chapter presents the trajectory and outcomes of the community-based climate education initiative "Decarbonize! 'Cause that's no small feat" developed in Galicia from 2017 onwards. Originally designed to address the climate emergency with adult populations, the project evolved through continuous pedagogical reflection, methodological adaptation, and interdisciplinary collaboration. Grounded in the theory of social representations, the initiative aims to reduce the social and psychological distance between climate change and everyday life, while fostering community resilience and action. Through participatory workshops, emotional engagement and collective decision-making; diverse groups —predominantly women and older adults— explored the links between climate issues and their daily lives. Over time, the initiative expanded to include intergenerational dynamics, gender-focused interventions, risk management strategies and the use of artistic narratives such as theatre and audiovisual media. The findings emphasize the importance of community engagement, affective pedagogies and utopian imagination in promoting critical and transformative climate change education that empower citizens and strengthen the social fabric necessary to address the climate crisis.

> *Ames, 2074. Years after the climate revolution, Biduído is filled with native trees. We hear birds. We see solar panels on the rooftops of the houses, many of which are traditional constructions. People working in the distance in community orchards...*

This is how the group of women who took part in the project *From Imbalance to Shared Dreams* began to imagine a desirable future to aspire to. This narrative exercise eventually crystallized into the collective creation of a script of the short film *Raíces. Da utopía en diante* (*Raíces. From utopia onwards*), starring the same group

---

M. Pardellas Santiago (✉)
Feitoría Verde S. Coop. Galega, Santiago de Compostela, Galicia, Spain
e-mail: miguelpardellas@feitoriaverde.com

J. J. Conde · P. Á. Meira Cartea
Social Pedagogy and Environmental Education Research Group (SEPA-interea), Universidade de Santiago de Compostela, Santiago de Compostela, Galiza, Spain

J. J. Conde et al. (eds.), *Practices in Education for Climate Emergency*,
https://doi.org/10.1007/978-981-92-2304-6_17

of women and involving their local community. In the transition from one stage to the next, a systematic educational process unfolded over many working sessions, in which the climate emergency, emotional bonds, everyday life and utopia became interwoven until the participating women threaded together a story of their own, where the threat of climate change and the ambition to move toward sustainable and desirable futures were combined in a coherent and optimistic manner.

The experience that led to the creation of this short film represents the final link in the development of a chain of socio-community-based educational actions that originated in the project *Decarbonize! 'Cause that's no small feat.* This environmental education initiative was designed to address the climate emergency with adult audiences, but its original approach and design have undergone a continuous evolution, guided by an ongoing process of evaluation and nourished by the constant dialogue between educational practice and theoretical reflection. This evolution has been essential to adapt the original methodology to the different social contexts in which it has been implemented and to allow it to adjust to the changing social perceptions of climate change and the policies responding to its threat potential.

1 First Stage: Design and Objectives

For over two decades, the Research Group on Social Pedagogy and Environmental Education (SEPA-interea) at the University of Santiago de Compostela has been developing a basic and applied research line under the general title of RESCLIMA Project (Conde et al., 2026). This research line primarily aims to understand how different population groups perceive and represent climate change and, based on the knowledge generated, to design and test effective socio-educational resources in addressing the threats arising from this issue. The ultimate goal is to support the role of education in the design and implementation of public policies aimed at reducing greenhouse gas emissions and minimizing human vulnerability to the consequences already underway due to climate change.

In addition to numerous recommendations and suggestions, the research carried out within the framework of RESCLIMA Project revealed a sense of disconnection between socio-educational research findings and the focus of most educational and communicative actions currently in use (Meira, 2017a; Meira et al., 2022). As a matter of fact, despite increasing evidence highlighting the limitations of scientific literacy initiatives that assume conveying the best available science on climate change will result in the necessary individual and collective changes (Azevedo & Marques, 2017; Carvalho, 2011; García-Vinuesa et al., 2023), this positivist approach continues to be dominant. In consequence, it overshadows others that prioritize cultural and political identity, emotional climate, ethics, and social interaction processes that mediate, hinder, or stimulate individual and collective transformation (Bentz, 2020; Hobson & Niemeyer, 2012; Lehtonen et al., 2019; Lombardi et al., 2013; Nigbur et al., 2010; Roeser, 2012).

To counter this inertia, the project *Decarbonize! 'Cause that's no small feat* was born around the idea of designing and testing an educational proposal to effectively incorporate the findings of the research line (Fig. 1). Several works have already detailed the early development of the project (Meira, 2017b; Meira et al., 2022; Pardellas & Meira, 2020). Additionally, the methodological guide (Pardellas, 2019) provides a detailed description of its goals and theoretical foundations, which is guided by three main objectives:

- To organize and support groups willing to contribute to the decarbonization and adaptation of their communities, increasing community resilience and facilitating the ecological transition toward low-carbon societies.
- To reduce the social, cultural, and psychological distance that separates the climate crisis from the everyday lives of most people and communities.
- To place climate change on the public agenda as a socially relevant issue for citizens, linking its threat potential to meaningful aspects where people can make synergistic social and environmental changes.

The methodological framework of *Decarbonize!* is built upon several theoretical pillars, among which the following stand out:

The need to adopt a socio-educational approach to address climate change: Since climate change emerged in the 1970s as a systemic sign of a global socio-environmental crisis, most response policies have focused on technological (e.g., combustion engine efficiency, renewable energy development, local emission reduction systems, carbon capture technologies) or economic solutions (e.g., fossil fuel taxation, disinvestment in polluting technologies, emission compensation mechanisms, subsidies for clean energy or carbon sink restoration projects) or a combination of both. Notably, the future scenarios explored in the IPCC reports rely primarily on biophysical, technological and economic variables, with minimal attention to social, psychosocial or cultural factors.

The partial, uncoordinated and superficial nature of these actions —limited to very specific areas and a few countries— has made mitigation strategies largely insufficient to halt climate change (Fernández Durán & González Reyes, 2015), with emissions continuing to rise over the past few decades. It is significant that the few

Fig. 1 Logo of *Decarbonize! 'Cause that's no small feat*

instances of emission reductions over the past 50 years have aligned with regional (e.g., Soviet Union's dissolution in 1992) or global economic crises (e.g., oil price spikes in the 1970s, the 2008–2009 financial collapse) or disruptive events like the COVID-19 pandemic in 2020 (Friedlingstein et al., 2025).

In this context, there is a growing call to integrate the social dimension —ethical, psychological, cultural, philosophical, pedagogical, political, etc.— into the design of adaptation and mitigation strategies. A clear example is the explicit way in which the social dimension is presented as an essential component in the IPCC's Fifth Assessment Report (Heras, 2016), a focus that is further developed in its Sixth Report (Creutzig et al., 2022). This means that, alongside economic factors —which are themselves social in nature— it is essential to also consider aspects such as governance, socio-cultural dynamics, education (both formal and informal), and communication, as well as the interactions among these elements. This expanded and comprehensive perspective is essential for designing more effective climate policies insofar as they are demanded, supported, and adopted by the diverse societies and human groups that bear varying degrees of responsibility for the causes of the climate crisis and that are more or less vulnerable to its consequences.

According to the IPCC (2018, p. 22), "Education, information, and community approaches, including those that are informed by indigenous knowledge and local knowledge, can accelerate the wide-scale behaviour changes consistent with adapting to and limiting global warming to 1.5°C." The report also emphasizes that "these approaches are more effective when combined with other policies and tailored to the motivations, capabilities and resources of specific actors and contexts," which aligns with the orientation of the community education initiative described here.

The urgency of taking social representations of climate change into account: It is possible to have extremely detailed information on certain social variables —beyond climate and technological data— yet still fail to effectively reach the population or to do so in a way that effectively motivates the necessary transformations. Information alone is not enough. This is where social representations become crucial; that is, understanding how the population perceives, interprets, feels, and evaluates climate change, and especially what implications it may have for their lives and their immediate environment. It is through the knowledge of these representations —shaped by intense social controversy— that more effective educational and communicative interventions can be designed to support the transition of individuals, communities and societies toward decarbonized lifestyles and modes of production and consumption (Takano et al., 2025).

From a perspective that bridges psychology and sociology, social representations are defined as a particular form of knowledge whose function is the collective interpretation of reality, the motivation of behavior, and the organization of communicative interactions among individuals within a society or social group (Moscovici, 1979). In other words, social representations theory offers an epistemological and ontological framework for uncovering how "common sense" knowledge is socially constructed, i.e. knowledge that aims to provide a socially shared meaning of the world, facilitate communication, and shape the identity of a social group.

Leveraging successful educational practices and approaches: As a transversal line of research, the analysis and evaluation of effective educational practices and approaches is one of the objectives of the RESCLIMA Project. Within this line of work, the decision was made to study —through a socio-pedagogical lens— the initiatives of the so-called newest social movements that emerged at the turn of the century. We refer here to degrowth movements, post-carbon cities, the slow movement, or transition town networks; all of which are indebted to earlier and contemporary social movements —such as the environmental movement, indigenous movements, the anti-globalization movement, etc.— but which have introduced innovative actions and initiatives that, in practice, function as methodologies for environmental-educational intervention with great potential for adaptation to other contexts, even if their creators and advocates do not necessarily view them as such (Pardellas et al., 2017).

Despite their different trajectories and theoretical-ideological foundations, the social activism of these neo-movements contains many elements of educational relevance, such as the use of future-oriented scenarios, the emphasis on participation and shared decision-making, the essential role of emotions in community processes, and the development of accessible and mobilizing strategies for relocalization.

2 Second Stage: Initial Steps (2017–2020)

Research into social representations of climate change has provided valuable insights into how citizens tend to perceive and evaluate this socio-environmental challenge. At the same time, the identification of successful educational practices offers concrete, replicable examples that help recognize the threat posed by climate change and respond through civic engagement. The convergence of these two lines of work forms the core of the intervention methodology behind *Decarbonize!*, a methodological proposal that seeks to connect an issue often perceived as abstract and remote —such as climate change— with meaningful aspects of people's daily lives, both individually and collectively.

The resulting sequence of activities consists of several training sessions in which adult and socially diverse participants are invited to analyze and reflect on the causes and consequences of climate change, and to identify and activate individual and collective actions to reduce greenhouse gas emissions. These actions range from lowering household electricity use and cutting energy bills to promoting sustainable transport and responsible food consumption, as well as initiating various activities that help make their communities better places to live.

Through various group dynamics, participants decide what actions to take in order to reduce their emissions in areas of their lives that are particularly meaningful. In doing so, the project seeks to activate several levers simultaneously: a shared responsibility for the problem, highlighting what can be achieved through collective effort; the organization of joint activities that resonate with participants' emotions and daily lives; mutual reinforcement of the actions to be undertaken, particularly on an emotional level; and a sense of community identity, acknowledging its limitations while developing its potential.

With these foundations, the project began its journey with two pilot experiences in the municipality of Santiago de Compostela, involving several user groups from the Municipal Socio-Cultural Centers. These initial trials provided essential insights for refining and consolidating the methodological approach, which was then more thoroughly tested in a subsequent and more extensive intervention carried out simultaneously in Santiago de Compostela and Donosti, with funding from the *Fundación Biodiversidad* (a Spanish public foundation). In this second phase, the educational strategy also included training municipal cultural service staff from both cities, who then independently implemented the work sequence with participating adult groups.

By the end of this first stage, the project had reached around 200 participants across Santiago and Donosti, working continuously with them through 8 to 10 sessions of approximately 90 min each. The participant profile was predominantly composed of women over the age of 65, involved in various activities at sociocultural and civic centers. Some of the key elements identified in the socio-educational intervention methodology included (Pardellas & Meira, 2020, pp. 36–39):

- **Group dynamics and participatory engagement as both means and end.** In contrast to typical interventions in most climate change education initiatives focused on individuals —or, at most, households—, the *Decarbonize!* methodology is rooted in group work and constructivist learning dynamics (García, 2004). The group engages in dialogue, analyzes information, reflects, and learns collectively in order to eventually make and implement decisions together. This participatory and educational process, centered on everyday situations and decisions, requires participants to internalize information about climate crisis and integrate diverse perspectives within the group.
- **Community as a foundation.** Group work consistently emphasizes the community scale as the setting for educational action. On one hand, the community to which the group belongs (neighborhood, urban district, village) becomes the shared social space where concrete, meaningful and feasible action can take place. On the other hand, a focus on the community level breaks with individualistic tendencies and reinforces collective identity, which has greater influence on people's beliefs and behaviors.
- **Acknowledging emotions.** Emotions play a crucial role in educational processes in general, and particularly in those related to climate change. As a politically charged issue involving conflicts of interests and values, the climate crisis can generate intense emotional responses —often negative— such as fear, helplessness or powerlessness. It can also lead to denial, indifference or skepticism, often encouraged by political and economic interests. For this reason, the intervention sequence pays special attention to participants' emotional responses to climate change, aiming to create —whenever possible— a positive and even enjoyable atmosphere that fosters collective emotional satisfaction. Group care and mutual support are also essential to creating this kind of environment.

- **Memory of everyday life.** When working with adult audiences, tapping into their collective "low-emission memory" serves as a powerful cognitive and emotional catalyst. Without romanticizing the past or ignoring its hardships, childhood and youth memories help link the emotional and ordinary aspects of life with the greenhouse gas emissions responsible for climate change. These shared memory exercises also strengthen group and community bonds, highlight their relevance for resilience, while reaffirming the value of collective celebration and leisure.
- **Systemic interpretation of reality.** Climate change encompasses energy crisis, individual transportation, biodiversity loss, meat consumption and consumerism, among others. Having clear, shared objectives across all sessions —such as decarbonization, adaptation and community resilience— gives the intervention sequence strong potential for fostering a systemic understanding of reality, where personal and vicarious experiences combine into a complex and interconnected mosaic. This systemic view is further supported by collective memory exercises and the inclusion of intergenerational dimensions, which also facilitate activities involving people of different ages.
- **Without fear and without nostalgia.** Fear and alarmism are poor companions when it comes to dealing with global —and seemingly abstract— threats like the climate crisis. The usual response to such projections is paralysis or apathy in the face of what seems inevitable. Instead, it is more effective to offer alternatives that are both feasible and desirable. Once again, collective memory can play a key role, as it holds examples of lifestyles marked by more frugal consumption and lower emissions. Still, nostalgic views of the past that overlook difficult realities must be avoided, especially in terms of quality of life and respect for basic human rights (labor rights, children's rights, women's rights, etc.). Learning from the past does not mean replicating all the experiences it entailed.
- **Engaging with uncertainty with rigor and quality.** Epistemic uncertainty is inherent to scientific practice and therefore part of the complexity of climate change itself. Communicating this uncertainty to lay audiences —whether regarding the attribution of specific weather events, medium- or long-term consequences, the effectiveness of proposed solutions, or interactions between human systems and the climate— can lead to doubts about the reliability of scientific knowledge and the level of consensus within the scientific community towards climate change. As Lewandowsky et al. (2015) suggest, absolute certainty is unattainable; yet uncertainty —which is perhaps one of the few certainties of climate change— can serve as a powerful motivator for action. Working through the inherent uncertainties in the threats posed by climate change can stimulate civic responses to the crisis; i.e. through simulation exercises and future-thinking practices that engage with both uncertainty and scientific rigor.
- **Resilience and conscious adaptation.** Even in "well-adapted" communities, achieving a level of resilience capable of overcoming climate risks is an almost impossible task in today's world. Exercises and strategies around resilience and adaptation must also acknowledge their limitations while enabling actions aimed at influencing decision-making processes at other scales (municipal, regional,

national, etc.). Although it is a difficult leap, connecting the local with the global is essential and must remain in focus.

3 Third Stage: Constant Adaptation (2020–2024)

At the beginning of 2020, several opportunities emerged to transfer the project *Decarbonize!* to other municipalities in Galicia, while, at the same time, its replicability in other regions of the Spanish state was also being considered. However, the arrival of COVID-19 disrupted and stalled this expansion process. Nevertheless, the need to adapt the educational approach to the new pandemic and post-pandemic conditions prompted a process of methodological evolution that would become a defining feature over the following years.

This process, in addition to aligning the project's foundation as an action research initiative, was catalyzed by the involvement of the cooperative *Feitoría Verde* in the development of the project. *Feitoría Verde* is a non-profit cooperative with a long track record in the field of environmental education and an ongoing collaborative relationship with *SEPA-interea* group, based on both academic and personal ties. As a result of its integration into the project, numerous changes were introduced into the session dynamics, with a stronger emphasis on the importance of group cohesion, while preserving the original sequenced structure. Below, some of the projects derived from *Decarbonize!* project and developed during this third stage (see Fig. 2) are described:

3.1 Exploring the Intergenerational Perspective

The first initiative of this new phase emerged from a project designed by *Feitoría Verde* as a proposal to address health-related issues among adolescents, adopting a holistic perspective. This experience, titled *Des-Equilibristas*, was presented as an adventure aimed at forging healthy connections between food, environment, emotions, neighborhood relationships and reference communities; an initiative

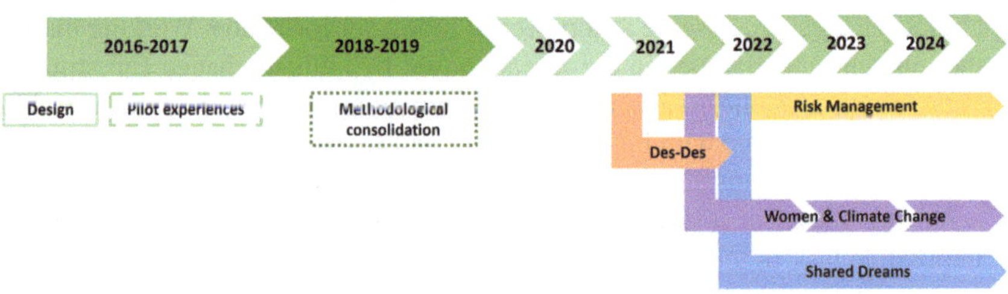

Fig. 2 Timeline *of Decarbonize! 'Cause that's no small feat* development

largely conceived in response to the needs that arose during the pandemic. Since one of the recurring issues identified in the *Decarbonize!* groups was the need to involve younger age groups —given that the majority of participants up to that point were older adults— an intergenerational proposal was developed under the name *Do Desequilibrio á Descarbonización* (*Des-Des*, it can be translated in English as: *From Imbalance to Decarbonization*). The aim was to bring together groups of older adults and groups of youth and adolescents to carry out joint actions.

After an initial phase of working separately —with groups of adolescents and young people on the one hand, and older adults on the other— a joint action was proposed aimed at decarbonizing their surrounding environment. In the co-designed planning of this intergenerational action, both the health crisis and the climate crisis were addressed simultaneously, with the goal of complementing the needs and interests expressed by both age groups, resulting in an action (in the neighborhood, the village, etc.) that integrated environmental and emotional education.

Under the framework of a public grant program for launching environmental activities in municipalities across the province of A Coruña, the methodological proposal of *Des-Des* was implemented in the Galician municipalities of Padrón, Ames, and Santiago de Compostela. However, it was only in the latter where it gained some traction, through occasional meetings between groups of young people and older adults. Still, this fell short of the initial aspiration to design and carry out joint actions framed in terms of community climate change mitigation or adaptation.

In the evaluation conducted by the organizing team (*SEPA – Feitoría Verde*), the need for a better adaptation the methodology to work with youth groups was identified, as well as the need to facilitate intergenerational bonding, which encountered significant resistance from both age groups. Personal interaction barriers and the emotional impact of the pandemic and post-pandemic context were also considered, as these posed additional difficulties to the already complex challenge of harmonizing different age groups, along with limitations in the timing of the intervention sequence, which did not align well with the slow processes required to build emotional bonds and solid commitments among participants.

Even while the *Des-Des* project was still ongoing, three new lines of action began to emerge, which would take on a leading role in the years to come, focusing on the themes of gender, the identification and management of climate risks and future narratives.

3.2 The Gender Perspective Emerges

From the earliest pilot experiences of *Decarbonize!*, the number of female participants in the groups was disproportionately higher than that of men. This gender imbalance persisted over the following years and eventually sparked a debate within the working team about the need to design an intervention methodology that would take the specific weight of this variable into account. As a response to this challenge, the project *Mulleres fronte á emerxencia climática* (*Women Facing the Climate Emergency*) emerged.

The literature on the intersection of gender and climate change is extensive and well-established. Most studies highlight the different ways in which women are affected by this issue, either due to the likelihood of their —already precarious and unequal— conditions worsening as a result of the most severe consequences of climate change, or through experiences in which women demonstrate a greater capacity for personal and community adaptation and resilience in response to those consequences (Iglesias et al., 2020; Puleo, 2019). In the accumulated experience with the *Decarbonize!* groups, both dimensions were present, but in an unbalanced way. The work around memory and a "low-emission past" also served to highlight the inequalities that women experienced during that earlier time of their life, and how these inequalities created situations of extreme vulnerability and suffering.

However, this experience of discrimination was also a reflection of tremendous resilience, as even under such adverse conditions, women identified and led both individual and collective practices to cope with circumstances marked by scarcity and hardship. These practices, as discussed in the group sessions of *Decarbonize!*, were valued as examples of what could be done in the present —or in the medium and long-term future— to face the challenges of climate emergency. In light of these considerations, an intervention sequence was developed in which the gender variable was prioritized through several dynamics. The focus was placed not only on analyzing inequalities and how they increase women's vulnerability to the potential impacts of climate change, but also on the traditional —and still ongoing— role women play in sustaining social cohesion and reinforcing community resilience in the face of any threat that disrupts it.

Since 2021, with public funding from the *Deputación da Coruña* through its grant program aimed at promoting gender policies at the local government level, the project has been implemented in the municipalities of Santiago de Compostela, Padrón, Ames, Brión, A Pobra do Caramiñal and Vedra. Depending on the groups involved and the degree of municipal engagement, in some cases it has been possible to carry out sustained work in the long term, unfolding across as many as three distinct phases.

All groups begin the project with first phase, in which the links between climate change and gender are collectively analyzed and then situated within the local and community context of each group. From there, depending on the possibilities of each setting and the decisions made by the participants, each group may choose between two intervention pathways: one focused on the development of a gender-sensitive climate adaptation plan —identifying assets and proposing actions— or the creation of an emotional map of their environment, highlighting unique elements of their surroundings (natural features, tangible or intangible heritage, social organization, etc.) from a gender perspective. In the evaluations carried out to date —especially in municipalities where a multi-year socio-educational sequence has been implemented— the outcomes have been very positive, particularly regarding the improvement in self-esteem among the groups and the women within them, as well as a strengthened belief in their capacity to act as key agents in energizing their communities and reinforcing resilience in the face of the climate crisis or other external threats.

Paradoxically, this social activation has not been accompanied by similar engagement from municipal administrations, revealing a disconnection in the development of adaptation strategies. As noted by research in this field (Arto et al., 2025; Satorras et al., 2020), public administrations —notwithstanding some exceptions in regions such as the Basque Country— are, in general, significantly behind in launching climate adaptation initiatives. Moreover, they show little interest in establishing constructive dialogue with organized civil society in order to align community-based efforts with local climate policies, where such policies exist. This reluctance is even more pronounced when the proposals involve cross-cutting elements —such as gender in this case— which are often difficult to fit into the rigid and compartmentalized distribution of competences across different levels of government.

3.3 *Climate Risk Management*

Although it may seem paradoxical, the experience related to climate risk management serves as an example that it is indeed possible to approach things differently from within public administrations. In 2021, the Environmental Education Service of the Government of La Rioja expressed interest in adapting and transferring the *Decarbonize!* project to the context of its autonomous community. The chosen approach for its implementation involved organizing specific training for the team of educators within this service, who would later carry out the intervention sequence, along with setting a schedule of regular meetings to monitor the sessions and track the progress of the participating groups.

The first rounds of pilot implementation demonstrated the strength and flexibility of the intervention methodology, showing that it could be successfully adapted —with only minor adjustments— to a new social, environmental and cultural context, as well as to the diverse social profiles of the participating groups. This success translated not only into an expanding dynamic for the project —with more groups becoming involved— but also into its mainstreaming: what began as a modest initiative within the environmental education service grew to establish connections with the Directorate-General for Civil Protection and with other initiatives in socio-cultural centers.

In this case, the vector of collaboration has been citizen-led climate risk management. Since the first participating groups expressed concern about how to respond to risks such as droughts, wildfires or floods, it was decided to design a session sequence based on the guidelines and recommendations of the National Climate Change Adaptation Plan (MITERD, 2020) issued by the Government of Spain, as well as the Guide for the Development of Local Climate Change Adaptation Plans (Feliu et al., 2015).

The main goal is to analyze risks and develop proposals to minimize them from a community-based perspective that can later be presented to the relevant authorities.

Despite the level of abstraction and the complexity of concepts such as risk, exposure and vulnerability, the response from the groups —and the support from local and regional administrations— shows that there is significant potential for improving the design of socio-educational actions that help prevent and reduce risks related to climate change. It is also important to acknowledge the competence and commitment of the environmental education service staff, who were able to internalize the methodological approach and its pedagogical and socio-environmental foundations, thus successfully integrating its objectives and adapting to the specific needs and dynamics of each participating group.

3.4 A Necessary Utopia

Following the momentum gained in 2021 —a year in which, as previously described, several decisive events shaped the evolution of the *Decarbonize!* project—, a new opportunity emerged. In response to the call for proposals *Componer Saberes* by *the Daniel and Nina Carasso Foundation*, *Feitoría Verde* and *SEPA* developed a project that, while maintaining the intergenerational focus, introduced theater and the creation of future-oriented narratives framed in socio-environmental sustainability as central and innovative components. For this initiative, the theater company Chévere —a major cultural reference in Galicia and recipient of the Spanish National Theater Prize in 2014— joined the project.

This new phase of the project was named *Do Desequilibrio aos soños compartidos* (*From Imbalance to Shared Dreams*) and was conceived as an experience aimed at designing and testing an intergenerational intervention methodology to support the collective construction of narratives about possible, sustainable and desirable futures, using theater as a tool for education and community engagement (Fig. 3). Using theater as an educational tool, whether inside or outside of formal learning environments, is far from a new methodological approach. However, in the specific field of environmental education, it remains an underutilized strategy, despite its proven versatility and effectiveness in addressing emotional issues and managing conflicts within a group or community (Eizaguirre, 2009).

Reflecting this idea, the SEPA group held a gala to mark the end of the 2018–2019 activities of *Decarbonize!* aimed at highlighting the growing protagonism of older and female audiences. This audience aligned perfectly with characters created by the theater company Chévere, known as the *Fillas Bravas*: three archetypes of empowered older women with a feminist outlook on reality, living in a rural Galician village. They perform songs using traditional instruments and share stories —either recovered from oral tradition or newly invented— infused with humor and wisdom, and capable of tackling almost any complex issue, such as climate change and the *Decarbonize!* program itself.

With the momentum of that event and funding from RESCLIMA Project, a further step was taken by participating in the creation and production of a specific play titled *As Fillas Bravas e o Mito de Casandra* (*The Brave Daughters and the Myth of Cassandra*). In this new theatrical performance, *As Fillas Bravas* draw from sources in mythology and Greek theatre, as well as from traditional Galician music, to shape their unique narrative about the climate crisis. The piece is, on the one hand, a play in which three older women try to understand climate change and reflect on how to confront it by applying the knowledge and experience gained over a long life in the rural world. On the other hand, it is a play with a gender perspective that presents empowered female characters, highlighting women who are assertive, autonomous, creative and actively involved in the life of their community.

Fig. 3 Promotional poster of the project *From Imbalance to Shared Dreams*

This foray into the world of theatre and the collaborative dynamic established with Chévere served as key foundations for the theatre-based methodological component of *From Imbalance to Shared Dreams*. However, even more important than the artistic element was the construction of action around future narratives. The working logic that underpins the need to develop alternative narratives stems from the recognition that human beings are, by nature, narrative beings. From the beginning of our presence on Earth, we have constructed stories; first oral, then written and now in multiple audiovisual formats. Narratives help us represent and explain the past and present and imagine and project futures, either to fear or to desire.

Today, in the context of a deeply concerning socio-environmental scenario, cultural constructs are dominated by dystopian and apocalyptic framings: books, TV series, digital games and films that project human futures without alternatives, where it seems easier to imagine the end of the world than the end of capitalism (Martorell Campos, 2021). In contrast, utopia-oriented discourses are largely absent or, at best, given secondary importance. The dominance of a nihilistic and pessimistic outlook on the future disrupts the expectations of continuous progress that shaped cultural narratives and expressions beginning with the Renaissance —indeed, it was Thomas More in 1516, on the threshold of that historical period, who coined the concept of utopia to name his ideal society.

Optimistic visions of the future continued to influence the modern imagination, embodied in the idea of progress nourished by the Industrial Revolution and the momentum of social struggles throughout the 18th, 19th, and early 20th centuries. However, in the final decades of the last century and the early years of the current one, the production of narratives about desirable futures has become exceedingly rare. Paradoxically, despite the absence of utopias, we are witnessing a proliferation of strategic prospective documents at various levels, reflecting a clear preoccupation with predicting the future. These documents often emerge from forecasting and projection exercises carried out by dedicated departments within governments, international organizations or corporations producing their own narratives and future scenarios.

In this contradictory context, in which a certain anthropological pessimism tends to outweigh utopian impulses, there is no shortage of strategic documents aiming to increase the ability to control what the future might look like. However, these foresight proposals suffer from a significant shortcoming: there are very few initiatives —whether from public administrations or civil society— that focus on the pedagogical dimension; that is, on the development of educational interventions aimed at constructing utopias rooted in present-day reality, while also exploring future scenarios that are not only possible, but also socially and environmentally just and desirable. Whether these desirable futures are more likely to become reality depends, to a large extent, on our capacity to imagine what they might look like and on our ability to stimulate the socio-educational and cultural processes needed to make them real.

Likewise, numerous studies emphasize the importance of the emotional dimension in building alternative social and cultural conventions with an ecosocial orientation

(Gutiérrez-Bastida, 2025). The growing concern over climate change is increasingly manifesting as eco-anxiety, linked to the internalization of increasingly negative expectations about the future —especially among younger generations (Franquesa, 2024)— and to the self-perception that nothing can be done, individually or collectively, to avoid a future that appears increasingly threatening.

Narratives aimed at offering solutions to mitigate or adapt to climate emergencies based on scientific criteria often emphasize the need for an appropriate and effective normative and legislative framework in the short and medium term. However, other elements must also play a key role in mobilizing committed climate action, namely education and communication (Caride et al., 2025). For all these reasons, it is urgent to develop educational interventions that are both theoretically sound and methodologically grounded, in order to stimulate imagination and creativity —above all, as a collective exercise in proactive hope— to envision alternative futures and work toward the realization of shared, desirable, just and sustainable utopias.

That said, one of the greatest challenges of this task —and of environmental education more broadly— lies in the need to establish a praxis that not only stands in opposition to the current paradigm, but more importantly, contributes to the creation of meaningful alternatives that inspire hope for a better world, particularly among younger generations. As scientific evidence increasingly shows a link between hopelessness, demobilization, and social nihilism (Bryan, 2020; Ojala, 2023), it becomes vital to promote a pedagogy of hope in relation to the ecosocial crisis and humanity's capacity to mitigate it. However, this must not be a naive or empty form of hope (Freire, 1997), disconnected from the structural causes and the material and social manifestations of the crisis. Social research on the use of overly optimistic messages —detached from reality— indicates a lack of correlation between the hope such messages attempt to convey and meaningful action. More effective are interventions grounded in "critical hope," which begin with the present reality and explore the possibilities of alternative scenarios and the pathways —or transitions— that might lead to their realization. Furthermore, it is important to emphasize that the relationship between hope and action can be reciprocal: action can also nourish hope (Ojala, 2023).

For all these reasons, the interventions carried out within the framework of *From Imbalance to Shared Dreams* project —with intergenerational groups in the municipalities of Padrón, Ames, and Pontevedra— focused on generating situated utopias, that is, utopias rooted in local experience and oriented toward a shared community future with an ecofeminist perspective.

- In the municipality of Padrón the project was implemented with a group of older adults associated with the Municipal Volunteer Office and a second-year high school class from the public secondary school *Camilo José Cela*. The work with both groups culminated in the activity *Xardín Utopía* (Utopian Garden), a collaborative sound-based intervention held in Padrón's historic garden (Feitoría Verde, 2023a).

- In Pontevedra, a group of 18 participants aged between 21 and 78 was brought together through the coordinated efforts of the city council, *Eduardo Pondal* neighborhood association and the University of Vigo. After an in-depth exploration of the city's past and present, the group focused its activity on producing *Future Diaries*, which captured the participants' visions of what a desirable Pontevedra might look like in the year 2073 (Feitoría Verde, 2023b).
- As for the municipality of Ames, the initiative had a distinctive gender focus and involved a group composed entirely of women of various ages (from 14 to 80), all members of the neighborhood association *O Reguiño*. Starting from their own lived experiences and their relationship to the causes and consequences of the ecosocial crisis, the group produced a radio program titled *Construír o futuro entendendo o noso passado* (*Building the Future by Understanding our Past*), centered on the community's perspective on this issue and the importance of traditional knowledge (Chévere, 2023).

Both in the final evaluations conducted with the participant groups and in those involving the municipal staff and elected officials —whether directly or indirectly involved in the project— it became clear that conveying a message as complex as that of a future utopia to the broader public posed a significant challenge. While the work carried out with the groups was positively assessed, a shared conclusion emerged: the need for a final product with a greater capacity to make an impact on society. In this regard, the audiovisual format was identified as having enormous potential; not only because of its ease of dissemination, but also due to its power to influence the collective imagination.

As a result, a second phase of the project was launched, consisting of the design and filming of a short film whose narrative recreates a utopian, sustainable, and desirable future based on the future scenarios developed during the first phase. To this end, and thanks to additional funding provided by the *Daniel and Nina Carasso Foundation*, the audiovisual production company *Illa Bufarda* was brought into the team. Their role was to facilitate the group creative process for script development and to oversee the filming and editing of the short film.

For various reasons related to the dynamics of the participants, it was decided to work exclusively with the women's group from Buduído (Ames municipality) for this second phase. This is a group with which *Feitoría Verde* has been collaborating for years, which is highly active in promoting the tangible and intangible heritage of their surroundings and known for a discourse that combines the affirmation of rural identity with the empowerment of the women who inhabit it. Following a new sequence of preparatory sessions, the scenes for the short film were shot during June and July 2024, featuring the women's group as the protagonists and with the participation of many community members as extras (Bufarda, 2024).

In the evaluation of this second phase, both the project team (*Feitoría Verde*, *Chévere*, *SEPA*, and *Illa Bufarda*) and the participants from the group of the village of Biduído expressed a very positive assessment of both the outcome and, above all, the process itself. The outcome was celebrated because the mere fact of successfully

Fig. 4 Promotional poster for the short film *Raíces. Da utopía en diante*

filming and publicly premiering the short film was an achievement in itself, especially considering it was a high-quality audiovisual production created with limited resources and featuring a group of women acting in front of a camera for the first time.

However, it was the process that had the greatest impact on the women, due to its ability to distill the complex ideas explored around climate emergency and utopia into a story they made their own. It also helped them overcome fears related to acting and participating in public events where they could share and reflect on their experience in the project. The impact of the short film —and of the experience it brought visibility to— on the wider community, particularly in the sites of *Raíces* and *Casalonga*, was significant (Note that Galician villages are usually composed of smaller population entities called *lugares*). The proof lies in the turnout at the pre-premiere screening, where more than 200 residents watched the film and later took part in a discussion with the actresses (Fig. 4).

Overall, the entire process served as a powerful catalyst for empowerment and for the discovery, by the women involved, of their own capacity for social agency both individually and collectively. After its premiere, the short film was submitted to several festivals, and in parallel, the organization of public screenings for various groups was launched, many of which included the participation of the protagonists in both the presentation and the post-screening discussion.

4 As a Final Reflection

From the accumulated experience of *Decarbonize! 'Cause that's no small feat* and its subsequent derived projects (*From Imbalance to Decarbonization*, *Women Facing the Climate Emergency*, *Climate Risk Management in La Rioja*, and *From Imbalance to Shared Dreams*), several general reflections can be drawn.

First, in light of the climate emergency and its increasingly evident impacts —both in terms of frequency and intensity— it is more important than ever to implement educational initiatives aimed at adult audiences, grounded in pedagogically sound designs that are adapted to the social profiles of the target populations and to the specific socio-cultural and environmental characteristics of the territorial contexts in which they are carried out.

Second, it is essential to support initiatives that are sustained over time and capable of fostering lasting and deepening connections among participants, between participants and educators, and with the communities of reference in which the initiatives are situated. Continuity is indispensable for individuals and communities to establish meaningful links to the climate emergency, both in terms of acknowledging their share of responsibility in its causes and demanding more ambitious policies to reduce greenhouse gas emissions, as well as participating in efforts to reduce their own vulnerability to the consequences of climate disruption, many of which are already occurring or will inevitably do so.

Third, it is advisable for these socio-educational initiatives to adopt an action-research logic, integrating participants into the design process itself within a community-based framework that strengthens their autonomous capacity for action, while keeping in sight the ultimate goal of advancing the construction of desirable futures in terms of social and environmental sustainability.

And fourth, it is necessary to harmonize the fourfold challenge of feeling, thinking, dreaming and acting; moving from present threats and risks toward the horizon of utopia and doing so through a form of collective educational action that is both intentional and organized. This pedagogical approach must enable people to realistically recognize and understand the current manifestations of the climate crisis —as well as those projected to unfold in increasingly proximate spaces and times— while also fostering hope as a driver to envision and realize alternative, possible and desirable futures from an ecosocial perspective, both on a personal and a community level.

Acknowledgments This research was supported by RESCLIM@TIEMPO project (PID2022-136933OB-C21), funded by MCIN/AEI/10.13039/501100011033 and by ESF Investing in your future. JJC acknowledges the State Research Agency (AEI) for his postdoctoral fellowship (Grant reference FJC2021-047345-I) funded by MCIN/AEI/10.13039/501100011033 and by the European Union NextGenerationEU/PRTR. SEPA-interea is a Galician Competitive Research Group (GRC) funded by Xunta de Galicia (Grant reference ED431C-2021/07).

Throughout its history, *Decarbonize! 'Cause that's no small feat* has had funding from various institutions: Santiago de Compostela City Council (2016-2019), Fundación Biodiversidad (2018-2019), Deputación da Coruña (2022-2025), Environmental Education Service of the Government of La Rioja (2022-2024), and Daniel and Nina Carasso Foundation (2021-2024).

References

Arto, M., Merelas, T., & Pardellas, M. (2025). *CLIMACTIVA. Iniciativas ciudadanas ejemplares. Integrar la adaptación y la resiliencia ante los riesgos climáticos en los estilos de vida.* Feitoría Verde.

Azevedo, J., & Marques, M. (2017). Climate literacy: A systematic review and model integration. *International Journal of Global Warming, 12*(3,4), 414–430. https://doi.org/10.1504/IJGW.2017.084789

Bentz, J. (2020). Learning about climate change in, with and through art. *Climatic Change, 162*(3), 1595–1612. https://doi.org/10.1007/s10584-020-02804-4

Bryan, A. (2020). Affective pedagogies: Foregrounding emotion in climate change education. *Policy & Practice: A Development Education Review, 30*, 8–31.

Bufarda, I. (2024, November 7). *Trailer: Raíces. Da utopía en diante [Video]. Vimeo.* Retrieved July 2, 2025, from https://vimeo.com/1027223561

Caride, J. A., Gutiérrez, J., & Meira, P. Á. (2025). Las narrativas climáticas en las agendas educativas: obre el valor del tiempo para la investigación pedagógica y las políticas ambientales. *Revista española de pedagogía, 83*(290), 83–103. https://doi.org/10.22550/2174-0909.4147

Carvalho, A. (2011). *As alterações climáticas, os media e os cidadãos.* Grácio.

Chévere. (2023, May 3). *Construír o futuro comprendendo o noso pasado [Podcast]. Ivoox.* Retrieved July 2, 2025, from https://go.ivoox.com/rf/107503853

Conde, J. J., García-Vinuesa, A., Bisquert, K. M., & Meira, P. Á. (2026). RESCLIMA: Two decades of research in educational and social responses to climate change. In W. Leal Filho, M. Sima, A. Lange Salvia, M. Kovaleva, & E. Manolas (Eds.), *University initiatives on climate change education and research* (pp. 2087–2118). Springer. https://doi.org/10.1007/978-3-031-59731-2_109

Creutzig, F., Roy, J., Devine-Wright, P., Díaz-José, J., Geels, F. W., Grubler, A., Maïzi, N., Masanet, E., Mulugetta, Y., Onyige, C. D., Perkins, P. E., Sanches-Pereira, A., & Weber, E. U. (2022). Demand, services and social aspects of mitigation. In IPCC (Ed.), *Climate change 2022: Mitigation of climate change. Contribution of working group III to the sixth assessment report of the intergovernmental panel on climate change* (pp. 503–612). Cambridge University Press. https://doi.org/10.1017/9781009157926.007

Eizaguirre, P. (2009). *El teatro como recurso para la Educación Ambiental. Boletín Carpeta Informativa del CENEAM.* Centro Nacional de Educación Ambiental. https://www.miteco.gob.es/content/dam/miteco/es/ceneam/articulos-de-opinion/2009_02eizaguirre_tcm30-163558.pdf

Feitoría Verde (2023a, March 25). *Xardín Utopía [Video].* YouTube. https://www.youtube.com/watch?v=dkMydDOIB5o

Feitoría Verde (2023b). *Diarios do futuro. Pontevedra 29 de abril de 2073.* https://doi.org/10.5281/zenodo.15795980.

Feliu, E., García, G., Gutiérrez, L., Abajo, B., Mendizabal, M., Tapia, C., & Alonso, A. (2015). *Guía para la elaboración de Planes Locales de Adaptación al Cambio Climático.* Oficina Española de Cambio Climático. Ministerio de Agricultura, Alimentación y Medio Ambiente.

Fernández Durán, R., & González Reyes, L. (2015). *La Espiral de la energía.* Libros en Acción.

Franquesa, T. (2024). *Cambio climático y ecoansiedad.* Ediciones Oberon.

Freire, P. (1997). *Pedagogía de la esperanza. Un reencuentro con la pedagogía del oprimido.* Siglo XXI Editores. (Original work published 1992)

Friedlingstein, P., O'Sullivan, M., Jones, M. W., Andrew, R. M., Hauck, J., Landschützer, P., Le Quéré, C., Li, H., Luijkx, I. T., Olsen, A., Peters, G. P., Peters, W., Pongratz, J., Schwingshackl, C., Sitch, S., Canadell, J. G., Ciais, P., Jackson, R. B., Alin, S. R., & Zeng, J. (2025). Global carbon budget 2024. *Earth System Science Data, 17*, 965–1039. https://doi.org/10.5194/essd-17-965-2025

García, J. E. (2004). *Educación ambiental, constructivismo y complejidad.* Diada Editorial.

García-Vinuesa, A., Andrade, M., & Meira, P. Á. (2023). La representación social del cambio climático en estudiantes universitarios brasileños: un estudio exploratorio-descriptivo en el marco de una investigación internacional. *Revista Brasileira de Educação, 28*, e280041. https://doi.org/10.1590/S1413-24782023280041

Gutiérrez-Bastida, J. M. (2025). De que hablamos cuando hablamos de educación ecosocial. *Revista de Educación Ambiental y Sostenibilidad, 6*(2), 2101. https://doi.org/10.25267/Rev_educ_ambient_sostenibilidad.2024.v6.i2.2101

Heras, F. (2016). La dimensión social del cambio climático en el Quinto Informe del IPCC. In P. Á. Meira, M. Arto, M. Barba, & M. Andrade (Eds.), *Investigar o cambio climático na interface entre a cultura científica e a cultura común.* Universidade de Santiago de Compostela. https://doi.org/10.15304/9788416954193

Hobson, K., & Niemeyer, S. (2012). What sceptics believe: The effects of information and deliberation on climate change skepticism. *Public Understanding of Science, 4*(22), 396–412. https://doi.org/10.1177/0963662511143045

Iglesias, L., Pardellas, M., & Gradaílle, R. (2020). Públicos invisibles, espacios educativos improbables: el Proyecto "Descarboniza! Que non é pouco…" como educación para el cambio climático. *Pedagogía Social. Revista Interuniversitaria, 36*, 81–93. https://doi.org/10.7179/PSRI_2020.36.05

IPCC. (2018). Summary for policymakers. In *Global warming of 1.5°C. An IPCC special report on the impacts of global warming of 1.5°C above pre-industrial levels and related global greenhouse gas emission pathways, in the context of strengthening the global response to the threat of climate change, sustainable development, and efforts to eradicate poverty* (pp. 3–24). Cambridge University Press. doi:https://doi.org/10.1017/9781009157940.001

Lehtonen, A., Salonen, A. O., & Cantell, H. (2019). Climate change education: A new approach for a world of wicked problems. In J. W. Cook (Ed.), *Sustainability, human well-being, and the future of education* (pp. 339–374). Palgrave Macmillan. https://doi.org/10.1007/978-3-319-78580-6_11

Lewandowsky, S., Oreskes, N., Risbey, J. S., Newell, B. R., & Smithson, M. (2015). Climate change denial and its effect on the scientific community. *Global Environmental Change, 33*, 1–13. https://doi.org/10.1016/j.gloenvcha.2015.02.013

Lombardi, D., Sinatra, G. M., & Nussbaum, E. M. (2013). Plausibility reappraisals and shifts in middle school students' climate change conceptions. *Learning and Instruction, 27*, 50–62. https://doi.org/10.1016/j.learninstruc.2013.03.001

Martorell Campos, F. (2021). *Contra la distopía. La cara B de un género de masas.* La Caja Books.

Meira, P. Á. (2017a). Creer o no creer en el cambio climático, ¿razón científica o sentido común? In I. M. Arto & P. Á. Meira (Eds.), *Resclima: aproximación ás claves sociais e educativas do cambio climático* (pp. 63–70). Aldine Editorial.

Meira, P. Á. (2017b). El Proyecto RESCLIMA: de la representación científica a la representación social del cambio climático. In P. Á. Meira, M. Arto, M. Barba, & M. Andrade (Eds.), *Investigar o cambio climático na interface entre a cultura científica e a cultura común* (pp. 31–50). Universidade de Santiago de Compostela. https://doi.org/10.15304/9788416954193

Meira, P. Á., Bisquert, K. M., Pardellas, M., & Iglesias, L. (2022). Le projet de recherche Resclima: réponses éducatives et sociales au changement climatique. *Éducation relative à l'environnement, 17*(2), 1–16. https://doi.org/10.4000/ere.9759

MITERD. (2020). *Plan Nacional de Adaptación al Cambio Climático, 2021–2030.* Ministerio de Transición Ecológica y el Reto Demográfico.

Moscovici, S. (1979). *El psicoanálisis, su imagen y su público* (N. M. Finetti, Trans.). Huemul. (Original work published 1961).

Nigbur, D., Lyons, E., & Uzzell, D. (2010). Attitudes, norms, identityand environmental behaviour: Using an expanded theory of planned behaviour to predict participation in a kerbside recycling programme. *British Journal of Social Psychology, 2*(49), 259–284. https://doi.org/10.1348/014466609X449395

Ojala, M. (2023). Hope and climate-change engagement from a psychological perspective. *Current Opinion in Psychology, 49*, 101514. https://doi.org/10.1016/j.copsyc.2022.101514

Pardellas, M. (2019). *Manual de instrucións. Proxecto "Descarboniza! Que non é pouco...".* SEPA-Interea, Universidade de Santiago de Compostela. https://doi.org/10.5281/zenodo.15800186

Pardellas, M., & Meira, P. Á. (2020). *Descarboniza! Que Non é Pouco... Una Experiencia Educativa para la Emergencia Climática.* Fundación Cristina Enea. https://www.researchgate.net/publication/339747224

Pardellas, M., Meira, P. Á., & Iglesias, L. (2017). Transition communities and the glass ceiling of environmental sustainability policies at three universities. *International Journal of Sustainability in Higher Education, 2*(18), 576–593. https://doi.org/10.1108/IJSHE-05-2015-0088

Puleo, A. H. (2019). *Claves Ecofeministas. Para rebeldes que aman la Tierra y los animales.* Plaza y Valdés.

Roeser, S. (2012). Risk communication, public engagement, and climate change: A role for emotions. *Risk Analysis, 6*(32), 1033–1040. https://doi.org/10.1111/j.1539-6924.2012.01812.x

Satorras, M., Lara, Á., del Moral, L., Oteros-Rozas, E., Berraquero-Díaz, L., Baró, F., March, H., & Ruiz-Mallén, I. (2020). *Compilación de iniciativas comunitarias de resiliencia urbana en Sevilla y Barcelona: la sociedad civil frente a los efectos del cambio climático.* Universitat Oberta de Catalunya. http://hdl.handle.net/10609/146788

Takano, H., Castillo, A., & Meira, P. Á. (2025). The educational potential of the theory of social representations applied to climate change: A critical literature review. *Revista Española de Pedagogía, 83*(290), 27–61. https://doi.org/10.22550/2174-0909.4097

Innovative Strategies to Approach Climate Literacy in the Classroom: Fostering the Bond Between Academia and Educational Institutions

Harumi Takano-Rojas, Marla Arce Pimienta, Nallely Geraldine Hernández-Almanza, Ximena J. Torres, Eugenia González del Castillo, Alicia Castillo, Juan Ramón Brambila-Topete, María Elena Méndez-López, and Alondra Velasco-Morón

Abstract This chapter explores a collaborative initiative aimed at promoting climate literacy and community resilience through environmental education in a public high school on the southern coast of Jalisco, Mexico. Rooted in over two decades of research and outreach with local communities surrounding the Chamela Biological Station, the project engaged students and educators through a series of activities centered on a newly installed weather station. These included exploratory workshops, a science fair, and a multi-session workshop that introduced students to meteorological and climatic concepts while equipping them with the technical and research skills needed to use the weather station and interpret the data it produces. The activities were designed to foster student engagement with local meteorological phenomena and to support the development of context-based research projects that were meaningful and relevant to their experiences. The integration of the school into UNAM's

H. Takano-Rojas · M. Arce Pimienta · N. G. Hernández-Almanza · X. J. Torres · A. Castillo · A. Velasco-Morón
Instituto de Investigaciones en Ecosistemas y Sustentabilidad, Universidad Nacional Autónoma de México, Morelia, Michoacán, México

H. Takano-Rojas (✉)
Posgrado en Ciencias de la Sostenibilidad, Universidad Nacional Autónoma de México, Ciudad de México, México
e-mail: atakano@cieco.unam.mx

E. González del Castillo
Instituto de Ciencias de la Atmósfera y Cambio Climático, Universidad Nacional Autónoma de México, Ciudad de México, México

J. R. Brambila-Topete
Escuela Preparatoria Regional de Cihuatlán, Módulo Miguel Hidalgo, Universidad de Guadalajara, Miguel Hidalgo, Jalisco, México

M. E. Méndez-López
CIHTI-Centro de Investigación en Ciencias de Información Geoespacial, Mérida, Yucatán, México

J. J. Conde et al. (eds.), *Practices in Education for Climate Emergency*,
https://doi.org/10.1007/978-981-92-2304-6_18

Programa de Estaciones Meteorológicas del Bachillerato Universitario (PEMBU) further positioned the station within a broader academic and institutional network. Drawing on these experiences, the chapter reflects on the importance of youth inclusion, context-specific pedagogical strategies, and the strengthening of ties between academic institutions and local schools as critical components for fostering climate awareness and community agency in the face of environmental change.

1 Introduction

Climate change is one of today's most complex challenges. This global crisis involves not only transformations of the climate system, but also profound social, economic and cultural changes that affect both human communities and the ecosystems that sustain them. Therefore, addressing it requires comprehensive, multisectoral and multiscale strategies that transcend traditional disciplinary boundaries (Kubisch et al., 2022; Makrooni et al., 2026).

In this context, education based on transdisciplinary approaches emerges as a critical tool to address the multiple levels of uncertainty and complexity that characterize the current era. These approaches seek to integrate scientific knowledge, but also to incorporate local, indigenous and multispecies perspectives that enrich the understanding of contemporary socio-ecological dynamics (Ferguson & White, 2023).

The transdisciplinary approach is characterized by its emphasis on the co-production of knowledge, where scientists, educators and communities collaborate to address concrete problems in local and global contexts. In educational settings, this allows not only for greater relevance and applicability of scientific knowledge, but also for a critical reconnection with students' everyday practices and socio-ecological realities, promoting a sense of agency and responsibility for climate action (Ferguson & White, 2023). Transdisciplinarity requires articulation between academic institutions (such as universities and research institutes) and school environments to foster transformative, relevant and contextualized climate education. Although many universities currently play a key role in training people with a sensitivity to the current crisis, and with the necessary tools to contribute to climate change mitigation (Filho et al., 2024), there are still challenges to be solved associated with the territorialization and dissemination of knowledge generated in academic environments.

Academic institutions have the potential to function as scientific, methodological and pedagogical support platforms for schools. Collaboration can materialize in school research projects, development of joint didactic materials, teacher training, community practices and local mitigation actions (Filho et al., 2024; Kubisch et al., 2022). One avenue for this dialogue is the development of articulated pedagogical models. Kubisch et al. (2022) propose a transdisciplinary education model in which students work together with scientists and social actors in their contexts to address problems related to climate change. This approach allows transcending traditional classroom boundaries by fostering knowledge co-production, critical thinking, and active engagement.

On the other hand, the articulation and dialogue between the scientific community and basic education teachers might build stronger pedagogical strategies for climate change education. In this sense, Makrooni et al. (2026) highlight the importance of teacher narratives as a tool for understanding how climate change is experienced and taught in school environments. These narratives are enhanced when there is action research and participatory methodologies, where the teacher not only receives knowledge, but also becomes a co-creator of contextualized pedagogical strategies.

Transdisciplinarity in science education presents itself as a promising path to foster transformative learning, which is not only limited to the acquisition of technical knowledge, but also cultivates critical competencies, such as systems thinking, ethical reflection and committed action for social and environmental change (Echegoyen-Sanz et al., 2024; Kubisch et al., 2022; Schirnding de Almeida & Boucsein, 2025). The linkage between academic institutions and school environments will be key to combine scientific rigor with cultural relevance, foster active and meaningful learning, and strengthen individual and collective capacities to face the challenges of climate change (Riley et al., 2024).

An example of this institutional engagement can be found in the long-standing efforts of the National Autonomous University of Mexico (UNAM) to support environmental research and education in tropical ecosystems. Over 50 years ago, UNAM established two biological research stations to study the country's major tropical ecosystems. The first, founded in 1969, is located in a humid tropical rainforest of the Los Tuxtlas region of Veracruz. The second, the Chamela Biological Station, was established in 1971 in the tropical dry forest of Jalisco. Both stations are situated near coastal rural communities, providing a unique opportunity for collaboration between scientists, educators, and local populations (Castillo et al., 2021).

For the past 25 years, our research group has carried out sustained work with the human communities surrounding the Chamela Biological Station. Initial studies focused on understanding local perceptions and perspectives on key topics such as regional productive activities (including agriculture, cattle ranching, fishing, and tourism), environmental challenges, biodiversity conservation, and the effects of governmental land-use planning and environmental protection programs (Castillo et al., 2005, 2009; Pujadas & Castillo, 2006). As the scope of our projects evolved, it became increasingly clear that it was essential to address issues of direct relevance to local populations, including the broader environmental phenomena that shape their lives and the ecosystems they inhabit. Among these issues, climate change has become particularly salient due to its growing impact on productive activities. Although local residents may not explicitly name climate change as a concern, they do recognize and express unease about observable changes—such as shifting rainfall patterns, increasing temperatures, and the growing frequency and intensity of hurricanes—which directly affect their daily lives and livelihoods (Takano-Rojas et al., 2023).

Indeed, a recent vulnerability assessment by Ramírez-Sánchez et al. (2024) identified the state of Jalisco as highly vulnerable to the impacts of climate change, while

Takano-Rojas et al. (2023) documented significant climatic shifts along the southern coast of Jalisco over the past four decades. These transformations highlight the need for inclusive and forward-looking strategies to address its impacts. In particular, it is imperative to meaningfully involve younger generations, who will face the long-term consequences of the climate crisis. This requires the promotion of climate literacy through innovative, context-sensitive educational practices that consider students' lived experiences and socio-environmental realities. Moreover, fostering community engagement and local agency is essential to ensure that responses to climate challenges are not only scientifically informed but socially grounded and collectively owned. Such approaches can help cultivate resilient communities equipped to act on and adapt to the ongoing transformations of their environments.

The next section outlines a series of activities carried out at a public high school on the southern coast of Jalisco, Mexico. These included exploratory workshops to assess students' relationships with their local climate and their understanding of weather and climate change; a science exhibition that linked climate change to local realities; the inauguration of a school-based weather station; and a set of hands-on sessions designed to train students in using meteorological data for their own research projects. The final section reflects on the impact of these initiatives on students, faculty, and researchers, and considers their broader implications for climate education and community engagement. We conclude with key insights and recommendations for future work in this field.

2 Building Climate Literacy on Campus

2.1 Context

Since 2017, with the aim of working closer with the communities surrounding the Chamela Biological Station, graduate and undergraduate students from the Laboratory of Communication for Sustainability have developed a series of research projects in partnership with students and faculty of *Preparatoria Módulo Miguel Hidalgo* (Miguel Hidalgo High School). These initiatives have addressed a range of topics, including the co-creation of environmental education materials with students (Arreola, 2018), education for the sustainable management of tropical dry forests (Vigil, 2021), environmental communication through digital platforms along the coast (Fernández & Fraire, 2024), and broader themes related to education for sustainability (Moreno, 2024).

Preparatoria Módulo Miguel Hidalgo is part of the Higher Secondary Education System of the University of Guadalajara. It offers a competency-based high school program structured into six semesters, each comprising a variety of subjects. Alongside academic coursework, the curriculum also includes vocational training in areas such as carpentry, electricity, and nursery garden management, which enrich the educational experience of students.

Located on the southern coast of Jalisco, Mexico, the school operates in a region that is not only vulnerable to the effects of climate change but is also increasingly impacted by organized crime. Young people in the area are among the main targets of this violence. This context differs significantly from that of urban high schools and presents a complex and challenging environment for educational work.

The activities described below were developed as part of the doctoral research project *Social representations of climate change in rural communities of the southern coast of Jalisco*, led by Harumi Takano-Rojas. During fieldwork, informal conversations with the daughters, sons, and nephews of interviewees revealed a generational gap in climate knowledge: while adults and elders possess deep, experience-based understanding of local climatic and meteorological conditions, this knowledge is not as prevalent among younger people. To further study this, several activities were developed in collaboration with *Preparatoria Módulo Miguel Hidalgo*, leveraging both an existing relationship with the school and its role as an educational hub for students from multiple local communities.

2.2 First Approach: Getting to Know Each Other

To better understand students' relationships with weather and climate, as well as their knowledge and perceptions of climate change, an exploratory workshop consisting of four sessions was carried out during October 2023. This workshop served as a starting point for identifying how young people interpret and engage with environmental changes, and provided valuable insights into the ways in which local and global climate narratives are understood—or overlooked—by the younger generation.

The workshop –developed with 25 students divided in two groups– consisted of both individual and group activities designed to explore students' knowledge, perceptions, and experiences related to weather, climate, and climate change. Group activities focused on assessing students' understanding of the local climate throughout the year—for instance, identifying the rainiest and driest months, the hottest and coldest periods, and whether they associate rainfall with events such as hurricanes. Other group exercises aimed to gauge students' familiarity with the concept of climate change, its causes, and its implications. Finally, some activities invited students to consider ways they could take action in response to climate change, both at the local and global levels.

Individual activities were conducted through a specially designed activity notebook, in which students were encouraged to reflect on their personal experiences with weather in daily life—for example, what they typically do or avoid doing on rainy or particularly hot days (Fig. 1). They were also asked to reflect on observable changes in the local landscape, such as how the beaches or surrounding vegetation has transformed over time, what might have caused those changes, and what actions could be taken in response. These reflections provided valuable insight into how young people relate to their environment and how they make sense of both immediate weather patterns and broader climatic shifts.

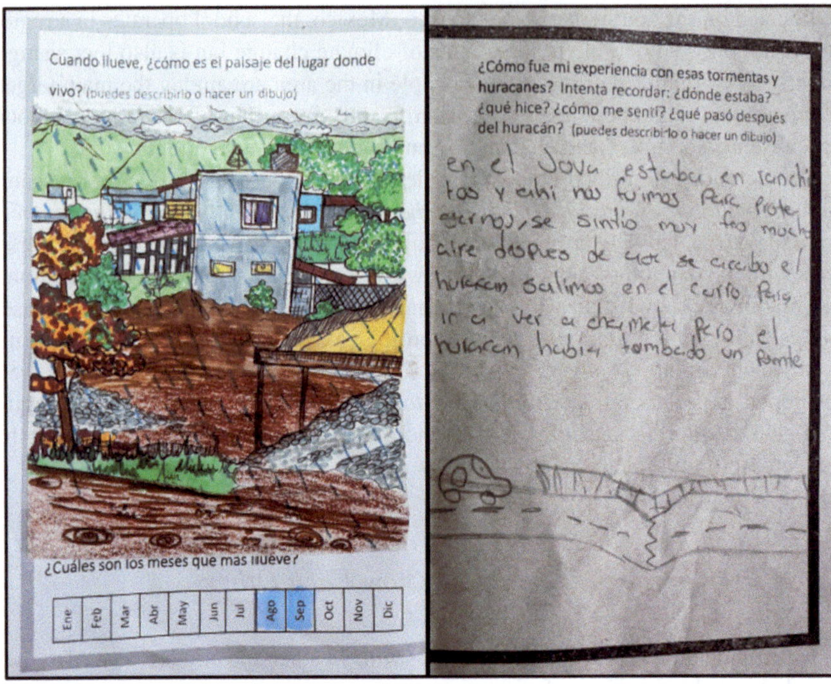

Fig. 1 Examples of responses in the activity notebook. On the left, a student's drawing depicts the landscape of her community during the rainy season. Muddy roads are shown and a river clouded with soil appears at the bottom of the image. The presence of both dry and green trees illustrates the seasonal transition. On the right, a student's written story recounts her experience during Hurricane *Jova*. She describes evacuating with her family to a nearby community for shelter and feeling frightened by the strong winds. After the storm, they were unable to return home because a bridge had collapsed

The results of the activities revealed several key insights. In the individual exercises, students generally struggled to recognize climatic cycles or patterns throughout the year, such as identifying the rainiest or hottest months. While group activities that encouraged discussion led to slightly more accurate responses, there remained a noticeable difficulty in identifying rainfall, hurricanes, and temperature patterns. Although students acknowledged that weather affects their daily routines—such as avoiding certain activities during heavy rain or extreme heat—this awareness did not appear to hold significant relevance or prompt deeper reflection.

Regarding climate change, most students were able to define the concept and identify its causes, including both biophysical and social factors. However, they rarely connected these global processes to their immediate surroundings or lived experiences. When discussing actions to address climate change, responses tended to focus on individual efforts related to waste and water management—such as recycling, separating trash, avoiding littering on beaches, and conserving water. These limited responses appeared to stem from feelings of being overwhelmed by the scale of the problem, as well as a sense of powerlessness due to a lack of resources or perceived ability to enact meaningful change.

These findings point to important challenges and opportunities for climate change education in rural contexts. While students demonstrate basic conceptual understanding of climate change and an awareness of environmental issues, the disconnection between global climate discourse and local realities limits their ability to recognize climate change as a process that directly affects their lives and communities. This underscores the need for educational approaches that not only convey scientific knowledge but also foster critical reflection on local environmental changes, encourage meaningful place-based engagement, and support students in envisioning themselves as agents of change.

2.3 Science Fair: *Olas de ciencia*

In light of the workshop results—particularly students' limited recognition of local climatic patterns and the disconnect between their understanding of climate change and its local manifestations—the installation of a weather station emerged as a valuable educational opportunity. It aimed to foster more direct, place-based engagement with the climate and to help students observe and interpret weather-related data in connection to their daily lives. To mark the inauguration of the station and reinforce these learning objectives, a science fair was organized at the school. This event was designed not only to celebrate the installation but also to provide a space for students to explore weather measurement instruments and the local dimensions of climate change, thereby strengthening the link between abstract concepts and their immediate context. The idea of holding a science fair was inspired by *Puertas Abiertas* ("Open Doors"), an annual outreach event at the Chamela Biological Station, initiated under the leadership of Station Chief Jorge Vega† (who served from 2007 to 2017) and Alicia Castillo. That event had responded to a similar need: to bridge the gap between scientific research and local communities, many of whose members had limited familiarity with the station and its work.

The success of *Puertas Abiertas* provided a strong model for science outreach that effectively connected academic knowledge with local communities, an approach that informed the design of the science fair at the high school. Over the years, *Puertas Abiertas* became a key event on the local social calendar. Attendance grew from 138 participants in its inaugural year to nearly 500 in its final edition. While early versions of the event required extensive outreach efforts, such as distributing posters and announcements in nearby communities, these were no longer necessary in later years as word of mouth and community interest sustained participation. The event consistently drew not only residents from surrounding communities but also visitors from more distant municipalities, some of whom traveled 2 to 3 h to attend.

During *Puertas Abiertas*, academic researchers and students presented informational posters, exhibited demonstrative materials such as plant specimens, seeds, live animals (e.g., lizards and snakes), and meteorological equipment. They also facilitated interactive activities, such as educational games and demonstrations on the life cycle of sea turtles, the importance of avoiding the captivity of parrots and parakeets, and watershed ecology—highlighting how pollution in upper regions affects water quality downstream and impacts the communities living along the lower reaches of the basin.

The experience proved to be of significant educational value not only for the visitors—ranging from young children to older adults—but also for the participating academics. Beyond the opportunity to become familiar with the facilities and to foster a welcoming and convivial atmosphere, the events consistently generated meaningful dialogues and mutual knowledge exchange. Visitors shared personal experiences and observations related to the terrestrial and marine ecosystems, while the researchers offered scientific explanations, answered questions, and, importantly, posed questions of their own to the local residents. As it is the inhabitants who engage with the natural environment on a daily basis, their insights were invaluable to deepening academic understanding of the region.

Building on the outreach model of *Puertas Abiertas*, *Olas de Ciencia* ("Waves of Science") –held in December 2023– adapted its core principles to the context of the high school, bringing science communication directly into the educational environment. Unlike *Puertas Abiertas*, which took place at the Chamela Biological Station, this new initiative was held on school grounds and centered specifically on climate and climate change. A particularly notable feature of *Olas de Ciencia* was the diversity of its participants. Alongside researchers and students from UNAM, the event brought together representatives from regional institutions such as the University of Guadalajara and the "Benito Juárez García" University for Well-Being. High school students, local residents, and members of the Careyes Foundation –a local nonprofit organization– participated not only as audience members, but also as presenters, sharing their knowledge, experiences, and perspectives on environmental change. In total, the event drew approximately 180 attendees.

The event proved highly effective in communicating a wide range of issues related to climate change, highlighting its connections to both local and global dynamics (Fig. 2). Importantly, the presentations addressed not only the biophysical dimensions of climate change—such as biodiversity loss, altered precipitation patterns, and changes in cloud formation—but also its social implications. These included the impacts of climate change on fisheries and cattle ranching, as well as a theatrical performance that explored the historical roots of the climate crisis.

The event also served as a platform to introduce the newly installed meteorological station, highlighting its relevance to various stakeholders and institutions potentially interested in utilizing the station's data for educational or outreach purposes. It also provided an opportunity to familiarize students and school faculty with the station and its objectives (Fig. 2). Additionally, the event fostered interactions among representatives of participating institutions, laying the groundwork for future collaborations. Indeed, 2 years later, the Careyes Foundation organized a similar event—*Cuidemos a la Costa* ("Let's Take Care of the Coast")—also held at the school facilities.

Fig. 2 Presentations during *Olas de Ciencia*. On the left, a researcher and a student lead an outreach activity on local flora for high school students. On the right, a researcher conducts a hands-on presentation about the components of a weather station. (Participants' faces have been blurred to protect their privacy)

2.4 Turning Knowledge to Practice: Installation And Maintenance of a Weather Station

In alignment with the goals set during the exploratory workshops and the *Olas de Ciencia* science fair, the weather station was introduced as a way to bridge the gap between students' everyday weather-related experiences and the broader climatic phenomena shaping their region. The station was envisioned not only as a tool for data collection but as an interactive educational resource designed to foster a deeper understanding of local weather patterns and their connection to global climate change. This initiative sought to empower students with the knowledge necessary to make informed decisions about environmental issues, encouraging both local and global action in response to the challenges posed by climate change.

A series of workshops were originally intended to take place shortly after the station's installation; however, they were conducted a year and a half later due to unforeseen administrative and organizational challenges. Despite this delay, the weather station remained a central component of the initiative for both the research team and the high school administration.

The automated weather station, a compact set of sensors and a processing console (Vantage Vue Wireless, Davis Instruments, Hayward CA) tracks variables such as temperature, humidity, wind speed and direction, and precipitation. It collects data in real time, which are then stored on a nearby computer. This system allows the collected data to be easily accessed for educational purposes. Located in a visible and accessible part of the school, the station was installed not only to function as an educational tool but to promote the active involvement of the school community. With its user-friendly design, the station was intended to be maintained by students and teachers, thus ensuring that it would remain a sustainable resource for climate education without relying on external expertise. Ensuring its appropriation by the school community was essential to guarantee its continued use and relevance beyond

Fig. 3 Workshop sessions on the weather station. On the left, participants engage in field identification of the weather station and its components. On the right, a participatory activity focuses on identifying the station's components. (Participants' faces have been blurred to protect their privacy)

the lifespan of the research project, which is constrained by the temporal scope of academic timelines.

The workshop was structured as a six-session program developed from a preliminary diagnostic that assessed students' and teachers' prior knowledge, perceptions, and potential interests regarding the weather station. Implemented with nearly 40 students over 4 weeks in March and April 2025, the sessions engaged students from various grade levels in both theoretical and hands-on activities (Fig. 3). Additionally, participants visited the Chamela Biological Station, where they explored its research infrastructure, including its meteorological monitoring systems. The workshop concluded with a final session in which students presented their work and the collaborative projects they designed, to other students and to the school community. These presentations attracted around 70 participants ranging from students, teachers, school administrators and maintenance personnel.

A key objective was to move away from conventional classroom dynamics. Each session was conceived as an open, interactive space that encouraged curiosity, collaborative learning, and experimentation. Rather than centering on lectures or assessments, facilitators prioritized participatory activities, discussion, and practical engagement, aiming to create an atmosphere that felt exploratory rather than evaluative.

The first session introduced the general structure of the workshop and key concepts in meteorology, while also creating a welcoming environment where students felt free to participate. The second session explored atmospheric variables, the function of a weather station, and its applications. The third and fourth sessions focused on the technical operation and maintenance of the station, as well as the practical uses of the data it generates. The fifth session introduced ecotechnologies as examples of how such data can inform responses to local environmental challenges.

The final session, titled *Laboratorio de Proyectos* ("Project Lab"), invited students to form teams and develop small-scale projects based on a local issue. These projects were designed to integrate meteorological data in addressing a question or hypothesis, reinforcing the relevance of climate monitoring to their own context. Through this process, the workshop sought to foster a sense of agency among students and to position them as active interpreters and users of climate information in a region particularly vulnerable to climate change.

The final projects developed by the students addressed a range of locally relevant issues, including the environmental and social impacts of Hurricanes *Jova* (2011) and *Patricia* (2015)—two highly destructive events that remain prominent in the collective memory due to the extensive damage they caused (Aguilar-Román et al., 2020). Other topics included the relationship between vegetation cover and temperature in the tropical dry forest, the effects of deforestation on local temperature patterns, the viability of rainwater harvesting as a sustainable strategy for domestic use, and the ways in which shifting rainfall patterns are affecting cattle ranching practices.

The range of topics selected for the final projects demonstrates that students were able to meaningfully connect the weather station to their own contexts and explore issues that were both locally relevant and personally significant. As facilitators, our main objective was not limited to technical instruction; rather, we sought to equip participants with tools that would support community empowerment and the development of collective resilience. Although the technological components initially captured students' interest, what they valued most was the opportunity to work collaboratively in settings outside the conventional classroom. All workshop sessions were held either in the library or in outdoor spaces, which contributed to a more dynamic and participatory learning environment. This departure from traditional educational routines not only enhanced student engagement but also facilitated broader involvement from the school community. Given the school's small size, our day-to-day presence allowed us to interact with teachers, staff, and students from different grades who, drawn by curiosity and proximity, began to take interest in the weather station and its uses—gradually integrating it into the school's everyday life (Cripps, 2024; González-Gaudiano et al., 2020).

Following the successful engagement fostered through the workshops, the station at *Preparatoria Módulo Miguel Hidalgo* has now become part of a broader institutional network: the *Programa de Estaciones Meteorológicas del Bachillerato Universitario* (PEMBU, "High School Weather Station Program"), a network established by Universidad Nacional Autónoma de México (UNAM) through the Institute of Atmospheric Sciences and Climate Change in the mid-1990s. PEMBU is a longstanding initiative that equips all highschools affiliated with UNAM in Mexico City with weather stations, facilitating climate monitoring and education. The program provides open access to real-time and historical meteorological data through its platform (https://www.ruoa.unam.mx/pembu/), fostering data-driven research across campuses.

Since 2012, PEMBU has also organized an annual contest in celebration of World Meteorological Day (March 23), where high school students—accompanied by a teacher or tutor—develop research projects using data from their own station or from the broader PEMBU network. As a result of the collective efforts around the installation and use of the station in Miguel Hidalgo, the school was formally incorporated into the PEMBU network (https://ruoa.unam.mx/pembu/mhgo-pembu/), and its students and staff have been invited to participate in the program's 15th edition in March 2026. This integration represents a significant step toward consolidating the weather station's educational potential and ensuring its continued use through institutional support and broader academic networks. Additionally, a key outcome was the creation of a technical-educational manual that not only explains the station's operation, but also situates its relevance by providing local weather and climate information, while offering a guide for high school students to design their own research projects (https://www.iies.unam.mx/laboratorios/socioecologia-comunicacion-sustentabilidad/monitoreo-climatico-y-educacion/).

Finally, it is worth noting that some of these activities were documented on *La costa que nos une* ("The coast that unites us"), an Instagram (@lacostaquenosune) and Facebook (https://www.facebook.com/groups/538656021999118/?ref=share&mibextid=KtfwRi) page created for outreach proposes related to environmental activities and events along the southern coast of Jalisco. This initiative stems from a previous project partially carried out at the same high school, which aimed to explore the use of social media as a tool for environmental communication (Fernández & Fraire, 2024).

3 Forecasting Futures: Reflections on Academic Collaboration and Educational Transformation

Our experience working with the high school proved to be an enriching learning experience, though not without its challenges. In this section, we organize our reflections around three main themes. First, we discuss key operational considerations that shaped our approach and were essential to the development of our work. Second, we reflect on how the partnership between an educational institution and academic researchers can strengthen climate change education and contribute to meaningful social transformation beginning within the classroom. Finally, we explore how the activities we implemented created opportunities for students to engage with weather and climate in a context-sensitive and participatory manner. We also highlight the emergence of unexpected yet valuable tools and outcomes that arose during these activities.

3.1 Laying the Groundwork for Collaboration

The project began as a relatively small initiative, but its continued development and eventual expansion were made possible by the strong commitment of all involved parties. From the beginning, there was a shared interest in sustaining and deepening the collaboration, which proved essential for its long-term success. One of the key factors enabling this growth was the presence of a supportive and coordinated team. This teamwork was crucial not only for addressing technical and logistical challenges, but also for securing sufficient financial resources and carrying out the activities effectively.

Equally important was the establishment of the project's objectives and scope in each stage. Although these evolved over time in response to emerging needs and opportunities, there was always a clear understanding of the project's goal and its limitations. From the beginning, the roles, responsibilities, and commitments of each participant were explicitly defined. This clarity facilitated the collaboration and ensured accountability. The fulfillment of these mutual agreements helped to build and reinforce trust among the participants, which was –and continues to be– a decisive factor in the continuity of the project. Looking ahead, we hope the school will have the necessary resources to continue these activities independently, using the weather station in ways that best suit their pedagogical and contextual needs. Hence, our role would shift to that of technical companions, offering support when required but encouraging full local appropriation and leadership.

Inter- and transdisciplinary initiatives have significant potential for addressing complex socio-environmental challenges, but they are also particularly vulnerable to failure when key resources are lacking. Limitations in time, funding, technical support or human capacity can easily undermine collaborative efforts. In our case, we noticed a disparity between the project-based, short-term timeframes of academic institutions, and the priorities and work rhythm of educational settings. These differences can result in gaps that hinder the continuity of joint efforts. In this context, the school's appropriation of the project becomes not only desirable but necessary. By cultivating students' and teachers' ownership through the integration of the weather station into the school's educational and institutional agendas, the project can continue to evolve in meaningful ways beyond the limits of external academic involvement. Such appropriation helps ensure that the initiative is not dependent on temporary external support, but instead becomes a part of everyday practices of the school community.

Although a variety of activities have already been carried out, the processes of appropriation and meaningful integration will require sustained, long-term efforts before their impacts become visible. A key aspect moving forward –and one that remains insufficiently addressed in our current practice– is the evaluation of how students' perspectives are transformed through their participation in these activities. Beyond acquiring knowledge, it is crucial to understand whether and how these experiences influence students' lives, their decision-making processes, and the way they frame and interpret local weather and climate related issues. Investigating these dimensions would allow us to assess the educational outcomes of the project, but also its broader social impacts. Developing mechanisms to trace and document such transformations over time remains a critical challenge and a necessary step toward deepening the relevance of such initiatives.

3.2 *Partnering for Transformative Climate Education*

Fostering a link between academic researchers and an educational institution was very valuable for multiple reasons. From the academic perspective, it was a meaningful opportunity to place scientific knowledge at the service of the communities that can benefit from it. This involves not only acting as a conduit for information, but actively learning how to make knowledge accessible and relevant by contextualizing it and connecting it to students' experiences and interests. In this regard, the *Olas de ciencia* event was particularly significant, as it facilitated direct engagement between researchers and local participants, thereby highlighting the social relevance of academic work.

On the other hand, high school authorities have emphasized the value of collaboration with institutions specializing in applied research. This partnership enables the creation of educational proposals that integrate practical methodologies alongside theoretical foundations. In a context of climate change and in a region highly vulnerable to its impacts, this becomes an urgent task. It is noteworthy that in a region where a biological station has been present for over 50 years, initiatives of this nature are not more frequently promoted or consistently integrated into local educational and community activities (Castillo et al., 2021).

The activities also created a space for students, teachers and researchers to develop a range of skills. Whether through gaining access to new knowledge, learning pedagogical tools, or developing communication and group management abilities, the diversity of activities fostered a dynamic learning environment for all participants. Moreover, the trust built through our sustained presence at the school, enabled the students to openly express their ideas and questions, laying the groundwork for the co-production of knowledge. This was apparent in the topics they chose for their final projects during the 2025 workshops, where they drew upon their own lived experiences -such as hurricane impact, rainfall patterns, or shifts in the local landscape- as legitimate subjects of research.

Perhaps one of the most valuable aspects of this collaboration was the mutual exposure of both communities –academic researchers and the school community– to different social and cultural contexts. Such encounters offer a meaningful opportunity to engage with people who live, think and act in ways that may differ significantly from our own. This kind of engagement broadens our understanding of reality, as it challenges assumptions and stimulates reflections. Exposure to different perspectives not only helps identify local challenges but also illuminates their connections to regional and global dynamics. In this way, these efforts do more than generate knowledge or serve as outreach activities, they cultivate capacities such as empathy, adaptability, curiosity, and critical thinking, which are essential for addressing the complex challenges of our time.

3.3 Towards Disruptive Educational Practices that Encourage Agency

Environmental and climate education extends beyond merely presenting data that illustrate the severity of the multiple crises humanity faces on this planet. It involves the collective construction of critical thinking that, in turn, enables learners to move from understanding the causes and consequences toward envisioning alternatives and cultivating hope in response to climate change. While there is an abundance of resources to support the transmission of information, the real challenge lies in guiding groups through processes that require the development of skills to navigate diverse emotional responses to often overwhelming content, which can evoke feelings of despair, impotence and hopelessness.

Schools are places where common ground can be established and where opportunities for meaningful engagement with the educational community can emerge. This includes not only students, but also teachers, maintenance staff, and administrative personnel. Although schools are well-suited to fostering creativity and collaboration, they have also historically and systematically functioned as institutions where hierarchical power relations are established. In these settings, "correct" or "valid" knowledge is often held in the hands of few, while the majority, typically younger individuals, are positioned as passive recipients who are discouraged from questioning or contributing to its construction.

Reflecting on our role as educational facilitators, we recognize our responsibility to create conditions that motivate and strengthen the agency of those with whom we build and share knowledge. To this end, developing participatory methodologies and shifting toward horizontal communication practices during our workshops have been essential to fostering an environment where individuals feel encouraged to share their knowledge, experiences, and emotions. These approaches have proven to be fundamental components of the learning experience.

Environmental and climate education is closely tied to the regions in which we live. Each territory has unique ecosystems, climate, flora, and fauna, as well as a unique history, social dynamics, and culture. In other words, each territory is a universe unto itself. Throughout this process, we witnessed the importance of taking into account the specific territories participants inhabit, especially in relation to the lived experiences and consequences of the climate crisis.

Moreover, this highly complex system is embedded within a broader framework: the Earth system. This perspective highlights the fact that we are interconnected both locally and globally through intricate ecological and social relationships. Climate and environmental issues provide us with a powerful lens through which to explore this interdependence, not only of the planet's biotic and abiotic elements but also of ourselves in relation to other human beings.

Considering this, we argue that environmental and climate education, when genuinely committed to the territories where it is developed and to the individuals that inhabit them, cannot be separated from climate justice. Educational content should critically address historical and ongoing processes of extractivism, genocide, displacement, climate migration, inequalities in energy production and consumption, land appropriation, and other structural injustices. Moreover, the methods used to approach these contents should align with these principles. Disruptive educational practices, in this sense, must be grounded in a strong environmental ethic and a sustained commitment to the communities with whom we engage.

Such involvement allows the objectives of our interventions to go beyond delivering content, ensuring that they remain relevant to the territories and communities with which we work. Furthermore, the learning environment we cultivate and the tools we employ are key to supporting young people in recognizing themselves as agents of change and in understanding their roles and potential impact at both local and global levels.

To date, the effectiveness of this initiative has largely depended on the strong commitment of the high school authorities and the sustained involvement of the graduate students leading the project, as well as the support of other university actors who have helped connect local efforts to broader academic and institutional contexts. These factors have been critical for building trust, securing resources, and maintaining continuity across activities. However, the more lasting value of these educational practices lie in the active, experiential engagement of the students themselves. By positioning them not as recipients of information, but as co-participants in the construction of knowledge, the initiative fosters a sense of ownership and relevance that is often absent in traditional educational models. In this approach, scientific knowledge is not delivered through a top-down process, but it emerges through participatory learning experiences rooted in the students' contexts, addressing their concerns.

Science can be disruptive when it is no longer perceived as distant, elitist, or inaccessible, particularly with regard to technology. Hence, processes of co-production of knowledge encourage students to reflect on their environment, recognize their capacity to be insightful, and connect local phenomena to wider social and ecological issues. These are essential elements if we are to imagine climate change education as a transformative process with the potential to catalyze meaningful social change. In light of this, we strongly encourage other academic institutions to embark on similar efforts to engage with the communities around them. Such effort is not only part of our collective responsibility to enhance the social relevance of academic work, but also a necessary response in face of the uncertainty and complexity of the times we are living in.

Acknowledgments We are grateful to all workshop participants and the school community who opened up their doors and made us feel welcome. We are also grateful to the researchers, students and local inhabitants who participated in *Olas de ciencias,* and to Enrique Azpra Romero, technical coordinator of PEMBU, and Omar López Antón, who aided us with logistical support to link the weather station to PEMBU's network. We are thankful to Guillermo Murray Tortarolo for his feedback and comments on this manuscript.

HTR is a doctoral student from *Programa de Doctorado en Ciencias de la Sostenibilidad*, UNAM, and received a fellowship (CVU 1004620) from *Consejo Nacional de Humanidades, Ciencia y Tecnología*. NGHA and XJT received a scholarship from *Programa de Creación de Becas a Estudiantes*, UNAM (Project number: 232 007). Funding to buy the meteorological station and develop this project was received from DGAPA UNAM (PAPIIT IN300422: *Interacciones ciencia-sociedad en el manejo costero: análisis de representaciones sociales, redes de conocimiento y comunicación*).

References

Aguilar-Román, E., Castillo, A., & Güiza, F. (2020). Vulnerability and risk management after Hurricane Patricia in a rural community on the Jalisco coast, Mexico. *International Journal of Disaster Risk Reduction, 45*, 101464. https://doi.org/10.1016/j.ijdrr.2019.101464

Arreola, L. (2018). *Generación de material educativo ambiental en conjunto con jóvenes de una preparatoria en la Costa Sur de Jalisco, México* (Tesis de licenciatura). Universidad de Guadalajara.

Castillo, A., Godínez, C., Schroeder, N., Galicia, C., Pujadas-Botey, A., & Martínez, L. (2009). El bosque tropical seco en riesgo: Conflictos entre uso agropecuario, desarrollo turístico y provisión de servicios ecosistémicos en la costa de Jalisco, México. *Interciencia, 34*(12), 844–850.

Castillo, A., Magaña, A., Pujadas, A., Martínez, L., & Godínez, C. (2005). Understanding the interaction of rural people with ecosystems: A case study in a tropical dry forest of Mexico. *Ecosystems, 8*, 630–343. https://doi.org/10.1007/s10021-005-0127-1

Castillo, A., Velasco-Morón, A., Arroyo-Arroyo, Y., Aranda-Fragoso, A., Aguilar-Román, E., Pérez-Escobedo, M., Tapia-Hernández, K., Coates, R., & Vega-Rivera, J. (2021). Two tropical research stations in Mexico: 50 years of contributions and challenges. *Environmental Challenges, 3*, 100037. https://doi.org/10.1016/j.envc.2021.100037

Cripps, E. (2024). School for sedition? Climate justice, citizenship and education. *Journal of Moral Education, 54*(1), 27–43. https://doi.org/10.1080/03057240.2024.2375805

Echegoyen-Sanz, Y., Morote, Á., & Martín-Ezpeleta, A. (2024). Transdisciplinary education for sustainability. Creativity and awareness in teacher training. *Frontiers in Education, 8*, 1327641. https://doi.org/10.3389/feduc.2023.1327641

Ferguson, J. P., & White, P. J. (2023). Science education in the Anthropocene: The aesthetics of climate change education in an epoch of uncertainty. *Frontiers in Education, 8*, 1281746. https://doi.org/10.3389/feduc.2023.1281746

Fernández, C. & Fraire, D. (2024). *Las plataformas digitales como estrategia de comunicación ambiental: una propuesta para la costa sur de Jalisco* (Tesis de licenciatura). Universidad Nacional Autónoma de México.

Filho, W. L., Sow, B. L., Perlin, A. P., Mbah, M. F., Costa, C. A. d., Safiul Azam, F. M., & Dinis, M. A. P. (2024). Addressing climate change education: Relevant contributions from universities. *International Journal of Sustainability in Higher Education*. https://doi.org/10.1108/IJSHE-11-2023-0542

González-Gaudiano, É. J., Meira-Cartea, P. Á., & Gutiérrez-Pérez, J. (2020). ¿Cómo educar sobre la complejidad de la crisis climática? Hacia un currículum de emergencia. *Revista Mexicana de Investigación Educativa, 25*(87), 843–872.

Kubisch, S., Krimm, H., Liebhaber, N., Oberauer, K., Deisenrieder, V., Parth, S., Frick, M., Stötter, J., & Keller, L. (2022). Rethinking quality science education for climate action: Transdisciplinary education for transformative learning and engagement. *Frontiers in Education, 7*, 838135. https://doi.org/10.3389/feduc.2022.838135

Makrooni, G., Zen, S., Azhar, K., et al. (2026). Once upon a time: Teachers' narratives on climate change education. *International Review of Education, 72*(1), 37–60. https://doi.org/10.1007/s11159-024-10114-5

Moreno, J. (2024). *Educación para la sostenibilidad en la Escuela Preparatoria Regional de Cihuatlán, Módulo Miguel Hidalgo en la costa sur de Jalisco* (Tesis de maestría). Universidad Nacional Autónoma de México.

Pujadas, A., & Castillo, A. (2006). Social participation in conservation efforts: A case study of a biosphere reserve on private lands in Mexico. *Society & Natural Resources, 20*(1), 57–72. https://doi.org/10.1080/08941920600981371

Ramírez-Sánchez, H., Fajardo-Montiel, A., García-Guadalupe, M., & Ulloa-Godínez, H. (2024). Climate change vulnerability assessment in Jalisco: A comprehensive analysis. *International Journal of Environment and Climate Change, 14*(6), 115–144. https://doi.org/10.9734/ijecc/2024/v14i64215

Riley, K., Jukes, S., & Rautio, P. (2024). Relational ontologies and multispecies worlds: Transdisciplinary possibilities for environmental education. *Australian Journal of Environmental Education, 40*, 95–107. https://doi.org/10.1017/aee.2024.23

Schirnding de Almeida, E., & Boucsein, B. (2025). Facing climate emergency: A transdisciplinary, cross-institutional approach for built environment higher education. *Cogent Education, 12*(1). https://doi.org/10.1080/2331186X.2025.2464284

Takano-Rojas, H., Murray-Tortarolo, G., Maass, M., & Castillo, A. (2023). Characterization, variability and long-term trends on local climate in a Mexican tropical dry forest. *International Journal of Climatology, 43*(11), 5077–5031. https://doi.org/10.1002/joc.8133

Vigil, L. (2021). *Educación para el manejo sustentable del bosque tropical seco: experiencia en una escuela preparatoria* (Tesis de licenciatura). Universidad Nacional Autónoma de México.

Teachers for Future in Spain

Miriam Campos Leirós, Carmen Gloder Ramos, and Laura Muñoz Gloder

Abstract This chapter explores the origins, evolution, and impact of Teachers for Future Spain, a educational movement aligned with the global Fridays for Future initiative. Comprised of educators across Spain, the organization advocates for climate justice and sustainability education through transformative pedagogical practices. The chapter situates the movement within broader discourses of environmental education, activism, and policy engagement, analyzing how Teachers for Future Spain mobilizes teachers as agents of change within and beyond the classroom. The chapter contributes to a growing body of scholarship on climate activism in education, emphasizing the importance of teacher-led movements in advancing socio-ecological transformation. Ultimately, Teachers for Future Spain exemplifies how educators can collectively respond to the climate crisis through activism, curriculum development, and community engagement, offering a model for sustainable action in the educational sphere.

1 Introduction

In 2019, two students from the University of Girona reached out to Miriam Campos Leirós to inform her that they were participating in a "climate strike" every Friday, inspired by a European movement that originated in Sweden, led by the student Greta Thunberg. The students shared their protest on *El Guiño Verde* (*The Green Wink*),[1] a blog created by teachers committed to sustainability. Following this publication, the movement began to gain momentum and spread. Soon after, *El Guiño Verde* documented the visit to Madrid of two Swedish students, classmates of Greta Thunberg,

[1] El Guiño Verde. Accessed 16 june 2026. https://educandoenmedioambiente.blogspot.com/

M. Campos Leirós · C. Gloder Ramos (✉) · L. Muñoz Gloder
Teachers for Future Spain, Gondomar, Spain
e-mail: teachersforfuturespain@gmail.com

J. J. Conde et al. (eds.), *Practices in Education for Climate Emergency*,
https://doi.org/10.1007/978-981-92-2304-6_19

who took part in a protest and visited schools engaged in climate action. Over time, the blog evolved into the collective *Teachers for Future Spain – Profes por el Futuro*,[2] a group of teachers concerned about the climate emergency and its consequences.

We began working under the umbrella of *Teachers for Future* with the goal of giving a voice to students in grades lower than the third year of ESO (*Educación Secundaria Obligatoria* - Compulsory Secondary Education) who, according to Organic Law 8/1985 (*Ley Orgánica 8/*1985*)*, are unable to participate in street demonstrations due to their age. However, through various channels, we became aware of their concern about the state of the climate crisis in our country. In addition, we were driven by the need to educate students from an ecosocial perspective, helping them understand that our actions have consequences and that every action, no matter how small or insignificant they may seem, add up and contribute to change.

At the same time, we sought to create a support network for all education professionals - teachers, professors, educators - who often feel isolated in their centers, carrying out work that is disregarded or undervalued. We believe it is necessary to create spaces that allow for meeting and collaboration, allowing educators to share experiences and the common commitment to transform education towards a more sustainable future. As a collective, we carry out concrete actions to transform school management, providing an eco-social approach and cross-disciplinary pedagogical approaches that favor critical and active thinking in favor of sustainability.

We are independent of any political party, religion or union, driven solely by our commitment to the planet, its inhabitants and the interdependent and eco-dependent relationships between them. Our collective includes members from all types of educational institutions—public, private, and subsidized—, across all educational stages, throughout the national territory, and in both formal and non-formal education. We are a collective that works on a voluntary basis and in a totally altruistic way.

Currently, our work has been growing, as the challenges related to climate change have not only increased, but also diversified. Our work seeks to respond to these problems and contribute, through education and classrooms initiatives, to the awareness and transformation of society. From *Teachers for Future Spain* and in collaboration with other associations and NGOs, we have developed materials such as "Climate Refugees", "Advertising and Consumption", "Fake News" or materials on climate denialism, democracy or technology. As teachers, we feel the need to address these new emerging challenges in a constantly changing world from within the classroom, thus helping the students understand this new reality while at the same time encouraging critical thinking and decision making on these new issues. All our educational materials are open and free to use, in order to support both teachers and the education of new generations.

[2] Teachers for Future Spain. https://teachersforfuturespain.org/acerca-de//

2 Origins

Our origins trace back to *El Guiño Verde*, the blog that gave rise to *Teachers for Future Spain.* It was founded as a result of various projects led by Miriam Campos Leirós with her students, aimed at showcasing the sustainability initiatives they were implementing in the classroom. As a result of their experiences and through social networks, many teachers from all over the national geography began to collaborate selflessly to spread and share educational experiences related to environmental education since 2017.

Later, following the *"School Strike 4 Climate"*[3] and the global *Fridays for Future* movement,[4] we investigated and discovered the existence of similar initiatives aimed at teachers. We decided to go a step further and establish *"Teachers for Future Spain - Profes por el Futuro"*. We chose to retain the name in English to emphasize our connection to an international network, which already included nodes such as Teachers for Future Germany,[5] Teachers for Future Italy,[6] Teachers for Future Canada[7] or Teachers for Future India.[8] Today, *Teachers for Future Spain* maintains contact with these international nodes, participating in communication groups for the coordination of work, specific collaborations and the organization of joint actions.

Our proposals arise from the individual work carried out in the classroom by any of our *Teachers*, whether in response to an issue present in our daily lives or in connection with key dates for environmental awareness. Our main goal is to share the resources that have proven useful in our own classrooms and make them available to other educators. All materials are shared under an open license, allowing them to be adapted to different classroom contexts, curricula, and student interests. One of our key objectives is to avoid working in isolation, as we are fully aware of the heavy workload teachers face, not only in daily classroom activities but also in lesson planning and administrative tasks (Fig. 1).

We strive to facilitate the implementation of sustainability-related activities and initiatives. This way, every teacher who develops an activity can make it available to others, allowing our classrooms to be filled with accessible and impactful initiatives for the entire educational community. We have two ways of sharing: on the one hand, we prepare our own materials that we believe may be of interest to the teaching community, which are free to download, free of charge and offered on our website; and on the other hand, we publish those good practices that are shared with us by e-mail or by other means, fostering a ripple effect and encouraging peer-to-peer learning.

[3] Juventud x el clima son la esperanza española en la lucha contra la crisis climática. Vice. Accessed June 16, 2026. https://www.vice.com/es/article/juventud-x-el-clima-cambio-climatico-espana/

[4] Fridays for Future. https://fridaysforfuture.org/

[5] Teachers for Future Germany. https://teachersforfuture.org/

[6] Teachers for Future Italy. https://www.teachersforfutureitalia.org

[7] Teachers for Future Canada. https://linktr.ee/Teachersforfuturecanada

[8] Teachers for Future India. https://www.facebook.com/teachersforfutureindia/

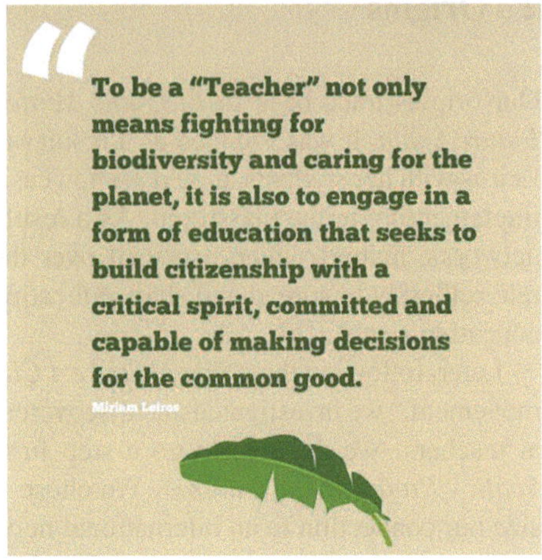

Fig. 1 The role of teachers as advocates for a sustainable future

3 Our Aim

Teachers for Future Spain was founded on the basis that education plays a fundamental role in addressing the environmental crisis. Our principles are rooted in the Tbilisi Declaration of 1977 (UNESCO, 1977) which established the foundation for environmental education, emphasizing that it must be a lifelong, interdisciplinary and holistic process. However, despite the decades that have passed since this declaration, progress has been insufficient. Today, the climate emergency is undeniable, and students continue their education in a world where no one can guarantee them a dignified future. Faced with the inaction of governments and the failure of countless climate summits, new movements advocating for environmental protection have emerged as a response to this critical situation.

At Teachers for Future Spain, we believe that the measures taken so far to protect the environment are inadequate. Therefore, we demand effective actions to curb global warming, as well as policies that are genuinely committed to sustainability. Education must be at the center of this change, as it is the foundation of all learning and personal development. However, we find that the current academic curricula, established by various educational administrations, fails to adequately address climate change, sustainability, and environmental education. We advocate for a comprehensive and transformative approach that integrates these issues meaningfully into the school system, preparing students to face the challenges of the future with knowledge, awareness, and commitment to action.

Therefore, we propose:

3.1 Naturalization of School Practices for Environmental Connection

Promoting pedagogical practices outside the classroom in the natural environment closest to the educational center, encouraging direct contact between students and their surroundings. We believe that only through proximity to their environment can students develop a sense of responsibility for their care. Only through a sense of belonging to nature do we cultivate ecological awareness.[9] In this regard, the school environment emerges as an inexhaustible source of teaching and learning elements for sustainability.

3.2 Ecosocial Education in Theory and Practice

In order to develop ecosocial education, we need a more ambitious school curriculum on sustainability issues, which gradually and transversally includes content in all areas of life (food, consumption, mobility, public space and forms of relationships, etc.). Moving towards an ecosocial education must result in a curriculum that connects the knowledge provided by the disciplines with everyday life, and this knowledge must arise from a dialogue between these disciplines.

Simultaneously, we consider that ecosocial education implies educating in another form of relationship, from care, good treatment in relationships, fostering horizontal and vertical collaboration, generating networks of help and association, that must be taught to create and care for, and awakening the critical spirit and initiative for the contribution to the common good. We cannot take care of the environment and not teach tools for respectful dialogue with our peers.

3.3 Promotion of Global Sustainability in Schools

We intend to put into practice the values of respect for the environment in the school day to day, encouraging contact with nature and the immediate environment. It should facilitate shared decision-making among the different members of the educational community: students, families, teachers, non-teaching staff, local councils and businesses, i.e., making the environment a network that contributes to education.

[9] Conciencia Ecológica. Accessed June 16, 2026. https://concepto.de/conciencia-ecologica/

4 Lines of Action

Since our creation and manifesto[10] in 2019, we have been evolving. Although we started with small individual activities tailored to the needs of each center, we soon realized that the different proposals could be organized into various lines of action, which we have been promoting in our annual activities. This is why we created a general program called "28,000 for the Climate",[11] referring to the approximately 28,000 educational centers of all types that exist today in Spain. Our main goal is to get all those centers to join our initiative and integrate eco-social education into their guidelines. This program is launched every year at the beginning of the school year, in September, with some modifications based on the experiences from the previous year or ongoing collaborations. Each edition usually has an honorary sponsor, and some of the figures who have collaborated with us in previous years are:

- 2021–2022 school year: Nacho Guadix, Head of Education and Children's Digital Rights at UNICEF Spain; and Fernando Valladares, researcher at the Spanish National Research Council (CSIC).
- 2022–2023 school year: Cristina Linares, professor at the University of Barcelona and member of the Office of Climate Change.
- 2023–2024 school year: María Neira, doctor and director of the Climate Change Office at the World Health Organization (WHO); and Carmelo Marcén, teacher and doctor in Geography.
- 2024–2025 school year: Isabel Moreno, physicist and meteorologist, known for being the weather presenter on various television programs.

The "28,000 for the Climate" program brings together all our lines of action, proposals, materials, and resources collected over the past few years. As part of the program, we also have an eco-social calendar, which is launched at the beginning of each school year and is designed to be displayed in classrooms. In it, you can find the main events of the year, accompanied by a QR code that links to the resources we have created on our website, specific to that day or campaign (Fig. 2)

All the proposals aim to contribute to the improvement of schools and a reduction in their environmental footprint. They are flexible initiatives that each school adapts to its circumstances. Additionally, there is no predetermined order: the idea is to start with the action that best suits the needs and possibilities of each educational community. In some cases, results can be seen quickly, while in others, changes require medium- or long-term work. However, all actions add up and contribute to the common goal of transforming education from a sustainable perspective. The different lines of action, which will be explained in detail in the sections that follow,

[10] Manifesto Teachers For Future Spain. https://teachersforfuturespain.org/manifiesto-teachers-for-future-spain/

[11] 28000 por el clima. Teachers for Future Spain lanza la nueva guía "28 000 por el Clima" para impulsar la acción ecosocial. EPALE. Accessed 16 June 2026. https://epale.ec.europa.eu/es/resource-centre/content/teachers-future-spain-lanza-la-nueva-guia-28-000-por-el-clima-para-impulsar

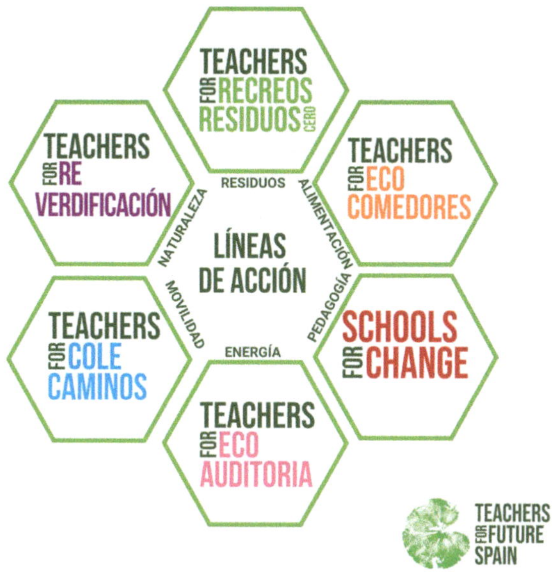

Fig. 2 Lines of action

are Recreos Residuo Cero—Zero Waste Recess, Ecocanteens, Schools for Change, Ecoaudits, Colecaminos—School Routes, and Green Spaces.

4.1 Recreos Residuo Cero - Zero Waste Recess

When the topic of sustainability, the environment, and climate change is addressed, especially in school and educational settings, the first idea that often comes to mind is the importance of teaching students how to recycle and dispose of waste in the appropriate bins. This idea is widely promoted, especially in campaigns by large companies aiming to engage with the educational sector, in the media, and through various local, regional, or national administrations. There is a widespread concern about educating future citizens in RECYCLING. As a result, many educational institutions had already begun implementing such practices in classrooms before the existence of Teachers for Future Spain, with separated waste bins and signage provided by companies or administrations.

However, at *Teachers for Future Spain*, we believe that the first "R" of sustainability should be REDUCTION. As teachers, we observed during breaks or recess that at the end of these periods, trash bins were filled with all kinds of waste from food packaging brought by students for lunch. This led us to start thinking about ways to reduce unnecessary plastic and packaging. The goal was, therefore, to reduce waste in the playground while simultaneously promoting healthy eating habits, such as encouraging students to bring fruit without plastic wrapping, opting for sandwiches instead of industrial pastries, or making homemade sweets as part of extracurricular and family activities. At the same time, we encouraged the use of reusable containers to bring lunch to school, avoiding plastic bottles, juice boxes, and single-use wrappers. This is how the Zero Waste Recess program (Recreos Residuo Cero) was born, which began experimentally in the 2017–2018 school year at a Galician school whose "Cooltureco"[12] project promoted both culture and ecology. Zero Waste Recess was taken on by *Teachers for Future Spain*, and from there, it was able to spread across the country, boosting both waste reduction and healthy eating.

This was the first program we launched to be implemented in classrooms, starting on a larger scale during the 2019–2020 school year. The initiative is very simple in its approach and easy to follow in all types of schools and educational stages. As previously mentioned, the aim is to reduce plastic waste in school lunches, while also improving nutritional habits and learning to recognize what waste is, the different types of waste, and their potential for recycling, if possible.

As with all our proposals, we try to make them cross-disciplinary activities with other subjects such as mathematics, natural or social sciences. One of the great advantages of this program is that it does not require any infrastructure or material costs for its implementation. It is not necessary for the whole center to do it at the same time, nor does it have to be started at the beginning of the school year. Each teacher has to evaluate when they can do it and what is the best moment for them, for their students and families. Experience has shown us that it usually started with a class, a level or a cycle and after a short time more classes were added until eventually the whole center joins.

For this purpose, we provide different resources to bring into the classroom, whether in early childhood, primary, or secondary education, as well as presentations to motivate and raise awareness among students, teachers, and especially families. All these resources are available on our website and they are all freely downloadable for everyone visiting the site. The different resources have been developed both by the group of Teachers' collaborators, as well as by different colleagues who have been adapting them to their center, to their reality and have shared them on our website, through the publication of experiences.

Some of the resources available on our website[13] to participate in the *Recreos Residuos Cero* program are summarized in the following table. These resources cover topics such as how to start and motivate the school community —teachers, students, and families— possible ideas for more sustainable and healthy snacks, and even how to compost the organic waste generated after recess (Table 1).

The success of this program became evident in 2019, when we collaborated with Comunidad #PorElClima[14] and #6DNow to raise awareness about the actions related to the Zero Waste Recess initiative taking place in our schools. To identify the schools that chose to join the movement, we provided banners for them to hang on their fences or façades, visibly showing their commitment on December sixth, coinciding with

[12] Cooltureco. Accessed 16 June 2026. https://cooltureco.blogspot.com/

[13] Zero Waste Recess Materials. https://teachersforfuturespain.org/materiales-recreos-residuo-cero/

[14] Comunidad #PorElClima. Accessed 16 June 2016. https://porelclima.org/

Table 1 Available resources for participation in the zero waste recess program

Articles	• Previous experiences on the Cooltureco Blog (2018–2019 academic year) • Greenpeace reports • Article by Ballena Blanca on the lies of 100% recyclables • General resources on the reality of waste • Organic waste compost
Presentations	• With questions on recycling, waste life and program proposals on how to reduce it • With waste-free lunch ideas • With information for families
Posters	• Murals about natural packaging • Explanatory posters to hang at the education center • Infographics with tips for waste reduction
Videos	• Video tutorial for sandwich carrier ideas to make it in class or at home • Rap for high school students
Extra material	• Zero Waste Recess Card, to reward students or classrooms that manage to reduce waste • Weekly or monthly waste registration models • Excel spreadsheets to calculate the amount of CO_2 and water saved • An App developed by the University of Salamanca for waste calculation

the 2019 United Nations Climate Change Conference (COP25)[15] held in Madrid. Thanks to this collective effort, we were able to give visibility to our challenge against climate change (Fig. 3).

Since then, our program has remained active, with an increasing number of educational centers joining the initiative. However, sustaining and expanding *Zero Waste Recess – Recreos Residuo Cero* is an ongoing effort that relies on keeping teachers, students, and families motivated and engaged. To support this, we highlight key dates in our ecosocial calendar to reinforce this commitment and share reminders through our social media channels.

We have a specific registration form[16] for our Zero Waste Recess initiative, as well as an updated map that continuously geolocates all the participating centers. The ability to visualize their location on the map is appealing to schools, as it allows them to feel part of a larger project. The sense of belonging and active participation have proven to be key factors for motivation. Currently (2024), the program has approximately 1600 registered centers, mostly located in Spain. However, educational institutions from Latin America, including countries such as Nicaragua, Mexico, Colombia, Chile, and Uruguay, as well as from the United States, also participate. Additionally, some Spanish centers in Europe, in countries like Portugal, Belgium, Italy and France, are also part of the initiative.

---

[15] COP 25. Nuestra primera semana en la Cop25 y la cumbre social. Teachers for Future Spain. Accessed 16 june 2026. https://teachersforfuturespain.org/nuestra-primera-semana-en-la-cop25-y-la-cumbre-social/

[16] Zero Waste Recess form. Accessed June 16, 2026. https://docs.google.com/forms/d/e/1FAIpQLSddRnEE5ndzes12slzdvD-37S2NSGvwqj65E6FfAl10Hi-gCg/viewform

Fig. 3 Official zero waste recess participation banner given to educational centers that took action during 2019

Over the years, we have requested the Ministry for Ecological Transition and the Demographic Challenge to recognize the participation of schools in the program through a certificate. Although this document does not grant credits or points for official matters, it does help reinforce the sense of contribution to the environment and to the Teachers for Future Spain movement. At Teachers for Future Spain, we handle the distribution of these certificates, and in return, we receive hundreds of messages of gratitude, which confirms the satisfaction of those involved in the program.

We would like to highlight that the *Recreos Residuo Cero* program was studied by the University of Salamanca in the publication titled "Whole School Approaches to Sustainability. Education Renewal in Times of Distress."(Wals et al., 2024). This study combined both quantitative and qualitative methods to evaluate the effectiveness of implementing a Whole School Approach (WSA) in waste management and promoting sustainability in school environments. This recognition reinforces the importance of the *Recreos Residuo Cero* program as a tool that helps foster sustainable habits within educational institutions. Through waste reduction during recess, the use of reusable containers, and raising awareness about the environmental impact of our consumption habits, the program contributes to the development of more responsible generations committed to the environment.

4.2 Eco-Canteens

Another area of action and improvement for a center seeking to reduce its environmental footprint is the school canteen. According to the study of Figueiredo et al., (2021), "universities play a crucial role in raising awareness about consumption choices, both individual and collective." We believe that this principle is equally applicable to other educational institutions, where promoting sustainable eating habits can have a significant impact on the school community. The meals for both students and teachers in an educational center must be taken care of in all aspects, from the nutritional, educational and environmental points of view. Beyond what the school mealtime represents for many student, not only in terms of nutrition but also as a fundamental part of daily life for many students, it is crucial to encourage habit changes and promote ecosocial education, both for children and for us as adults. While the most desirable and recommended options are often accessible to only a few, small actions can make a significant difference. Achieving an eco-canteen is a long process that also involves external agents: the educational administration that regulates the service, the catering company providing the meals, the suppliers, and the cafeteria staff, among others.

Depending on the different educational administrations, the rules and regulations are different when it comes to managing and supervising a school canteen. However, in all of them, the management team plays a crucial role, whether in selecting the company that provides such services to the center or when carrying out an evaluation of the quality of the services. For this reason, at Teachers for Future Spain, we propose a series of improvements to school meal services, such as incorporating menus with seasonal and locally sourced products, thereby promoting local commerce. We also advocate for the inclusion of organic foods (produced without pesticides) and whole-grain options, while encouraging the introduction of vegetarian or vegan choices, or at the very least, implementing a meat-free day once a week in school menus. Additionally, we promote the incorporation of traditional dishes from other countries when their environmental impact is low, particularly in schools with international students, as a way to foster inclusion.

Menu planning is an activity that not only can, but should be addressed in the classroom, integrating both healthy eating principles and sustainability. However, the most crucial aspect is ensuring that school cafeteria menus align with the recommendations of the Harvard Plate, also known as the Healthy Eating Plate.[17] This model, developed by the Harvard T.H. Chan School of Public Health, serves as a visual guide for balanced and nutritious eating. It suggests that half of the plate should consist of fruits and vegetables, one quarter should include whole grains, and the remaining quarter should contain protein sources. Additionally, it emphasizes the use of healthy oils and the consumption of water, coffee, or tea as primary beverages while also encouraging an active lifestyle.

[17] Harvard plate. Accessed June 16, 2026. https://www.health.harvard.edu/staying-healthy/healthy-eating-plate

4.3 School For Change

Schools for Change is an initiative that aims to turn schools into agents of change in their immediate surroundings. It is a program that seeks to mobilize teachers to transform educational centers and neighbourhoods through project-based learning (PBL) and service-learning methodologies (ApS). These projects are carried out with students in the final years of primary school or secondary school.

To transform society or undertake educational actions with transformative impact or intent, collaboration must be the starting point. The educational community is a key collective for transmitting values, commitment, and shared responsibility, as well as for promoting critical awareness and participation. Students' active participation in society as engaged citizens, rather than passive agents, is only possible by going beyond the walls of the classroom, permeating into the community, showing them their local environment, and asking them what they would do to improve it. This gives them the opportunity to express their opinions, participate, and transform their social and environmental surroundings on a small scale.

In order to implement an educational process of awareness, knowledge, critical thinking, and transformation, the entire educational community is needed: teachers, students, families, non-teaching staff, and administrations, working together on networked projects. To transform, there must be an interest in the environment, a deep knowledge of the surroundings: my school, my playground, my neighbourhood. It is essential to develop a sense of belonging to foster empathy. Students must be able to shift from the concept of individual to community. This is an approach rooted in holistic pedagogy, based on the sense of belonging to a group and the intention to improve or contribute. It is about developing the motivation and will to undertake voluntary action.

Transformation can start within the school itself, by modifying spaces like playgrounds to make them more welcoming and inclusive. This can involve creating spaces for coexistence and renaturalizing the environment by incorporating plants, trees, and green spaces instead of concrete. Introducing features like a butterfly oasis or a pollinator garden can serve as a starting point. However, giving students the opportunity to think about the playground they want and supporting them in the process of deliberation and decision-making is the core of the participatory and transformative education approach of S4C: Schools4Change.

On the other hand, if we want to go out into the nearby environment, we can look for ways to improve the urban planning of the neighborhood. The way cities are planned, designed and built is not an irrelevant issue. These are decisions that essentially affect our living and coexistence spaces, the development of our daily lives, the quality of life and, of course, the integral development of students. That is why it is so important to develop a culture of participation that allows all our voices, including those of children, to be heard and taken into account.

4.4 Ecoaudit

Another area of action of Teachers for Future, focuses on everything related to consumption at both individual and collective levels, aiming to raise awareness of how much we consume and how we go about it. We must keep in mind the need to move towards degrowth, waste reduction, and the promotion of responsible consumption that aligns with our goal of fostering eco-social education, environmental awareness, and a connection with nature.

To achieve this, we must analyze our consumption habits both individually and at the institutional level. One of the best tools for conducting this analysis is the "ecoaudit", which allows us to become aware of our consumption patterns based on concrete data. Although each school has its own characteristics and must operate according to its context, we recommend conducting data collection, or the ecoaudit, at three defined time points: the first, second and third trimester. This tool has been designed as an improvement plan, allowing for the correction of unsustainable practices and facilitating the implementation of a more immediate action plan to address real issues.

The ecoaudit model is designed to be carried out by all sectors involved in our educational community, even if not everyone can respond to every parameter. It is intended to be conducted by the school management team, teachers, service staff, students (starting from the final cycle of primary school, with some support), and families through the Parents' Association (AMPA). This approach ensures a greater diversity of perspectives, more data, and encourages the participation of the entire community in the changes we seek to implement in our schools. It takes into account the following parameters, with different questions that allow for an assessment of the state of the educational center, such as:

- **Center Parameters:** Is there a sustainability coordinator? Is there a selective collection system for paper and cardboard? Are refillable and eco-friendly writing materials used?
- **Consumption Parameters:** What is the water consumption in the center? How many photocopies are made per month? What type and amount of energy is used? Are large containers and refillable materials (such as markers) used?
- **Cafeteria Parameters:** Is the kitchenware reusable? Is the dishwasher used only when full to avoid water waste? Are suppliers local? Do ingredients come in large containers? Is the food offered seasonal?
- **Gardening and Green Space Parameters:** Is there a bicycle parking area? Is drip irrigation used? Is there a school garden? Does the center have birdhouses? Are there pollinator-friendly gardens?
- **Classroom Parameters:** Do students participate in activities that involve direct contact with nature? Are field trips to nearby natural environments organized? Is waste reduction promoted?
- **Administrative Parameters:** Are there safe school routes for students to reach the educational center? Is maintenance carried out on facilities to improve insulation and save energy?

Once the data has been analyzed, it will be up to the educational center to decide which actions to take and to what extent, allowing for follow-up through the same eco-audit. Like all the materials available on our website, this document can be freely downloaded in different co-official languages and includes additional resources to support decision-making and the implementation of actions.

4.5 Colecaminos - Schoolroutes

Another initiative developed by our collective focuses on promoting active and sustainable mobility, with the goal of creating safer and more accessible school environments. This initiative, called "Colecaminos - SchoolRoutes", seeks to encourage walking, biking, or using public transportation, therefore reducing the ecological footprint and improving the quality of the school environment.

It is not just about reserving space for bike racks or teaching road safety, but about promoting a shift in mindset, breaking away from the car-centric model of cities, and reclaiming streets and public spaces in the neighbourhood for children, while also reducing children's autonomy and contributing to sedentary lifestyles. For this reason, we aim to stimulate sustainable school mobility initiatives that encourage contact with their everyday environment through non-polluting modes of transportation, such as walking or cycling.

One of the objectives of the LOMLOE (Spanish Education Law of 2020) for primary education, the objective "n) Develop healthy, everyday habits of autonomous active mobility..." (Ley Orgánica 3/2020) de medidas urgentes para la protección de los derechos económicos y sociales de la infancia y adolescencia en España) refers, among other things, to how our students get to school each day. This is why we propose conducting a survey among students to understand how they commute. In this way, we can assess these habits and work towards making them increasingly active through various tools and proposals to raise awareness of this need. Once the data has been collected, we propose working on the concept of the School Route. While school routes can take various forms, in the educational field they should be understood as a resource that invites collective reflection on a problem that concerns society as a whole: a large part of childhood has lost the street as a public space, even for everyday activities such as going from home to school. The change we need is cultural; it is not just a list of routes and infrastructure.

A very interesting project is the *Pedibus*,[18] introduced in some cities like Melilla, which allows students to walk safely from their homes to school, without the company of their parents. Groups of children are accompanied by volunteer adults from the neighbourhood, who act as "drivers" of the *Pedibus*. These volunteers wear reflective vests and identifying signs, ensuring their visibility and recognition by the students.

[18] Pedibús. Iniciativas para ir andando al colegio. EROSKI Consumer. Accessed 16 june 2026. https://www.consumer.es/medio-ambiente/pedibus-iniciativas-para-ir-andando-al-colegio.html

The routes are marked on the streets and are also supported and monitored by members of the local police.

Another way to promote active mobility among our students was carried out in 2023, in collaboration with *Clean Cities*[19] and *Ecologistas en Acción*.[20] We developed "Mobility Saddlebags",[21] a set of materials aimed at promoting active and independent mobility in primary education. These kits include manipulable materials such as stories, street models, theater proposals, materials to create a "metrominuto" (a map of walking routes), and comics. The mobility saddlebags are itinerant and began circulating through the schools that requested them during the 2023–2024 school year. For all interested teachers, the materials are available for download on our website.

4.6 Space Greening

Having a green space in the center contributes to the development of environmental awareness. By bringing the natural environment closer to the school setting, learning opportunities are multiplied, and the understanding of the importance of nature, its conservation and care is fostered. However, in the current climate emergency, it is also necessary to implement strategies for climate change adaptation and mitigation and to improve the environmental quality of the educational center.

As Jean-Jacques Rousseau stated, "The best school is the shade of a tree" (Vela Castillo, 2019). Greening the schoolyard, renaturalizing spaces, or creating a nature corner within the educational center has become a priority. Children need to play with soil and mud, dig holes, trace paths, search for insects, and hide among the bushes. We are living in a time when the term *Nature Deficit Disorder*, coined by Richard Louv in his book *Last Child in the Woods* (Louv, 2008), is widely discussed. Although not classified as a medical disorder, there is evidence that contact with nature reduces stress and improves attention. Schools should actively create more opportunities for children to reconnect with nature.

A school garden, a green space, or even a small nature corner provides countless learning opportunities, contributes to the preservation of biodiversity, and allows students to develop various curriculum topics outdoors, in direct contact with their environment. Furthermore, creating a greener school also involves reclaiming underutilized spaces and materials, as well as undertaking the essential task of adapting to and mitigating climate change. Among the various green spaces within the center, the school garden stands out for its immense value as an educational resource. Through hands-on activities in the garden, many aspects of the curriculum can be explored in a practical and experiential way: the physical environment, weather changes, plant

[19] Clean Cities España. https://spain.cleancitiescampaign.org/quienes-somos/

[20] Ecologistas en Acción. https://www.ecologistasenaccion.org/quienes-somos/

[21] Alforjas de Movilidad. Teachers for Future Spain. Accessed 16 June 2026. https://teachersforfuturespain.org/alforjas-de-movilidad/

life cycles, the water cycle, local food and products, consumption cycles, farming systems, soil conservation, the rational use of natural resources, the organic matter cycle, and the discovery of new plants and vegetables, among others.

The lack of space and natural soil is no longer an obstacle to creating a school garden or a green area. Raised beds and planters made from recycled materials can be used, as well as green walls and rooftop gardens with hydroponic systems, expanding the possibilities for greening educational centers. Additionally, all these initiatives can be complemented with composting. On our website, educators can find numerous experiences shared by colleagues who have worked on these initiatives, from starting a school garden to creating a vertical garden or beginning composting. There are also resources available to integrate these activities across different subjects in the classroom. Furthermore, we collaborate with various institutions, such as *Composta en Red*,[22] offering training webinars for teachers.

5 Our Internal Organization

As we mentioned at the beginning, we are an organization independent of any political party, administration, union or religion. We do not receive any type of subsidy, membership fees or form of income. Our sole motivation is our commitment to working toward ecosocial education, one that raises awareness about the state of our environment, the climate emergency we are experiencing, and the challenges it poses for our society. Our approach is always centered on collective well-being, promoting actions that engage students: initiatives that emerge from their ideas and experiences. We want children and teenagers to be aware of the world they live in and to strive to change it by transforming their daily habits, from consumption to diet. At the same time, we aim to foster critical thinking, ensuring that educators are not merely figures of authority who impose solutions, but rather facilitators who provide students with the tools to find their own solutions, propose alternatives, and explore new ways of doing things.

Like everyone in Teachers for Future, whether part of the organizing team, collaborators, or simply supporters, we are "classroom teachers." We understand what it means to be in a classroom, not only in terms of managing a large and diverse group of students, but also dealing with the overwhelming amount of bureaucratic work that comes with teaching. That is why we recognize that proposing activities or actions that require additional effort can be challenging to implement in the classroom. For this reason, all our programs and proposals are designed to be flexible. Teachers can integrate them into their classrooms whenever they see fit and feel ready, without the pressure of adhering to specific start and end dates. We do not impose mandatory objectives or require reports. We believe that any teacher who chooses to bring our initiatives into their classroom is already motivated by the climate emergency and

[22] Comenzando con el Compostaje Escolar. Composta en Red. https://www.compostaenred.org/comienzo-compostaje-escolar/

our goal is simply to ensure that the ecosocial aspect of education, something we urgently need, is also included in their teaching.

What we do ask our colleagues is to share with everyone what they have done in their classrooms that have worked well, so we can continue to build and expand our resources. Each school is a unique environment and every classroom has its own characteristics and diversity, which may not align with another teacher's experience. That is why each teacher is free to modify, expand, or scale back what they consider appropriate for their context. Everyone is free to act and work in the way that best suits their reality. Our goal is for these actions and activities to be incorporated into the classroom in a cross-curricular manner, integrating with the curriculum being taught. We are fully aware that this is possible and have many examples to prove it. Connecting subjects such as mathematics, language, social sciences, and natural sciences is entirely feasible, just as integrating art and physical activity is.

In terms of internal organization in our collective, we have a coordination team and the "territorial nodes". The coordination team includes, on the one hand, the figure of the general coordinator, who assumes the functions of presidency regarding the signing of collaborations and other bureaucratic tasks. On the other hand, the team consists of a small number of members who make proposals for actions and/or materials, and work in communication networks, on the web or in databases, as well as in the coordination of the different programs and projects. In addition, we have "territorial nodes" that receive the proposals and organize themselves in a more or less autonomous way. These groups designate a person for communication, called a "collaborator", who is in charge of transmitting information to the nodes, assuming different tasks as needed or as they arise. Regular meetings are held to make joint decisions, promote actions or define tasks. Lastly, there is the bulk of Teachers for Future Spain, which are all teachers who, without wishing to participate actively in the organization, do take on board the proposals and take them to the classrooms and promote them.

In other words, the functioning is based on sociocracy. On one hand, decisions that affect the organization of work or have a political character are taken with the consent of the coordination group. There is consent if there is no reasonable objection, but if there is an objection, the person or group (referred to as regional groups or nodes) works to modify it and it is taken back into consideration for approval. Operational decisions, that is, those concerning daily tasks, are made autonomously, always with the premise of promoting the organization and its well-being, making the functioning smoother and more efficient.

In Teachers for Future Spain the coordination team assumes most of the work, but each person belonging to the above-mentioned "territorial nodes" is a member that can solve specific problems, and these territorial groups or nodes can make their own decisions with the consent of the members of their circle. Decisions of a political nature or affecting the organization of the association must have the consent of the coordination team, as mentioned above. Unlike sociocracy as it is traditionally known, there are no elections for candidates; instead, the process involves the nomination of a person who takes on the role of communication and other voluntary tasks. Each individual, regardless of their territorial node, can offer themselves to take

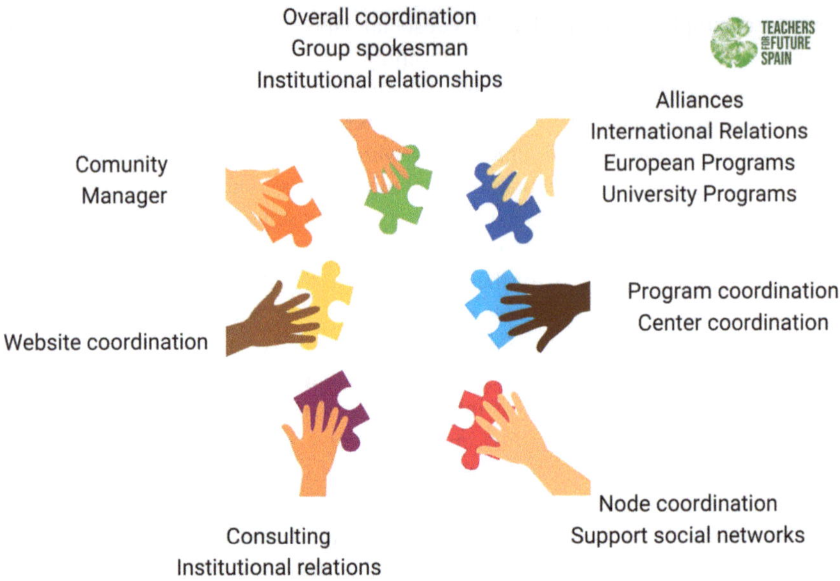

Fig. 4 Organigram showing the internal organization of Teachers for Future Spain

on work functions by communicating with the coordination team. If the workload is high, ongoing over time, and involves significant decisions regarding the functioning of the association, that person may join the coordination team (Fig. 4).

Being part of Teachers for Future is very simple: we have a membership manifesto in which we set out our ideas, which can be joined by those who wish to do so; but it is also possible to join the actions and register the school in one of our programs or our proposals. This is the most common way to join us. It is not necessary for the whole educational center to sign up, but reality has shown us that there is a contagious effect on the rest of the colleagues, and they tend to join the program. The important thing is to start doing something in our centers and classrooms. To those enrolled in our programs we send them information about things that are coming out on our website, new proposals or campaigns and other sources of interest. There is also the possibility to subscribe to our online blog to receive news and information about the activities we do, without having to join any program.

Teachers for Future exists throughout Spain and in many other places across Europe, North, Central and South America. In Spain, the regional nodes are getting stronger every day and in addition to sharing more local information, meetings are starting to be held among their members to share experiences and carry out more local actions. At the national level, we have already held three congresses, the first two in Madrid and the most recent one in Pamplona. We have been working for five years in favor of ecosocial education in our classrooms and our ultimate goal is to be present in the more than 28,000 educational centers that exist in our country.

6 Conclusions and Future Work

To conclude, we want to state that we are a movement that, ideally, should not have had to exist, because that would mean that a sustainable education is already present in the classrooms. Sadly, that is not the case. The day that all 28,000 schools in Spain adopt this approach to education, Teachers for Future Spain will have achieved its greatest goal, and our existence will no longer be necessary.

In the meantime, we are very proud that in just five years, and without any kind of subsidy, external support, or financial contributions, we have managed to reach so many schools and that our social networks and website are full of visits and contributions born from the collaborative work of educators who dream from the positive side of education, with actions for and by the climate. We will continue working and participating through volunteering, expanding our network across Spain, and seeking to engage schools, teachers, and especially students, to achieve a complete ecosocial education.

References

Figueiredo, C., Azeiteiro, U. M., García-Vinuesa, A., & Carvalho, S. C. (2021). Campus decarbonization: Students' perceptions for reducing meat consumption in a Portuguese university. *Sustainability, 13*(11), 6048. https://doi.org/10.3390/su13116048

Ley Orgánica 3/2020. (2020). de 29 de diciembre, de medidas urgentes para la protección de los derechos económicos y sociales de la infancia y adolescencia en España. Boletín Oficial del Estado, núm. 340, de 30 de diciembre 2020. https://www.boe.es/buscar/act.php?id=BOE-A-2020-17264

Ley Orgánica 8/1985. (1985). de 3 de julio, de Derechos y Libertades de los Alumnos y del Derecho a la Educación. Boletín Oficial del Estado, núm. 159, de 4 de julio de 1985. https://www.boe.es/buscar/act.php?id=BOE-A-1985-12978

Louv, R. (2008). *Last child in the woods: Saving our children from nature-deficit disorder*. Algonquin Books.

UNESCO. (1977). *Intergovernmental Conference on Environmental Education, Tbilisi, USSR, 14–26 October 1977: Final report (ED/MD/49)*. UNESCO. https://unesdoc.unesco.org/ark:/48223/pf0000032763

Vela Castillo, J. (2019). «La mejor escuela es la sombra de un árbol». Jimena Menéndez-Pidal, Fernando Higueras, el Colegio Estudio. *Boletín de la Institución Libre de Enseñanza, 113-114*, 69–90. https://www.fundacionginer.org/boletin/bol_113_114.htm

Wals, A. E. J., Bjønness, B., Sinnes, A., & Eikeland, I. (Eds.). (2024). *Whole school approaches to sustainability: Education renewal in times of distress*. Springer International Publishing. https://doi.org/10.1007/978-3-031-56172-6

Building Sustainability from the Ground Up: Community Experiences from the Environmental Culture Centers of Mexico City, Living Spaces for Citizen Transformation

Mariana Muñoz-Galván and Alejandra Atzin Ramírez Hernández

Abstract The Environmental Culture Centers of Mexico City's Secretariat of the Environment (SEDEMA) represent a key strategy to strengthen critical environmental education, foster biocultural connections, and drive socio-ecological action in urban settings. Through a diverse and immersive program of cultural and environmental education activities, these spaces promote ecological restoration, territorial awareness, and community participation. Initiatives such as the *School Garden*, *Gardens for Life*, *Bat Night*, and the *Axolotl Route* address both the conservation of endemic species and the appreciation of traditional knowledge, creating meaningful eco-social experiences. The Environmental Culture Centers also serve as hubs for training and territorial governance, facilitating co-educational processes and participatory methodologies focused on urban regeneration and the stewardship of shared commons. The approach of the Environmental Culture Centers goes beyond simply transmitting knowledge. It fosters co-creation of understanding, strengthens the sense of belonging, and builds urban resilience through participatory, intergenerational methods. In doing so, the Environmental Culture Centers establish themselves as incubators of alternatives for a fairer, more resilient city that values its biocultural heritage, inspiring new ways of inhabiting and transforming Mexico City.

1 Introduction

In a world marked by multiple crises (climatic, social, geopolitical, and cultural), educational processes with territorial and community-based approaches take on an urgent and transformative meaning. Confronting technocratic or decontextualised

M. Muñoz-Galván (✉) · A. A. Ramírez Hernández
Environmental Culture Directorate, Ministry of the Environment of Mexico City, Mexico City, Mexico
e-mail: mariana.munoz.sedema@gmail.com

A. A. Ramírez Hernández
e-mail: alejandra.atzin.sedema@gmail.com

J. J. Conde et al. (eds.), *Practices in Education for Climate Emergency*,
https://doi.org/10.1007/978-981-92-2304-6_20

approaches, environmental education rooted in Latin American thought proposes pathways built from people's knowledge, dialogue, affection, collective work, and a shared commitment to the commons (Leff, 2020; Martínez Ordoñez et al., 2024).

In a city as complex, vibrant, and biodiverse as Mexico City, discussing culture and environmental education is neither a luxury nor an optional extra, but an urgent necessity. Environmental culture, understood as the way people relate to nature through their knowledge, values, practices, and everyday decisions, provides a fundamental basis for imagining and constructing new forms of coexistence—both with our natural surroundings and with one another (UNESCO, 2021). It is about changing not only what we do, but how we understand our existence on this shared planet.

Within this context, the Environmental Culture Centers (ECC) stand out as strategic and deeply necessary spaces. They are not merely occasional visit sites but true citizen laboratories where education, collective action, and ecological and social regeneration are interwoven (González-Gaudiano, 2022). These centers bring together children, young people, women, adults, and elders to learn, engage in dialogue, and reconnect with the ecosystems that sustain the city. Here, it is not only the soil that is cultivated, but also consciousness, solidarity, and the possibility of imagining fairer and more sustainable futures (Ross & Raworth, 2019; Sauvé, 2004b).

The ECC are strategically located within Mexico City's protected natural areas—in Ajusco, the Sierra de Santa Catarina, Xochimilco, and Chapultepec Forest—allowing residents to reconnect with their biocultural heritage. Culture and "sentipensares" (feelings and ways of thinking) transform as people interact with the land, territory, endemic species, local knowledge, and traditions, directly confronting the real challenges of their environment. A key feature of centers such as Ecoguardas, Yautlica, and Acuexcomatl is that they are situated in historically peripheral zones, outside the city's cultural core. This positioning makes them crucial spaces for bringing sociocultural activities closer to neighbouring communities traditionally excluded from the central cultural offering.

The mission of the ECC goes beyond simply transmitting knowledge. They aim to create new ways of inhabiting and perceiving the city, recognising nature and people as part of the same living fabric. These spaces employ active methodologies that foster situated learning, collective inquiry, and critical reflection on contemporary urban challenges. The diversity of voices and actors converging in the ECC—including environmental educators—enriches activities and nurtures participatory management processes that build alliances between institutions, local communities, and social organisations.

Through these experiences of learning and action, participants in the ECC contribute to rethinking how to inhabit and transform the city. This collective approach drives an education based on territory, sensitive to context, and open to social innovation, where creativity and intergenerational dialogue spur change. Narratives highlighting local stories and connections to nature promote new forms of citizenship and ecological responsibility. By recovering urban agricultural practices, restoring degraded spaces, and celebrating biocultural diversity, the centers foster territorial belonging and collective ownership. This integral vision makes the ECC seedbeds of alternatives to tackle the city's pressing challenges. By co-creating knowledge and strengthening community bonds, these spaces sow hope and resilience for present and future generations.

The ECC are public spaces that foster environmental culture. Through educational activities such as workshops, interpretative walks, school gardens, exhibitions, and community actions, they offer meaningful learning experiences that challenge excessive consumption and promote care and respect for nature (Freire, 1998; Kolb, 1984; Toledo & Barrera-Bassols, 2009).

Ultimately, the relevance of these centers transcends the local sphere, contributing to global commitments on climate change, biodiversity, environmental justice, and the Sustainable Development Goals (UNESCO, 2021). In a world facing socio-environmental crises, these spaces provide local responses that build ecological and resilient citizenship, weaving a network of global actions seeking cultural transformation in the face of environmental collapse (Rockström et al., 2023).

In sum, the ECC of Mexico City embody the hope for a profound and necessary change in our relationship with nature and with the city itself. They are beacons illuminating pathways for collective learning, resilience, and care, inviting us to imagine and build fairer, more sustainable realities. Their daily work shows that it is possible to transform the present from the local level, weaving networks of solidarity, knowledge, and action that transcend generations. In these spaces, environmental education ceases to be mere discourse and becomes a living practice, capable of germinating alternatives to the challenges of our time and planting seeds of the future in the heart of the city.

2 Descriptions of the Environmental Culture Centers

2.1 Acuexcomatl Environmental Culture Center (Xochimilco, Mexico City)

Located in the south of Mexico City, within the Xochimilco borough, whose environmental importance was recognised when it was included on UNESCO's World Heritage list in December 1987 and designated under the Ramsar Convention for the conservation and wise use of all wetlands (Wake, 2008)—the Acuexcomatl Environmental Culture Center stands as a living tribute to the sacred relationship between water, land, and indigenous communities. Its Náhuatl name, meaning *Jar from which water springs*, embodies a profound mission: to act as a bridge between ancestral knowledge and today's sustainability challenges.

This center is far more than an educational space; it is an ecological refuge where chinampa wisdom meets contemporary science. Surrounded by canals, chinampas, and unique biodiversity, Acuexcomatl welcomes children, young people, women, and adults to experience transformative learning. Environmental education here takes shape through hands-on workshops on traditional agriculture, chinampa cultivation, responsible water management, food production, and soil restoration, as well as ecological restoration days and interpretive tours that reveal the complex web of life in wetlands. One of the region's most emblematic species is the *Ambystoma mexicanum*, which in the wild remains on the brink of extinction in its last natural habitat: the lakes of Xochimilco and Chalco (Wake, 2008). At Acuexcomatl, this species is promoted as an umbrella species, aiming to raise awareness of ecosystem loss in the region and the social and environmental consequences that follow.

Every activity seeks to strengthen Xochimilco's biocultural identity and empower the community to care for its territory. Beyond environmental training, the center promotes the recovery of indigenous knowledge on the use of native plants, community water management, and traditional festivities that link spirituality with respect for nature.

Community participation is a fundamental pillar at Acuexcomatl. Local residents, producers, and students actively collaborate on conservation projects, species monitoring, and the defense of common goods in the face of urban threats and climate change. The center is, in essence, a seedbed of environmental guardians and a living laboratory where social and ecological innovation thrives hand in hand with collective memory and citizen creativity.

The Acuexcomatl Environmental Culture Center envisions a future where resilience, environmental justice, and the dignity of indigenous peoples are part of a more harmonious and sustainable city. It promotes environmental education from a biocultural perspective, integrating ancestral knowledge, soil and water management, and traditional farming techniques with modern science and the challenges of climate change (Toledo & Barrera-Bassols, 2009). Through activities such as agroecology workshops, composting with earthworms, the revival of chinampa-based knowledge, and educational tours of the wetlands, the center teaches new generations the importance of water and biodiversity.

2.2 *Environmental Culture Center Ecoguardas (Ajusco Forest, Mexico City)*

The Environmental Culture Center Ecoguardas is located within the Ajusco conservation area, inside the Ajusco Forest, a protected region that forms an integral part of the southern forest zone of the city. This area is crucial for aquifer recharge and biodiversity preservation.

Ecoguardas is a sanctuary for biodiversity and environmental education. Among pines, oyameles, and oaks, this space offers training for educators, capacity-building for citizens and companies, and immersive tours that highlight the value of nature. It also promotes activities focused on bats, birds, pollinating insects, fungi, and endemic species. The center fosters citizenship committed to the care of shared natural goods and the protection of life in all its forms.

Ecoguardas promotes the exchange of knowledge between long-standing and new residents of the region. Its activities include talks on the importance of the Ajusco, identification of endemic flora and fauna, and sustainable practices for the use of natural resources. This center consolidates itself as a platform to build ecological citizenship and defend natural heritage, combining community efforts with public conservation policies and urban resilience initiatives.

2.3 Environmental Culture Center Chapultepec (Chapultepec Forest)

The Chapultepec Environmental Culture Center is the newest of all ECCs and is located in Chapultepec Forest. The center combines the ecological richness of the oldest urban forest in Latin America with a contemporary cultural and educational proposal. This space promotes interaction between nature, culture, and the city from a humanistic perspective (Secretaría del Medio Ambiente de la Ciudad de México, s.f.). It offers workshops, exhibitions, sensory tours, and artistic events so visitors can experience the forest as a dynamic ecosystem that fosters its care and regeneration.

The center integrates history, art, science, and nature, relying on environmental education as a tool for social and ecological change (Toledo & Barrera-Bassols, 2009). Through workshops, exhibitions, guided tours, and volunteer days, it encourages intergenerational dialogue and knowledge about urban natural spaces. Its programs address topics such as water management, urban conservation, waste reduction, and sustainable lifestyles (Altieri & Nicholls, 2012). It also promotes citizen participation in bird monitoring, the creation of pollinator gardens, and community orchards, strengthening ecological resilience and public use of the forest (IPCC, 2022).

The educational approach is integral and multidisciplinary, encompassing the recovery of cultural heritage and sustainable urban practices. Through art, science, and environmental awareness, it seeks to create a network of citizens committed to protecting natural and cultural heritage, laying the foundation for a more habitable city in harmony with its environment (Martínez-Alier, 2004; Ostrom, 2011).

2.4 Environmental Culture Center Yautlica (Sierra de Santa Catarina, Mexico City)

The Yautlica Environmental Culture Center is located in the Iztapalapa Borough, the most populous in Mexico City. This area faces multiple social challenges that hinder sustainable development, including high poverty rates, low levels of education, and limited and unequal access to public services. Furthermore, Iztapalapa has historically been marked by a social stigma that deepens conditions of exclusion and inequality.

Yautlica is located in a xerophilous scrubland area (Secretaría del Medio Ambiente de la Ciudad de México, 2024) and is surrounded by extinct volcanoes in the Sierra de Santa Catarina. Its name comes from the Nahuatl word Yautlica, meaning *place of fog*. This center is an educational space that seeks to reconnect the community with its local history and natural environment. The Yautlica Environmental Culture Center stands as a beacon for the communities of the Sierra de Santa Catarina, promoting knowledge, connection to the land, and the construction of a fair, inclusive, and resilient environmental culture.

It promotes environmental culture through dialogue, active listening, and the creation of safe and supportive spaces that encourage community action. The center offers children, youth, and families educational experiences on local ecological processes, the identification of native plant and animal species, and the importance of conserving soil and water in the area.

Activities include guided trail walks, urban agriculture workshops, ecological restoration days, and cultural gatherings that highlight local traditions. The center fosters community participation in projects to reclaim public spaces and encourages intergenerational collaboration to transmit knowledge across generations, strengthening social bonds.

Yautlica promotes citizen participation to develop a critical perspective on environmental challenges in eastern Mexico City. The center is a space for environmental education and social transformation, where every activity plants hope for a fair and sustainable future. It promotes a culture that values traditional knowledge and the community's role in caring for the land. Its work is crucial in a marginalized area, driving territorial justice and social regeneration.

3 Community Projects of the Environmental Culture Centers

The Environmental Culture Directorate of the Secretariat of the Environment is responsible for coordinating the educational and cultural environmental strategies promoted in the ECCs. In addition to their vocation and territorial history, projects and programs are implemented in alignment with sectoral and binding programs of Mexico City's environmental policy. For the purposes of this text, we describe three projects that maintain a community essence and align with Mexico City's environmental policy and global challenges. These projects are: (a) promoting agroecological gardens, (b) *Gardens for Life* with native and pollinator-friendly plants, and (c) educational and environmental culture activities in the ECCs (public schedule).

3.1 *My Garden, My Right: Agroecological Gardens in the City*

In Mexico City, where urbanization grows rapidly and green spaces are shrinking, it is urgent to strengthen the connection between new generations and nature. Faced with this challenge, the Utopia School Project, *My Garden, My Right: Planting Life in the City*, promoted by the Environmental Culture Directorate of the Secretariat of the Environment (SEDEMA), proposes an innovative and transformative strategy: the participatory installation of agroecological school gardens in primary and secondary schools. This initiative seeks not only to promote sustainable practices but also to affirm the right of children and adolescents to a healthy and dignified environment, recognizing environmental education as a fundamental tool for imagining and building more just and resilient futures (Secretaría del Medio Ambiente de la Ciudad de México, 2024; UNESCO, 2021).

The creation of agroecological school gardens is much more than an agricultural activity within an educational environment. It is a pedagogical, ecological, political, and poetic proposal that allows awareness to be planted from the roots. Cultivating a school garden means learning as a community, exploring nature's life cycles, understanding the importance of living soil, and discovering how urban biodiversity sustains life even in highly urbanized contexts (Toledo & Barrera-Bassols, 2015). It also strengthens emotional bonds with the earth and contributes to the development of environmentally responsible, participative citizens aware of today's socio-environmental challenges (Sauvé, 2004a).

In this context, school agroecology becomes a powerful tool for social transformation. The COVID-19 pandemic sharply revealed the fragility of food systems and the urgent need to strengthen food sovereignty, especially in urban contexts (Altieri & Nicholls, 2020). School gardens allow students to understand the value of producing food sustainably, learn where their food comes from, and acquire practical life skills. They also become living laboratories where scientific, traditional, and community knowledge converge, generating meaningful, situated learning (Capra & Luisi, 2016; Leff, 2002).

The project plans to start by installing agroecological gardens in at least 20 public schools in the capital, combining teacher training, student participation, technical support, and engagement with families and school communities. Through this strategy, schoolyards become spaces of integral learning where not only food but also emotional bonds, socio-emotional skills, environmental awareness, and a sense of belonging are cultivated (Morin, 1999). Schools thus become vital nodes for ecological and cultural regeneration, where planting is an act of justice and care for present and future generations.

From a rights perspective, this project also aligns with Article 4 of the Mexican Constitution, which recognizes the right of every person to a healthy environment for their development and well-being, and with the Environmental Education for Sustainability Program of Mexico City (PEAS-CDMX), which prioritizes strengthening education for sustainability from an intersectional, participative, and territorial perspective (Gobierno de la Ciudad de México, 2021). It also links with the United Nations 2030 Sustainable Development Goals, particularly Goal 4 (quality education), Goal 11 (sustainable cities and communities), and Goal 13 (climate action) (ONU, 2015).

My Garden, My Right is, above all, an affirmation of life amid concrete. It is the recognition that planting in schools is no small act: it is planting possible futures, climate justice, food autonomy, and environmental awareness. Each seed buried is a lesson of hope, an opportunity to rebuild broken ties with nature, and a tangible way to care for life, starting from childhood.

3.2 Gardens for Life: Sowing the Commons in the Territory

Gardens for Life is one of the central projects promoted by the Secretariat of the Environment of Mexico City and implemented by the Environmental Culture Centers and environmental educators. It is a project that promotes the revegetation of the city through the creation of pollinator gardens with native flora from the Valley of Mexico. Using nature-based solutions, it encourages an educational, ecological, and community approach. Gardens are recognized as important spaces for learning, participation, and regeneration. In this context, gardens with native plants are considered common goods, meaning shared spaces maintained collectively and dedicated to the well-being of all forms of life. The implementation of this project allows environmental educators to work directly in neighborhoods, schools, collectives, and public spaces. Moving away from the traditional classroom, learning takes place in the environments where we live: plazas, medians, schoolyards, street markets, housing complexes, parks, and sidewalks all become places to share knowledge.

Gardens for Life is carried out in coordination with organizations such as CORENA (Commission of Natural Resources and Rural Development) and CONABIO (National Commission for the Knowledge and Use of Biodiversity), agencies dedicated to promoting biodiversity conservation and sustainable use in Mexico. These institutions provide technical support, training, and native plant species adapted to the ecological conditions of each area. The project focuses on using native plants, promoting 23 species that are fundamental for conserving and restoring urban biodiversity. These species are adapted to local climate, soil, and ecological dynamics, making them strategic allies against climate emergency impacts and biodiversity loss. They sustain complex ecological relationships: they provide food and shelter for pollinators, birds, and insects; help regulate microclimates; favor water infiltration; and strengthen ecological connectivity between urban nature fragments (Martínez-Camacho et al., 2021). These gardens are also educational and cultural spaces: they allow children, teachers, and neighbors to connect with local ecological knowledge, the rhythm of the seasons, manual labor, and the botanical memory of local communities. They are also gardens of words, where stories are told, ancient plant names are recovered, identities are sown, and emotional bonds with the territory are cultivated.

The proposal includes several garden formats: school gardens, community gardens, neighborhood plots, in public or private spaces, transforming them into micro-reserves of life and collective appropriation. The aim is not to impose a single design but to adapt each garden to the specific cultural, ecological, and social context of each community, recognizing its diversity and inherent flexibility.

These gardens are common goods with political and educational dimensions: they promote the right to a healthy environment, build the city from the ground up, and foster local governance based on care, reciprocity, and participation. They uphold the commons, regenerate the territory, and plant seeds for possible shared futures.

3.3 Environmental Education and Culture Activities in the ECCs (Citizen Program)

The Environmental Culture Centers of SEDEMA have become dynamic spaces for learning, encounters, and connection with nature. Their seasonal programming includes immersive experiences and participatory workshops that bring citizens closer to ecosystems, traditional knowledge, and biocultural environmental practices. Among the most notable activities are interpretive tours focused on bats, amphibians, and wild bees, designed to highlight the biological diversity of the region and the challenges it faces in conservation (Toledo & Barrera-Bassols, 2009).

Camps dedicated to studying bats and fireflies offer unique nighttime experiences, allowing participants to understand natural cycles and emphasizing the importance of preserving healthy natural environments free from light pollution (Meira, 2013). Likewise, creative environmental visualization workshops encourage children and young people to imagine possible futures from artistic, technological, and ecological perspectives (Jickling & Sterling, 2017). Programming also includes activities centered around the nurseries and school gardens of the ECCs as agroecological solutions for pest management, where sustainable alternatives are shared to strengthen food sovereignty and soil health, based on collective knowledge and community experimentation (Alcántara Nieves & Larroa Torres, 2022; Altieri & Nicholls, 2020). Courses and workshops are also offered on mushroom cultivation, terrariums, plant pressing, medicinal plants, urban trees, hummingbirds, and more.

Finally, every month the Environmental Culture Directorate and the ECCs organize educational activities based on the natural and biocultural contexts of Mexico City. A highlighted activity is *Bat Night*, a nighttime tour to learn about this species essential to urban ecosystems, promoting its observation and protection. Another important activity is the *Axolotl Route* at the Acuexcomatl ECC, where participants explore the Xochimilco wetlands and learn about the conservation of the axolotl, an endemic species and symbol of resilience.

4 Community Experiences that Transform: Social Innovation in the Environmental Culture Centers of Mexico City

4.1 Exemplary Intergenerational Participation: The Impact of the "Ave Fénix" Collective and Scout Children on Environmental Culture

Throughout their history, the centers have functioned as catalysts for community and various groups. The experiences have been diverse: from short visits by schools or companies to sustained processes of cohesion and collective strengthening. A key element of the ECCs is their ability to engage a diverse audience, from children to the elderly, which gives their intergenerational reach great relevance in times of climate emergency and intergenerational challenges.

Children and older adults are among the most vulnerable to climate change and biodiversity loss (Rascón Villanueva & Cervantes Rendón, 2022), making their participation in socio-environmental issues not just necessary but essential. The ECCs have developed successful and representative community experiences with these groups. At the Acuexcomatl Environmental Culture Center, there is a well-established group of older women known as the *Ave Fénix* collective, and another group formed by children from a local Scout group. The women's group began with 15 participants and, over time, grew to 192 women. What started as occasional workshops evolved into a meaningful experience. These women, with traditional knowledge about medicinal plants, water saving, and food preservation, found in the center a place to share their wisdom and learn new practices related to recycling, composting, biodiversity protection, renewable energy, gender, entrepreneurship, and mental health.

Community garden work not only allowed them to reconnect with the earth but also to strengthen ties among themselves, regain a sense of purpose, and build emotional and even historical support networks in a region of high biocultural value. By participating in awareness campaigns and activities with schools and families, they became local role models, passing on environmental values to younger generations and fostering intergenerational connections from a local perspective.

Including older women in environmental culture activities not only combats isolation and strengthens their self-esteem by giving them an active community role but also harnesses their time and willingness to drive sustainable projects. Their participation positively impacts their families, promoting responsible practices at home, and improves their physical, emotional, and social health. Recognizing their contributions promotes intergenerational justice, valuing their experience in building a sustainable future. The *Ave Fénix* collective's experience, growing from 15 to 192 members aged 65 to 90, demonstrates that their inclusion brings not only ecological but also social and emotional benefits, making them key agents in fostering more just and resilient communities.

4.2 *Chalchiuhtlicue Collective: Women Guardians of Territory and Biodiversity*

The Chalchiuhtlicue Collective emerged thanks to the initiative of a group of women from the San Miguel Teotongo neighborhood, near the Yautlica Environmental Culture Center. Starting in 2019, they began building a network of community work focused on agriculture and environmental protection. What initially started as a desire to share knowledge gradually transformed into a living environmental education project with a gender perspective, rooted in local ecological knowledge and the recovery of the human-nature bond.

In 2020, the collective was recognized by Cornell University's Lab of Ornithology, receiving a symbolic grant through a call dedicated to Latin America and the Caribbean. This opportunity marked the beginning of a project that combines bird observation and appreciation with revaluing the natural and cultural heritage of eastern Mexico City, particularly Iztapalapa. Due to pandemic restrictions, the first activities were held virtually. When public spaces began reopening safely, the ECC Yautlica became a refuge: a place of trust, safety, and freedom for the women of the collective. Here, they could gather without fear, in conditions of respect and sisterhood, to walk, observe birds, learn together, and build community around nature. These experiences led to deep reflections. Members of the Chalchiuhtlicue Collective highlighted the lack of safe, harassment-free spaces for women to study, enjoy, and work in nature, particularly in urban contexts marked by violence. They raised their voices to emphasize the urgency of developing environmental education programs that incorporate care, sisterhood, and emotional well-being as essential elements of fieldwork. During this process, they also built connections with other women's collectives, such as *Amazilias*, a network of female birdwatchers promoting gender-sensitive field practices. These exchanges strengthened the collective's work, enriching it with new perspectives, partnerships, and methodologies sensitive to both territory and women's experiences.

With bird monitoring efforts and the collaboration of 298 identifiers, they created a database in *iNaturalist* containing 92 species and 1689 observations of wild birds at ECC Yautlica. This information updates biological records and highlights the diversity of birdlife in the area compared to other urban zones.

The work of the Chalchiuhtlicue Collective demonstrates that when safe, respectful, and supportive spaces are available for women in nature, new forms of knowledge, care, and community can flourish. Their experience is living proof of how environmental education and culture can become transformative practices when they arise from the territory, collective action, and active listening to its inhabitants.

4.3 Urban Evolution Collective: Community, Territory, and Socio-environmental Defense

The *Urban Evolution Collective* was born as a citizen initiative in the Sierra de Santa Catarina, in the eastern area of Mexico City. In this region, communities such as Potrero, El Paraíso, and La Poblanita face challenges derived from their position on the boundary between the boroughs of Iztapalapa and Tláhuac, resulting in institutional gaps and problems with access to services and safety. In this context, community organization has been key to building networks that improve quality of life and defend collective well-being. Within this process, the Yautlica ECC has become an essential ally. Its proximity and openness to the community have turned it into a meeting point for reflection, shared learning, and collective action. Thanks to this connection, educational and cultural environmental activities have been strengthened, combining local knowledge, territorial activism, and ecological regeneration.

Several initiatives have been promoted, including *I Defend the Sierra de Santa Catarina*, which encourages walks, environmental activism, and the creation of a native plant guide; collaboration with *TryCities* to foster youth participation in peripheral areas; and work with *Colectiva La Trashumante*, focused on food, agroecology, and folk arts. In all these efforts, ECC Yautlica has provided space, safety, technical advice, and educational support. These processes have not only advanced environmental care; they have also strengthened pride in the territory, neighborhood identity, and a sense of belonging. Children, youth, women, and residents of all ages have actively participated in reforestation days, clean-up campaigns, environmental monitoring, and training sessions that renew their connection with the land.

Urban environmental problems are not limited to pollution; they stem mainly from the deterioration of the social fabric that sustains life in cities. In peripheral areas, where roots to the territory are often recent, it is younger generations who are weaving that fabric, building community through environmental awareness and collective organization. The experience of the *Urban Evolution Collective* shows that community action, when recognized, supported, and strengthened, has the power to transform vulnerable territories into spaces of dignity, ecological justice, and shared hope. From the ECC Yautlica, we celebrate and thank these communities for trusting us to walk together toward a fairer, more sustainable, and deeply human future.

4.4 Seeds of the Future: Children and Their Transformative Role in the ECC

Children's participation in the ECC has also been highly significant. What began 3 years ago (2022) as a simple collaboration involving 10 scouts and the ECC has progressively evolved into a solid community work program that currently involves more than 600 scouts from four provinces. Initially, the scouts participated sporadically in specific activities, such as clean-up days, workshops, and environmental actions. Over time, this evolved into a deeper, sustained collaboration with the ECC. Recognizing shared values and goals, the need arose organically to formalize this relationship through a joint work program with long-term planning, assigned responsibilities, and an annual activity calendar. From occasional visitors, the scouts became active allies in key projects such as reforestation, maintenance of urban gardens, educational workshops, and environmental awareness campaigns.

After 3 years of continuous engagement, the experience has been transformative for the members of the four scout provinces involved—not only in terms of environmental learning but also in collective work, perseverance, and understanding the real impact that intergenerational and community collaboration can achieve. One of the most significant lessons has been understanding that real change does not happen instantly but is built through constant efforts and small actions that, combined, lead to tangible results. This journey has left a lasting mark on every participant, reaffirming their vocation as agents of change committed to building a more just, supportive, and sustainable world.

5 Conclusions

The Environmental Culture Centers (ECCs) of SEDEMA have become essential spaces for socio-environmental transformation in Mexico City. Their sustained work—delivered through continuous activities, innovative pedagogical approaches, and a strong connection with local ecosystems—embodies a living model of critical environmental education, community action, and territorial stewardship.

Over recent years, the ECCs have built a network of meaningful learning experiences for thousands of participants: children, youth, teachers, public officials, families, organized communities, and citizens in general. Their monthly, diverse, sensitive, and context-specific programs maintain an active link between society and nature, fostering ongoing processes of environmental awareness, civic participation, and ecological commitment. Among their most emblematic initiatives are interpretive walks, Bat Nights, the Axolotl Route, agroecology workshops, thematic camps, art and environmental visualization experiences, and training courses for environmental educators. Each of these opens pathways for understanding the territory through an integrated lens, one that acknowledges the interdependence of species, knowledge systems, and landscapes.

Programs such as *Gardens for Life* have expanded this vision to public, school, and community spaces, connecting ecological knowledge with urban restoration, collective care, and education for peace. This nature-based solution approach also carries a biocultural dimension: it values both biodiversity and cultural diversity, expressed through inter-institutional agreements, collaborative training processes, and socio-environmental governance strategies actively promoted by the ECCs.

The ECCs are not merely physical venues; they are living nodes where relationships are nurtured, knowledge is planted, species are protected, and possible futures are collectively built. They represent a determined commitment to a fairer, more resilient city that cherishes life, where every daily action—whether a walk, a planting, or a conversation—becomes an opportunity to reconnect with the Earth.

In times marked by climate emergency, biodiversity loss, social fragmentation, and weakening community bonds, the ECCs offer a tangible alternative: learning alongside nature, rooted in the territory and within the community. Their experience reinforces the understanding that environmental education is not a complementary activity but a central axis for imagining and shaping new ways of inhabiting the planet.

References

Alcántara Nieves, N., & Larroa Torres, R. M. (2022). La multifuncionalidad de los huertos urbanos en la Ciudad de México. *Espiral (Guadalajara), 29*(83), 187–229. http://www.scielo.org.mx/scielo.php?script=sci_arttext&pid=S1665-05652022000100187&lng=es&tlng=es

Altieri, M. A., & Nicholls, C. I. (2012). Agroecology scaling up for food sovereignty and resiliency. In *Sustainable agriculture reviews* (pp. 1–29). https://doi.org/10.1007/978-94-007-5449-2_1

Altieri, M. A., & Nicholls, C. I. (2020). Agroecología y resiliencia al cambio climático: principios y prácticas [Agroecology and resilience to climate change: principles and practices]. *Agroecología, 8*(1), 7–20. https://revistas.um.es/agroecologia/article/view/182921

Capra, F., & Luisi, P. L. (2016). *The system's view of life: A unifying vision.* Cambridge University Press.

Freire, P. (1998). *Pedagogía del oprimido.* Siglo XXI Editores.

Gobierno de la Ciudad de México. (2021). *Programa de Educación Ambiental para la Sustentabilidad de la Ciudad de México (PEAS-CDMX) 2021–2024.* Secretaría del Medio Ambiente de la Ciudad de México.

González-Gaudiano, E. (2022). *Educación ambiental crítica: Contribuciones para una transformación ecosocial.* Fondo de Cultura Económica.

IPCC. (2022). *Sixth assessment report.* Intergovernmental Panel on Climate Change. https://www.ipcc.ch/ar6/

Jickling, B., & Sterling, S. (2017). *Post-sustainability and environmental education: Remaking education for the future.* Palgrave Macmillan.

Kolb, D. A. (1984). *Experiential learning: Experience as the source of learning and development.* Prentice Hall.

Leff, E. (2002). *Saber ambiental: sustentabilidad, racionalidad, complejidad, poder.* Siglo XXI Editores.

Leff, E. (2020). *Ecología política: epistemología, racionalidad ambiental y modernidad.* Siglo XXI Editores.

Martínez Ordoñez, L. M., Martínez Ordoñez, M. P., Auquilla Andrade, E. G., Miranda Armijos, W. A., & Riera Palacios, G. M. (2024). Etnoeducación y Saberes Ancestrales: Preservación Cultural a Través del Sistema Educativo En Ecuador. *Ciencia Latina Revista Científica Multidisciplinar, 8*(5), 8437–8456. https://doi.org/10.37811/cl_rcm.v8i5.14250

Martínez-Alier, J. (2004). *El ecologismo de los pobres: Conflictos ambientales y lenguajes de valoración.* Icaria Editorial.

Martínez-Camacho, E., Rivera-Cantú, A., & Ramírez-Corzo, N. (2021). Jardines para la biodiversidad urbana: Diseño y restauración con flora nativa. *Revista Mexicana de Biodiversidad, 92*, e923184. https://doi.org/10.22201/ib.20078706e.2021.92.3184

Meira, P. (2013). Elogio de la educación ambiental. *Revista Iberoamericana de Educación, 61*(1), 7–18. https://doi.org/10.35362/rie6112013

Morin, E. (1999). *Los siete saberes necesarios para la educación del futuro*. UNESCO.

ONU. (2015). *Transformar nuestro mundo: la Agenda 2030 para el Desarrollo Sostenible*. Naciones Unidas.

Ostrom, E. (2011). *El gobierno de los bienes comunes: La evolución de las instituciones de acción colectiva*. Fondo de Cultura Económica.

Rascón Villanueva, A. E., & Cervantes Rendón, E. (2022). Vulnerabilidad social y clima extremo en estudios de América Latina. 2000-2019. *Tlalli. Revista De Investigación En Geografía, 8*, 6–32. https://doi.org/10.22201/ffyl.26832275e.2022.8.1801

Rockström, J., Steffen, W., Noone, K., Persson, Å., Chapin, F. S., III, Lambin, E. F., Scheffer, M., Folke, C., Schellnhuber, H. J., Nykvist, B., de Wit, C. A., Hughes, T., van der Leeuw, S., Rodhe, H., Sörlin, S., Snyder, P. K., Costanza, R., Svedin, U., Falkenmark, M., & Foley, J. A. (2023). Planetary boundaries: Exploring the safe operating space for humanity. *Ecology and Society, 14*(2), 32. https://doi.org/10.5751/ES-03180-140232

Ross, F. (2019). Kate Raworth—Doughnut economics: Seven ways to think like a 21st century economist (2017). *Regional And Business Studies, 11*(2). https://doi.org/10.33568/rbs.2409

Sauvé, L. (2004a). Una cartografía de las corrientes en educación ambiental. *Revista de Educación Ambiental, 4*(2), 7–34. https://laboratoriogrecia.cl/wp-content/uploads/2016/04/Lucia-SauveEAM3.pdf

Sauvé, L. (2004b). Una cartografía de corrientes en educación ambiental. En M. Sato e I. Carvalho (Orgs.), A pesquisa em educação ambiental: Cartografias de uma identidade narrativa em formação. Artmed.

Secretaría del Medio Ambiente de la Ciudad de México. (2024). Índice de biodiversidad urbana (IBU) 2024 [PDF]. Gobierno de la Ciudad de México. Recuperado el 3 de agosto de 2025, de https://www.sedema.cdmx.gob.mx/storage/app/media/DGCPCA/ibu/IBU2024.pdf

Secretaría del Medio Ambiente de la Ciudad de México. (s. f.). Conoce el Centro de Cultura Ambiental Chapultepec. Gobierno de la ciudad de méxico. Recuperado el 6 de agosto de 2025, de https://www.sedema.cdmx.gob.mx/comunicacion/nota/conoce-centro-cultura-ambiental-chapultepec

Toledo, V. M., & Barrera-Bassols, N. (2009). Saberes ambientales y sustentabilidad. *Revista Iberoamericana de Educación, 49*(3), 1–12. https://doi.org/10.35362/rie4931926

Toledo, V. M., & Barrera-Bassols, N. (2015). *Agroecología y saber local: Hacia una nueva racionalidad*. Siglo XXI Editores.

UNESCO. (2021). *Reimaginar juntos nuestros futuros: Un nuevo contrato social para la educación*. UNESCO. Recuperado el 5 de agosto de 2025, de https://unesdoc.unesco.org/ark:/48223/pf0000379707

Wake, D. B. (2008). Flying an amphibian flagship for conservation. *Conservation Biology, 22*(5), 1405–1407. https://doi.org/10.1111/j.1748-1090.2008.00044.x

Youth Advisory Climate Councils: The Case of Galiza (Spain)

Xabier Riádigos-Couso, Antonio García-Vinuesa, Rita Gradaílle-Pernas, and Inés Merayo-Fernández

Abstract This chapter explores the experience of the Youth Advisory Climate Councils in Galiza (Spain), established within the framework of the European project *People & Planet: A Common Destiny* (2020–2024). The initiative aimed to foster youth participation in local decision-making processes related to sustainability and climate change. Through participatory methodologies, the Youth Advisory Climate Council enabled young people aged 15–35 to design and implement concrete actions, including awareness campaigns, community-based initiatives, and policy proposals addressed to local governments. Drawing on documentary analysis and semi-structured interviews with project coordinators and facilitators, the study reveals positive impacts on youth participants, facilitators, and municipal stakeholders. However, it also identifies challenges such as generational diversity, maintaining long-term engagement, limited resources, and the need for greater institutional coordination. Despite these obstacles, the Youth Advisory Climate Council are presented as innovative and disruptive socio-educational structures that promote youth empowerment and facilitate their integration into local governance. The chapter concludes by emphasizing the importance of sustained institutional commitment to ensure the continuity of such initiatives as effective responses to the climate emergency. It also highlights the need for future research incorporating youth perspectives and insights from other project partners.

1 Introduction

Anthropogenic Climate Change (CC) constitutes a global challenge, the impacts and manifestations of which are increasingly visible and acutely experienced at the local level. Scientific evidence demonstrates that the adverse effects of CC will disproportionately affect younger generations, particularly those living in contexts

X. Riádigos-Couso (✉) · A. García-Vinuesa · R. Gradaílle-Pernas · I. Merayo-Fernández
Social Pedagogy and Environmental Education Research Group (SEPA-interea), Universidade de Santiago de Compostela, Santiago de Compostela, Spain
e-mail: xabier.riadigos.couso@usc.es

J. J. Conde et al. (eds.), *Practices in Education for Climate Emergency*,
https://doi.org/10.1007/978-981-92-2304-6_21

of risk and social, economic, cultural, or geographical vulnerability (Save the Children, 2021; Thiery et al., 2021; United Nations Children's Fund, 2023). Indeed, this demographic perceives it as one of the most pressing socio-environmental issues (Dunne & Bijwaard, 2021).

The United Nations Committee on the Rights of the Child, in General Comment No. 26, recognises the right of children to a clean, healthy and sustainable environment. It also highlights that the intensifying and prolonged negative effects of CC seriously undermine the fulfilment and protection of these rights. In response, the Committee urges governments to fulfil their obligations by adopting climate emergency policies that ensure the protection of children's rights.

Among the measures recommended to States for safeguarding children's rights in the face of CC threats is the need to include children and adolescents in decision-making processes, and to develop policies and actions aimed at climate adaptation and mitigation. This perspective underscores the significance of children's and young people's agency and participation in matters related to the climate emergency.

The urgency and severity of the crisis is not lost on the most vulnerable —and least responsible— populations, notably indigenous communities and young people (Prosser et al., 2020), who demand to be actively involved in the socio-cultural transformations required. With regard to youth movements, the *Fridays For Future* initiative has become a global point of reference, partly due to generationally relatable public figures such as Greta Thunberg. These individuals have inspired young people from diverse regions, morally committed to addressing the climate emergency (Misch et al., 2021), to participate in the design and implementation of climate action strategies and policies (Parsons et al., 2023). This movement represents a paradigmatic shift that challenges adult-centric notions of environmental ideologies and worldviews, encouraging everyday activism and sustainable lifestyles (Firinci, 2022).

The prominence of these global, counter-hegemonic youth movements has served as an international catalyst, despite criticism from various media outlets seeking to delegitimise them on the basis of age (Bergmann & Ossewaarde, 2020). Their achievements contribute to the recognition of children and young people as human rights defenders and as creators of alternative, dissident futures (Escorial et al., 2022; Valdivia et al., 2024).

Among the many actions and initiatives developed by children and young people, this chapter focuses on the creation and operation of local participatory spaces that have become key tools for youth involvement in decision-making processes. In particular, it explores Youth Advisory Climate Councils (YACCs) in Galiza as disruptive socio-educational initiatives seeking to respond to the climate emergency. As stated in the *Young European Citizens Declaration* (People & Planet: A Common Destiny, 2023), Youth Advisory Climate Councils are structured groups that aim to engage European youth in promoting sustainable development at various territorial levels (local, national, and global), with a particular emphasis on climate action and water scarcity.

To examine the creation of these councils, understand their functioning, and assess their scope and achievements, this study draws on two primary sources of information:

1. *Documentary sources.* Institutional documentation from the European project *People & Planet: A Common Destiny* and from the Galician Fund for Cooperation and Solidarity (FGCS), the body responsible for coordinating the project in Spain, has been consulted. In addition, materials, outputs and initiatives generated by participants of four YACCs in Galicia, as well as those by partnering social entities over the project's 4 year development period, were recorded and analysed (see Table 1).
2. *Primary sources.* Data from a semi-structured group interview were also recorded and analysed. Two key informants participated, both of whom were significantly involved in the coordination and implementation of the YACCs: the coordinator of the Galician Fund for Cooperation and Solidarity and an educator from the Dalle Que. Dalle Cooperative Society (a Spanish institutional equivalent to an NGO), the latter contracted as a professional facilitator for the YACCs in Galiza (Spain).

2 Youth Participation as a Socio-Educational Response to the Climate Emergency

The climate emergency has emerged as a scientific, political, and social reality over the past decade. Within the scientific community, more than 15,000 scientists from 158 countries have signed a climate emergency declaration, warning of the risks posed to humanity, primarily as a result of anthropogenic disruption to the climate system (Ripple et al., 2020). At the societal level, the *Climate Emergency Declaration* initiative (https://climateemergencydeclaration.org/) reports that over 2366 jurisdictions worldwide —including the Spanish Government (2020)— have declared a climate emergency, representing over one billion people. In part, these declarations respond to the fact that attention to the risks and threats associated with the emergency has largely focused on global and regional responses, underestimating the potential for local action (Salvia et al., 2023).

In Spain, the declaration encompasses various areas of social, educational, industrial, cultural, and economic governance, among other organisational aspects of the country. It proposes action lines and recommendations aimed at the decarbonisation of both the economy and society (Spanish Government, 2020). Within this framework, education is positioned as a key pillar in the transition towards a decarbonised and just society, highlighting the importance of full civic participation, as stated in point 2:

> Strengthen existing participatory mechanisms and structurally ensure citizen engagement in climate change decision-making through the establishment of a Citizens' Climate Assembly,

Table 1 Project partners and countries

Country	Organisations and/or Local Authorities
Cape Verde	Municipality of Maio
Germany	City of Baden-Wurttemberg
	FINEP
	Municipality of Dornstadt
Ireland	SLÌ - Waterford Sustainable Living
Italy	FELCOS Umbria
	WeWorld
Netherlands	LBSNN - National Town-Twinning Netherlands-Nicaragua
	Municipality of Maastricht
	Municipality of Zoetermeer
Poland	Buy Responsibly Foundation
Portugal	Conexão Lusófona – Youth association
	FUEL Publicidade
	IMVF - Instituto Marquês de Valle Flôr
	Municipality of Loures
	RICD - Rede Intermunicipal Cooperação para o Desenvolvimento
Romania	APSD – Agenda 21
	Brasov County
Spain	Fondo Galego de Cooperación e Solidariedade (FGCS)

Note Information extracted from website of people & planet (https://dearprogramme.eu/project/people-planet-2/)

> whose composition will reflect the principle of gender balance and include the participation of young people. (p. 5).

This point addresses several priority lines of action: integrating climate emergency considerations across all public policies, including those at the local level; reinforcing synergies with rural development strategies; promoting circular economy initiatives and waste legislation; and fostering a culture of consumption aligned with the challenges posed by the climate crisis. Likewise, citizen participation is emphasised as key to identifying vulnerable groups and avoiding social inequalities, ensuring collective decision-making and equal conditions across all sectors of the population.

In this sense, child and youth participation constitutes an essential tool for achieving the proposed goals, since not only contributes to giving young generations a voice but also equips them with the necessary skills for negotiation and informed decision-making. As such, they become vital and meaningful agents in the transformation of their own communities, since —in the face of the risks of an unstable climate— their responses provide added valuable contributions.

3 People & Planet: A Shared Future in a European Project

In 2020, within the framework of the 2030 Agenda, the European initiative *People & Planet: A Common Destiny* (hereafter *People & Planet*) was launched with a dual purpose: to raise awareness among European youth about the severe consequences of climate change, enhancing their agency to drive transformations towards a more sustainable world; and to encourage municipal governments to promote public policies aimed at sustainable development through a *glocal* (global and local) perspective.

The project—funded by the European Commission's Development Education and Awareness Raising (DEAR) programme— was structured around two converging lines of action aimed at supporting the transition towards decarbonised societies (Spanish Government, 2020). The first line consisted of a series of activities targeted at young citizens, providing them with digital tools to enable them —as future agents of change— to communicate and raise awareness among their peers. The second line focused on capacity-building and skills development for local decision-makers, with the aim of empowering local authorities to reinforce their role in promoting good practices and sustainable lifestyles within their communities.

During the implementation period (2020–2024), a total of 19 organisations participated in the project, representing eight EU Member States (Germany, Italy, Ireland, the Netherlands, Poland, Portugal, Romania, and Spain) as well as Cabo Verde (see Table 1). The initiative was coordinated by the Câmara Municipal de Loures (Portugal).

The 2022 global campaign #WaterOfTheFuture, under the slogan "We Are Not Camels", stands out for its relevance and impact. It was created with a dual objective: to raise awareness among European youth citizens about water scarcity, and to equip them with tools to change their everyday behaviours and address this issue within local, national, and global public agendas. The campaign generated a range of activities aligned with the Sustainable Development Goals (SDGs), including street actions, advisory councils, activism labs, surveys, and exchange programmes, among other initiatives[1] (People & Planet: A Common Destiny, 2024).

In the report produced by the People & Planet project partners, entitled *Acting locally, impacting globally – Compendium of good practices for promoting partnerships between youth and local authorities for climate action and sustainability* (People & Planet: A Common Destiny, 2024), Youth Advisory Climate Councils were identified as essential local structures to advance the following goals:

1. To meaningfully engage young people in climate action.
2. To raise awareness, foster understanding, and enrich youth knowledge about climate issues and sustainable lifestyles.
3. To provide platforms for the direct participation of children and young people in policy design and decision-making through local co-governance.

[1] Accessed February 24, 2025, from https://www.linktr.ee/waterofthefuture

4. To offer resources, ideas, and practical materials on climate change and sustainability to various municipal socio-educational stakeholders, including local authorities, associations, and civil society organisations committed to promoting sustainability at the local level.

4 Youth Advisory Climate Councils of Galiza

In the Spanish state, the People & Planet project was implemented exclusively in the region of Galicia, coordinated by the Fondo Galego de Cooperación e Solidariedade (FGCS). The FGCS is a non-profit association established in 1997, composed of Galician local authorities. It currently includes 114 municipalities and four provincial councils. Its core mission is to promote international cooperation from a local perspective, through projects in various countries —primarily within the Lusophone sphere— that respond to the needs of its partners (municipalities, associations, or other local entities). These initiatives cover areas such as water supply, gender equity, education, and health, among others. The FGCS also engages in humanitarian aid, supports Galician NGOs working abroad, and conducts global citizenship education and awareness-raising activities for its member municipalities.

To achieve the goals of *People & Planet* effectively, the FGCS adopted an innovative and diverse socio-educational approach, delivering more than 70 activities aligned with three main areas: climate action, artistic creation (*artivism*), and education for global citizenship. Notable initiatives included performance workshops, guerrilla communication, sustainable cooking, and urban sketching for the revalorisation of natural spaces; training sessions on the 2030 Agenda; exchange programmes with political and technical staff from FGCS member administrations; as well as audiovisual competitions, virtual exhibitions, and awareness-raising mural projects in eight Galician municipalities focused on SDG 6 (Clean Water and Sanitation) and SDG 13 (Climate Action). Among all these initiatives, this study focuses on the creation of Youth Advisory Climate Councils (YACCs) as a disruptive, timely, and necessary socio-educational strategy to promote democratic and equitable responses to the climate emergency.

The Youth Advisory Climate Councils (YACC) represent one of the core activities proposed within the People & Planet project. The FGCS selected four member municipalities to establish these Councils in Galicia: A Guarda, As Neves, Oleiros, and Teo (see Fig. 1). This selection was based on several criteria, including each municipality's level of engagement with the FGCS (e.g. participation in assemblies, timely submission of required documentation), their suitability for the proposed activity, and their existing work in promoting youth participation. Each YACC was directly linked to the municipal departments responsible for social services, education, or youth affairs.

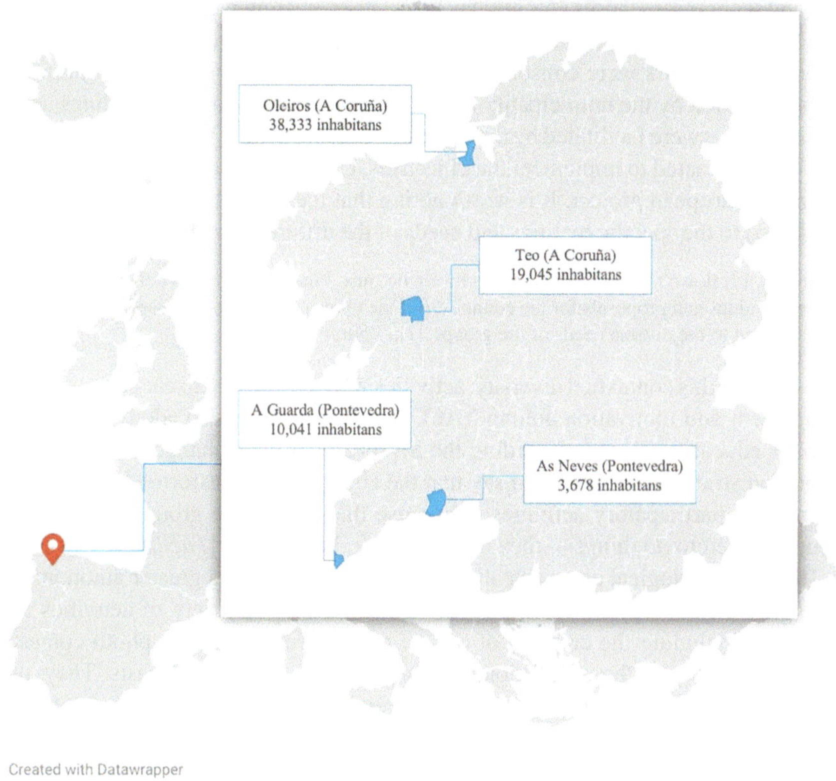

Fig. 1 Municipalities participating in the people & planet in Galiza

4.1 Youth Advisory Climate Councils: Foundations and Actions

In June 2021,[2] the four Youth Advisory Climate Councils (YACCs) were established as participatory bodies composed of 25 young people aged between 15 and 35.[3] However, until October 2024, other members joined intermittently. The process for recruiting young people was varied, with each municipality employing different outreach strategies: from posters in educational centres (As Neves, Teo), announcements via municipal social media such as Instagram, Facebook or the Municipal Youth Information Office (A Guarda and Teo), to public presentations of the project at youth houses and fairs (Oleiros), and promotional activities at entrepreneurship

[2] The first year was dedicated to the implementation of the People & Planet project, the development of other common activities, and the tendering process for the facilitation of the YACC in Galicia.

[3] This criterion was adapted on different occasions, respecting the interest expressed by various young people to be part of the YACC (different age than the one stipulated or residing in areas bordering the selected municipalities).

fairs and secondary schools (A Guarda, Teo). Other existing participatory structures, such as UNICEF's local councils for children and adolescents (Teo), were also used.

Once the YACCs were consolidated, monthly 1 h meetings were held in public spaces provided by the municipality (e.g. Youth Houses, Town Hall buildings, etc.). These sessions were facilitated by the educator from the *Dalle Que Dalle* cooperative, who was contracted to implement the objectives, educational content, and actions set out in the European project. It is worth noting that the original proposals needed to be adapted to the specific realities and needs of the different groups in each territory:

> At first, I followed the handbook quite closely because I had a contract to deliver training, create guides and proposals for the councils. But the plan ended up being tossed out and we readjusted to the diverse needs of the groups. [*Facilitator*]

To respond to this contextual diversity, activities were redesigned to encourage greater engagement and motivation among YACC members, while also addressing young people's educational needs regarding the SDGs and climate change mitigation and adaptation strategies: "You could say that the content was transformed, mostly into other, more participatory activities —because that's what the groups wanted. The groups wanted to *do* things— they didn't want to just listen." [*Facilitator*].

This methodological approach allowed the YACC members greater autonomy in designing, organising, implementing and disseminating a variety of activities and initiatives, including the creation of escape rooms, treasure hunts, photo contests, clean-up events, and the production of regular climate-themed podcasts. They also held meetings with municipal environmental representatives to discuss sustainability initiatives implemented by local governments and climate mitigation and adaptation measures developed by the YACCs:

> A great added value of these activities was that they were proposed by the participants themselves or tailored to their own image and preferences. They ended up leading and running them. This was very positively received wherever we presented the initiatives. [*Coordinator*]

To develop and deliver these activities, logistical support was often provided by the local councils, thus facilitating communication between young people and local government: "In most cases, when possible, the councils supported the logistics, and we managed to establish the kind of contact between youth and the municipality we were aiming for — even if it wasn't as strong as we'd hoped" [*Coordinator*]. To encourage internal communication among participants, internal consultations were held in each YACC through WhatsApp groups. However, as these groups included municipal representatives, communication was not as fluid as expected:

> If the WhatsApp group is very serious and only serious topics are discussed, the young people disengage. I think it's fine for silence, for archiving, and they open it when they're tagged, when someone @mentions them, and it buzzes, because otherwise, nothing happens. And if it has a more 'hihi-haha' theme, it's the girls who disconnect, because they are working, and amidst all the WhatsApp messages, they get lost and then tell me: 'Can you message me when there's a meeting or when I need to do something? Because with so many WhatsApp messages, I got lost.' So, communication doesn't end up being as smooth as it should be. [*Facilitator*]

With regard to external communication, difficulties were also noted in building stronger links between the councils and young people. A lack of effective communication and dissemination channels was cited: “We’ve discussed this before — the communication channels between the council and young people either don’t work well or aren’t the right ones. I don’t really have a solution either, but we’ve clearly identified the problem” [*Facilitator*]. This issue was compounded by young people’s disinterest in official institutional social media, which led to a lack of awareness of municipal activities: “Why should I follow the council’s Instagram? Politics isn’t really my thing. It’s like a stain on their cool Instagram feed full of top people… It doesn’t fit, so they miss out on activities” [*Facilitator*]. To help bridge this gap, the YACC facilitator served as a link, providing two-way communication between the councils and the municipalities: “When it was the young people leading the activities, the facilitator would act as a bridge, explaining to the council what the event was about, when it would take place, sharing the poster, checking if anything was needed...” [*Coordinator*]. The Galician Fund for Cooperation and Solidarity (FGCS) also helped promote YACC actions and activities through its own social media channels and press releases:

> At the Galician Fund, we shared everything the facilitator sent us. When there was a trip, for example, we wrote a press release — and the municipalities loved it, they liked to show off and say: ‘A girl from our town is presenting in Brussels…’ and they’d share it too. [*Coordinator*]

One key aspect in the creation and functioning of the YACCs was the participatory evaluation process. These were designed to improve the internal dynamics of the advisory councils and support the educational goals of the project. Beyond the formal evaluations required by the project (feedback surveys, session assessments, attendance tracking), the young people gave the facilitator open and honest feedback about the sessions, activities and actions, often suggesting alternative approaches to improve participation, learning, involvement, and group cohesion:

> They would just tell me straight up — no filter (laughs). If they didn’t like something, or if they loved it or thought it was awful, they’d say so. There was no filter because they got on well with me and they’d say: ‘We really don’t feel like doing this’ or ‘Why are we indoors on a day like this?’ That taught me to always bring backup options — sometimes I had three different activities ready in my pocket. [*Facilitator*]

4.2 *Key Actions of the Youth Advisory Climate Councils*

Since the establishment of the YACCs, actions have been designed, planned, and executed by their members, addressing their interests, concerns, and local realities. The following actions stand out:

- Beach and green space clean-ups.
- Creation of murals and awareness-raising posters.
- Participation in youth and volunteering fairs.

- Production of awareness-raising videos, management, and posting on social media.
- Screening of the documentary "*A auga que trouxeron*".
- Request for a water fountain at an educational centre.
- (In)formative meetings of the councils at the regional level (to meet and engage in topics such as the creation of an environmental association, consumerism and the textile industry, as well as talks with "*Ecologistas en Acción*").
- Design of a guide on good practices for individual and collective environmentally respectful lifestyles, presented at the Lisbon meeting in June 2024.
- Development of climate action proposals for local governments (Table 2).

Regarding the success of the actions carried out, both the coordinator and the facilitator of the YACCs highlight: the youth's satisfaction with the usefulness, visibility, and impact of the actions in their local contexts, the encouragement of creativity, and the freedom to express themselves without feeling judged in these participatory spaces. However, they also identified obstacles that prevented some members from participating in specific actions. For example, participation in meetings and actions with other councils in the project was hindered by the language barrier (English), which emerged as a deterrent to involvement: "However, I must say that they found it hard to get excited. There were certain complexities, like having to speak in English or other things that seemed a bit strange" [*Facilitator*]. They also mentioned the challenge of conflicting schedules, particularly for the most active members: "We have several people who sign up, but their lives are too busy. It's the usual thing — the most involved profiles, or those who want to participate in more activities, eventually get saturated" [*Facilitator*]. Finally, the issue of family consent was raised as a key factor in facilitating the participation of members under 18 in activities outside their municipalities:

> In the case of trips, talking to other organisations (like Erasmus Plus...), sometimes it's hard to gather participants from this age group. In some cases, even if the young person wanted to go, it was the parents or family who pulled back. Either they weren't convinced by us or our way of working. [*Coordinator*]

4.3 Youth Advisory Climate Councils' Proposals for Local Governments on the Climate Emergency

To delve into the proposals developed, their social and environmental impact over the 3 years of the YACC's operational period, the following is a summary (Table 2) of the proposals made and the evidence of success in their respective municipalities.

Table 2 Relationship between initiatives addressing the climate emergency and degree of implementation in the municipalities

Municipality	Proposals	Evidence of sucess
As Neves	1. Energy communities. 2. Green routes and bike lanes. 3. Second-hand local market. 4. Increase in recycling bins. 5. Reforestation with native species. 6. Youth participation. 7. Expansion of community spaces.	1. Road safety: traffic calming, reduced speed limits, pedestrian priority, and fruit tree planting. 2. Extended opening hours of community spaces and expanded book offerings. 3. Hiring of a youth centre facilitator. 4. Design of a wildfire prevention plan.
Teo	1. Improve public transport: new routes and increased bus frequency. Development of a geolocation app. 2. Greater variety of recycling bins and more waste bins in rural areas. 3. Awareness campaigns against pollution and climate change. 4. Offer fiscal benefits or other types of economic rewards. 5. Community surveys on environmental issues.	1. Inclusion of some proposals in the terms of the new municipal waste contract. 2. Request to the bus company for the implementation of the geolocation app. 3. Improved public transport: more frequent services and connections. 4. Installation of compost bins for organic waste. 5. Campaigns for proper waste separation and reduction of consumption.
Oleiros	1. School gardens and chicken coops. Talks with workers in the primary sector and fishing industry. 2. Climate-neutral city: clean energy, the European New Bauhaus construction model, or energy communities. 3. Protect the forest: planting native species, beekeeping, outdoor sports. 4. Support youth entrepreneurship with advice and reduced bureaucracy.	1. Creation of gardens in kindergartens, primary schools, and secondary schools. 2. Development of the Urban Agenda, including public policies on energy efficiency. They report having the YACC's collaboration in its preparation.
A Guarda	1. Forest fire management: planting native species and aromatic plants, beekeeping, livestock farming, among others. 2. Water: sewage system, wastewater treatment plants, beach cleaning, awareness campaigns, etc. 3. Urban space: green routes, urban gardens, vegetable planters, or environmental murals, among others. 4. Sustainable festival: workshops, art, sports, food, competitions, market, among others. 5. Increased youth participation.	1. Progress in fire prevention: creation of a green ring, removal of invasive species, or hiring a new municipal worker responsible for these areas, among others. 2. Currently studying other YACC proposals. 3. Willingness to receive new proposals from YACC.

Note Information obtained from the meeting minutes between YACCs and municipal governments in November 2024

4.4 *Impact of the Youth Advisory Climate Councils Initiative on Education and Society*

Over the 3 years that the YACC have been operational, their actions have contributed to the socialisation of the climate emergency across the four municipal contexts. The interview with the individuals responsible for coordinating and facilitating the YACC revealed the positive impact of the initiative on three social actors: youth, the facilitators, and municipal agents.

4.4.1 Impacts on Youth

Participation in the climate councils has fostered certain transformations in the attitudes, skills, and habits of the young individuals involved. Although some of these changes are expected to manifest in the medium or long term, significant progress has been observed in the expansion of knowledge and critical thinking regarding the climate emergency and other socio-environmental issues: "I believe they learned a great deal about the climate emergency, about measures, and most importantly, to have a critical spirit and analyse things a little before accepting everything that comes through social media" [*Facilitator*]. In particular, social media posts have served as a key tool for the dissemination and consolidation of these learnings, with one facilitator highlighting:

> As they linked the Instagram account of the climate council to their own profiles, they published everything we did and shared it. [...] At least, participating in the climate councils opens doors or windows for them to access information on other topics. [*Facilitator*]

In terms of daily habits, modifications attributable to the campaigns promoted under the project have been identified, such as the use of reusable bottles to reduce single-use plastic consumption. Furthermore, there has been a recorded increase in aspects related to socio-environmental awareness and their agency in community life through dialogue with local authorities:

> Attitude changes in youth regarding participation in activities and encouraging them to interact with the town hall. Not as an entity in a 'political' sense, but as a government that legislates, makes actions, and does things that affect day-to-day life. So why not give your opinion, propose things that you need or think are necessary for daily life in your village, town, or whatever. [*Facilitator*]

Finally, some participants have expanded their involvement in climate action, with one coordinator stating: "There were already people who joined other European projects" [*Coordinator*].

4.4.2 Impacts on the Role of the YACC Facilitator

The experience of facilitating the YACC has had a substantial impact on the personal and professional development of the facilitator, who highlights both positive aspects and challenges encountered throughout the 4 years of the project: "Positively, many things, from personal growth to professional development. There were also moments over these 3 years when I felt down because I no longer knew how to get things moving" [*Facilitator*]. The facilitator values the inclusive and horizontal nature of the YACC, as it fostered interaction between themselves and the young members of the council, as well as with others interested in participating in the actions promoted:

> In the end, youth are not as closed off as they are often portrayed; they are welcoming. It happened in the climate councils that very diverse people joined, and overall, the group always welcomed everyone, no matter how strange or special they seemed, or the odd opinions they may have had. There were no difficulties in this regard. I also take this as a positive experience. I've learned a lot that youth are very flexible. […] They see you as older, but they are inclusive. It's like 'you are older, but come on, let us explain how things work, how we want them, how we like things to function, how we want the activity, how we think it's right'. [*Facilitator*]

Among the challenges identified, the facilitator notes the need to adapt content and activities to the group's interests, the generational gap between the facilitator and the youth, and the difficulty in maintaining participants' motivation at certain points in the process. Confronting these challenges enabled the redefinition of socio-educational strategies, the management of frustration, and the development of greater empathy with the sensitivities, priorities, and interests of the youth: "Sometimes we think 'these kids have nothing to do, why don't they answer me when they're on their phones all day?' But their lives have complexities you don't understand" [*Facilitator*]. Moreover, the facilitator emphasises the youth's capacity to take responsibility, generate ideas, and commit to climate action. Thus, they stress the importance of "actively involving youth in the entire participatory process to increase their commitment and achieve the project's objectives" [*Facilitator*].

4.4.3 Influence on Municipal Agents

The final evaluation carried out in the four municipal governments by the project coordinator suggests positive progress in the perception of youth participation, particularly with regard to decision-making and involvement in their municipalities. In this regard, the evidence of success (Table 2) reinforces the need for local authorities to be accountable for the proposals put forward by the YACC. These results suggest a progressive transformation in local governance, where youth are beginning to take on a more active role in shaping policies addressing the climate emergency at the municipal level:

> I can't speak on behalf of the municipalities, but I believe they realised that it is important to involve youth. In the final meetings, I insisted a lot that when they had further advancements on the proposals that they made, when they wanted to implement something, please involve them. You can incorporate the voice of youth. In one case, for example, when they were working on the urban agenda (Oleiros), they suggested to the colleague managing the urban agenda that the participatory process should not only involve neighbourhood associations but also take into account the perspective of youth. We insisted a lot on this and hope that, as a result of the councils, they now recognise that this participation is important" [*Coordinator*].

5 Conclusions, Challenges, and Lessons Learned

The development of the YACC has allowed the identification of a series of obstacles that influence their implementation and drive dynamic processes to adapt and respond effectively to the objectives for which they were created. Some of these challenges pose future difficulties that require coordinated efforts to advance youth participation in addressing the climate emergency at the local level.

On the one hand, there are inherent challenges related to the developmental stages of the participants due to the wide age range (15–35 years) of the young people involved. This age difference entails a great diversity of interests and expectations, which affects the level of motivation and commitment to participation and highlights the correlation between youth training and guidance from adults. Furthermore, the lack of short-term results and rewards, due to the scheduling and sequencing of activities, as well as the need for greater commitment and responsibility in some initiatives, also present challenges for their achievement or even lead to the abandonment of some members. Thus, ensuring the continuity of the YACC, achieving effective management, and facilitating the transition to new recruits are significant challenges to maintaining their activity and operation. Another identified obstacle, specifically related to certain actions, has been the reluctance of some participants to publicly expose or "label" themselves in the digital content associated with the activities carried out within the YACC for social media platforms. Finally, there is the limited capacity to access funding and available resources—both material and space—which conditioned their operation, and the outcomes achieved. Nevertheless, the YACC represents a suitable initiative with great potential for promoting child and youth participation in the municipalities where it was implemented.

However, to guide the future implementation of other YACC and improve the operation of those currently in place, it is necessary to identify certain challenges. In an increasingly fast-paced world —both climatically and socially— there is a need to coordinate the agendas and daily schedules of YACC members to plan initiatives that respond to their needs, interests, and demands. Therefore, it is crucial to understand the context (rural/urban) of the different participating localities, the residential locations of the participants, and the transportation opportunities, as these are some of the factors that will decisively influence the political participation of young people. Similarly, the family context will contribute to the degree of acceptance of the initiative, as certain family education models constrain the autonomy and sense of agency of their children, fostering distrust, fear, and detachment. Lastly, for the success of the YACC to be ensured, there must be a favourable disposition from local authorities, facilitating communication and coordination between local authorities, the individuals responsible for and members of the initiative, and other social and educational agents in the municipality (schools, families, associations, etc.). In this regard, the implementation of the YACC resulted in greater collaboration among various actors, with notable new collective actions, such as reforestation with native trees (As Neves) within the framework of the 1Planet4All programme by *Ayuda en*

Acción, meetings with other youth structures from Spain (Andalusia and Madrid), and participation in climate action forums by UNICEF Spain.

6 Study Limitations and Prospects for Future Research

This research focused on the analysis of the perceptions of the adults responsible for coordinating the European project and facilitating the Youth Advisory Climate Councils (YACC) in Galicia. Despite the potentialities, some limitations have also been identified, notably regarding information bias, as it solely reflects the adult perspective without being contrasted with that of the participating youth. Furthermore, the study has concentrated on a single partner of the People & Planet project, which restricts the representativeness of the findings.

In terms of foresight and continuity, the next steps of the research will include the collection of perceptions from other key actors in the project. First, the stories and experiences of the youth members of the YACC in Galicia will be gathered through online questionnaires; a method that allows for efficient outreach to a wide number of participants. However, to gain deeper insights and analyse subjective nuances, in-depth interviews will be conducted with those young people who wish to participate voluntarily. In this regard, it is worth noting that in Portugal, interviews have been scheduled with the coordinator and the facilitator of the YACC, in addition to online questionnaires being sent to the youth. In the other participating countries, online questionnaires will be directed at both facilitators and youth members of the YACC.

The aim of this proposal is to obtain a more comprehensive understanding of the participatory dynamics, challenges, and best practices of these spaces, fostering a holistic view of the impact and functioning of the YACC. This line of research could be complemented with broader qualitative studies, incorporating ethnographic methodologies or longitudinal studies that would allow for an analysis of the evolution of youth participation as a disruptive socio-educational response to the climate emergency.

Acknowledgements This research was supported by Resclim@tiempo project (PID2022-136933OB-C21), funded by MCIN/AEI/10.13039/501100011033 and by "ESF Investing in your future". This work is also carried out within the framework of the Spanish Ministry of Universities' Grants for University Teacher Training (FPU21/00667).

References

Bergmann, Z., & Ossewaarde, R. (2020). Youth climate activists meet environmental governance: Ageist depictions of the FFF movement and Greta Thunberg in German newspaper coverage. *Journal of Multicultural Discourses, 15*(3), 267–290. https://doi.org/10.1080/17447143.2020.1745211

Dunne, A., & Bijwaard, D. (2021). *Pan-European survey. Main multi-country report*. Ipsos Belgium. https://www.suedwind.at/bildungsmaterial/pan-european-survey-a-main-multi-country-report/

Escorial, A., Mini, M., & Ponce, C. (2022). *Los retos que la degradación medioambiental y el cambio climático plantean a la infancia [The challenges that environmental degradation and climate change pose to childhood]*. Plataforma de Infancia España. https://www.observatoriodelainfancia.es/oia/esp/documentos_ficha.aspx?id=7912

Firinci, T. (2022). Youth's everyday environmental citizenship: An analytical framework for studying interpretive agency. *Childhood, 29*(4), 495–511. https://doi.org/10.1177/09075682221107750

Misch, A., Kristen-Antonow, S., & Paulus, M. (2021). A question of morals? The role of moral identity in support of the youth climate movement Fridays4Future. *PLoS One, 16*(3), e0248353. https://doi.org/10.1371/journal.pone.0248353

Parsons, M., Bhor, G., & Crease, R. P. (2023). Everyday youth climate politics and performances of climate citizenship in Aotearoa New Zealand. *Environment and Planning E: Nature and Space, 7*(3), 1436–1460. https://doi.org/10.1177/25148486231208205

People & Planet: A common destiny. (2023). *Young European citizens declaration*. Accessed Febraury 16, 2025, from https://www.imvf.org/wp-content/uploads/2023/03/yec-water-declaration-people-and-planet.pdf

People & Planet: A common destiny. (2024). *Acting locally, impacting globally: Compendium of good practices for promoting partnerships between youth and local authorities for climate action and sustainability*. Accessed Febraury 16, 2025, from https://wateroffuture.org

Prosser, G., Bonilla, N., & Arboleda-Ariza, J. (2020). Educación para el cambio climático y saberes ambientales en declaraciones del alto segmento de la COP25 [Education for climate change and environmental knowledge in statements from the high segment of COP25.]. *Revista Mexicana de Investigación Educativa, 25*(87), 873–899. https://comie.org.mx/v5/sitio/wp-content/uploads/2020/11/RMIE_87

Ripple, W. J., Wolf, C., Newsome, T. M., Barnard, P., & Moomaw, W. R. (2020). World scientists' warning of a climate emergency. *Bioscience, 70*(1), 8–100. https://doi.org/10.1093/biosci/biz152

Salvia, M., Reckien, D., Geneletti, D., Pietrapertosa, F., D'Alonzo, V., Hurtado, S. D. G., Chatterjee, S., Bai, X., & Ürge-Vorsatz, D. (2023). Understanding the motivations and implications of climate emergency declarations in cities: The case of Italy. *Renewable and Sustainable Energy Reviews, 178*, 113236. https://doi.org/10.1016/j.rser.2023.113236

Save the Children. (2021). *Born into the climate crisis—Why we must act now to secure children's rights*. Save the Children International. https://resourcecentre.savethechildren.net/document/born-climate-crisis-why-we-must-act-now-secure-childrens-rights/

Spanish Government. (2020). *La Declaración del Gobierno ante la Emergencia Climática* [The Government Declaration on the Climate Emergency]. Spanish Government. https://www.miteco.gob.es/content/dam/miteco/es/prensa/declaracionemergenciaclimatica_tcm30-506551.pdf

Thiery, W., Lange, S., Rogelj, J., Schleussner, C. F., Gudmundsson, L., Gudmundsson, L., Seneviratne, S., Andrijevic, M., Frieler, K., Emanuel, K., Geiger, T., Bresch, D., Zhao, F., Willner, S., Büchner, M., Volkholz, J., Bauer, N., Jinfeng Chang, J., Ciais, P., Dury, M., François, L., Grillakis, M., Gosling, S., Hanasaki, N., Hickler, T., Huber, V., Ito, A., Jägermeyr, J., Khabarov, N., Koutroulis, A., Liu, W., Lutz, W., Mengel, M., Müller, C., Ostberg, S., Reyer, C., Stacke, T., & Wada, Y. (2021). Intergenerational inequities in exposure to climate extremes. *Science, 374*(6465), 158–160. https://doi.org/10.1126/science.abi7339

United Nations Children's Fund. (2023). *The climate-changed child: A children's climate risk index supplement*. UNICEF. https://reliefweb.int/report/world/climate-changed-child-childrens-climate-risk-index-supplement-enar

Valdivia, A., Ibáñez-Canelo, M. J., Palenzuela-Fundora, Y., & Cruz-Martínez, Y. (2024). Futuros y educación en los activismos juveniles socioambientales [Futures and education in youth socio-environmental activism]. *Revista Latinoamericana de Ciencias Sociales, Niñez y Juventud, 22*(1), 1–27. https://doi.org/10.11600/rlcsnj.22.1.6001

A Final Reflection

Environmental Education at the Limits: Accelerating and Deepening Educational Responses to the Climate Emergency

Pablo Ángel Meira Cartea

Abstract This concluding chapter offers a critical reflection on the trajectory and limitations of Environmental Education in the face of climate emergency. The author argues that Environmental Education has often failed to confront structural drivers of ecological crisis, aligning too closely with the growth-oriented paradigm of sustainable development. In response, it calls for a disruptive reorientation positioning Environmental Education as a counter-hegemonic praxis. The text highlights four guiding imperatives for rethinking the field: a moral duty to reduce suffering, an ontological awareness of planetary limits, a strategic recognition of human eco-dependence, and a political commitment to democratic sustainability.

Environmental Education (EE) is a discipline with a distinctive identity and trajectory. Making use of the sociological concept of field, coined by Pierre Bourdieu, we have always argued that EE is a trans-field or sub-field located at the interface between the educational and environmental fields. Qualified as such, it has another important peculiarity: in both fields EE usually plays a subsidiary or secondary role, which contrasts with the relevance and performative transcendence usually granted to it in practically all international agreements and treaties related to the educational or environmental fields. In its characterization as a trans-field, EE constitutes a necessarily hybrid domain of pedagogical knowledge and educational praxis. As such, it is a practical example of the aspiration of post-normal epistemology to construct representations of reality that are inter- or even transdisciplinary, that aim to transform it, and that engage in dialogue —bridging the social and natural sciences— with other cultural forms of understanding and interacting with the world and within it.

Despite what more conventional foundations may suggest, the object of EE is not the biophysical environment; that is, in any case, its indirect object. The object of EE is to transform the social and symbolic relationships we establish among ourselves and from which we value, transform, distribute, and metabolize the natural

P. Á. Meira Cartea (✉)
Social Pedagogy and Environmental Education Research Group (SEPA-interea), Universidade de Santiago de Compostela, Santiago de Compostela, Spain
e-mail: pablo.meira@usc.gal

J. J. Conde et al. (eds.), *Practices in Education for Climate Emergency*,
https://doi.org/10.1007/978-981-92-2304-6_22

resources provided by the beautiful, complex, and fragile planet we inhabit. As I often tell my students to challenge their naturalistic representations of EE and the environmental crisis, this task is more closely related to economics, sociology, anthropology, politics, or ethics than to ecology or biology. As the climate crisis demonstrates, an economy that is not also ecological or an ecology that is not also social, may lead us toward civilizational collapse.

Environmental Education in its contemporary sense began its journey three-quarters of a century ago. We must recognize it: this is not a story of success. Perhaps we have won some battles on the cultural and symbolic fronts, and also at the local scale, but we are losing the global war of human readjustment to the planet's biophysical limits. The world of 2025 is more unsustainable —also more unjust and unequal— than the world of the 1960s and 1970s. This failure is not the responsibility of EE; but EE, especially hegemonic EE, does bear some share of responsibility in this failure, because it ignores the structural causes of the environmental crisis, fails to acknowledge that our world is finite, and does not understand that environmental sustainability cannot be achieved without social sustainability.

In its most critical approaches, Environmental Education has been and continues to be a countercurrent praxis, with a marked countercultural and counter-hegemonic orientation. History ordained that its institutionalization coincided with the neoliberal tsunami of the 1980s and 1990s, which ushered in what we may label the global market economy. In the 1970s, there was an aspiration —admittedly naïve— to situate the environmental issue among the pillars of human well-being. Such an aspiration entailed recognizing the planet's biophysical limits in its capacity to provide resources for humanity and to absorb the impacts of our activities; but it also required a fair and equitable distribution of those resources and impacts.

However, the very model of production and consumption that has brought us to this impasse distorted those aspirations by linking environmental sustainability and human development to the imperative of economic growth. Such is the fallacy that underlies the precept of sustainable development, a conceptual framework conceived in the workshops of the World Bank in the 1980s, which still dominates environmental and development rhetoric, as expressed in the UN's 2030 Agenda. We are not opposed to the values that this document seeks to embody, but the reality of the results achieved after a decade of implementation reveals its great weakness: it is impossible to halt the degradation of the biosphere —which enabled the proliferation of our species over the last 10,000 years— if the growth of economic magnitudes associated with production and consumption, and with capital reproduction, continues to be accepted as a condition without which human development is deemed impossible. More precisely, if target 1 of SDG-8 on "Decent Work and Economic Growth," which seeks to sustain per capita economic growth at normal annual rates and at 7% in the least developed countries, is fulfilled, it will be impossible to meet the targets set in the environmental SDGs. This reality is evidenced by the scarce achievements regarding the targets set for SDG-13, "Climate Action," linked to the temperature limitation goals established in the Paris Agreement (2015). It must be stressed that the first goal set in Paris, limiting the average atmospheric temperature increase to 1.5 degrees Celsius by the end of this century, had already been surpassed in 2025, and all projections indicate that the second target established in the city on the Seine —limiting the increase to 2 degrees Celsius by the end of the century— will be exceeded during the next decade.

Following the logic that associates growth with development, since EE began its trajectory in the 1960s, global GDP has multiplied by 100. Yet inequality in the distribution of the income generated by that increase has also grown, and several planetary boundaries have been exceeded, such as climate stability, biodiversity loss, soil depletion, and the availability of sufficient water resources (Richardson et al., 2023). The reality is that we are reaching the limits, and this scenario of overshoot, collapse, or socio-environmental proto-collapse forces us to reconsider the identity, mission, and strategies of Environmental Education. Disruptive pedagogical and social work to facilitate the transition toward a horizon of sustainable and just degrowth is possibly the last opportunity to frame the praxis of Environmental Education realistically and ethically coherently.

The initiative for this publication arises from our deep concern over the runaway trajectory of the climate crisis and over the weakness and irrelevance of the educational responses articulated thus far. As we anticipated in defining EE as a subsidiary trans-field of both education and environment, and despite the evidence of the climate crisis, education occupies a secondary role in climate policy, just as climate occupies a secondary place in educational policy. This imbalance can be observed at different spatial and institutional scales.

When we decided two decades ago to place the challenges of climate change at the center of our scientific and academic activity, we were not well understood.[1] Why grant such relevance to this issue compared to so many others that express the misalignment between human societies and the biosphere? Why prioritize educational responses to the climate crisis when so many pressing human challenges are associated with poverty, justice, equity, or even with other environmental misalignments?

Time, unfortunately, has proven us right.

We always explain it with an analogy we consider quite eloquent. The pool of humanity's problems is overflowing and continues to accumulate challenges for contemporary education. But anthropogenic climate change is not just another problem, for it entails a structural fracture of the pool itself. How we respond to that fracture will determine whether it is possible to adequately address the other environmental and human issues floating within it.

And it is not only a structural problem because of its systemic biophysical implications; it is also so —this consideration appearing implicitly or explicitly in all the educational experiences that comprise this monograph— because its ultimate causes

[1] We refer to the RESCLIMA Project, within which the initiative of this publication is framed. RESCLIMA is the acronym for *Educational and Social Responses to Climate Change*. This university research and transfer program, promoted by the Research Group on Social Pedagogy and Environmental Education of the University of Santiago de Compostela (SEPA-interea), took its first steps a quarter of a century ago. For a more detailed account of the initiatives it has developed, see the website https://resclima.info/ or the essay *RESCLIMA: Two Decades of Research in Educational and Social Responses to Climate Change* (Conde et al., 2026).

point directly to the model of production and consumption initiated with the Industrial Revolution and exacerbated in the current phase of global market capitalism. It is staggering to think that since Environmental Education began its journey in the 1960s and 1970s, greenhouse gas emissions have tripled, and the only recorded annual decreases were not the result of the implementation of the meager mitigation policies agreed upon in the global governance system, but of the conjunctural crises of the prevailing economic model, the disappearance of the Soviet Union, or a disruptive event such as the COVID-19 pandemic. Indeed, the already fragile global governance system associated with the work of the UN, which produced the Kyoto Protocol in 1997 or the Paris Agreement (2015), currently in force, appears increasingly weakened and incapable of articulating effective policies to reduce — or even flatten— greenhouse gas emissions, which continue to grow year after year, rebounding after the relative collapse caused by the paralysis of the world economy during COVID-19.

Perhaps this is why we procrastinate so much in climate policy responses. And perhaps this is also why the field of Environmental Education —and Education without any qualifier— still struggles to place the climate crisis at the center of its priorities.

Given this context of climate and socio-environmental emergency, there is a question laden with eco-pessimism that multiplies in the territories of Environmental Education: **are we still on time?** Perhaps it is not the right question —or at least not the most pertinent— for those of us who act as educators. Recalling the legacy of Paulo Freire, we concur with the precept that there can be no pedagogy without hope; but we are also aware that hope must be realistic and, above all, proactive. Thus, we prefer to reframe the question to ask ourselves: **what are we still on time for?** The search for answers to this question is at the origin of this monograph.

It is not easy to address, but we dare to suggest **four keys** for beginning to explore possible answers from pedagogical theory and educational praxis. Readers of this monograph will have been able to observe that these keys have been taken into account, in whole or in part, in the design and development of the educational experiences it includes.

The first invokes a moral imperative: we are still on time to avoid a significant share of human and non-human suffering that could be caused by a biosphere outside the climatic parameters that have made it habitable for our species; a suffering that is and will be especially intense among social groups most vulnerable to its imbalances.

The second key is ontological: through educational praxis we must cultivate an awareness of limits and of the need to live within that safe and just space which the forgotten Cocoyoc Declaration located in 1974 between the internal limits of satisfying basic human needs and the external limits of a finite planet. A sustainable and just degrowth is a requirement for transitioning toward a viable humanity.

The third key is strategic: human beings remain as eco-dependent as the first hominids were, or as any living being is that inhabits this pale blue dot floating on the edge of the Milky Way. We doubt that any hyper-billionaire can terraform Mars and cultivate vegetables on its barren surface scorched by cosmic radiation. The reality is that the more than 8 billion human beings alive today, and all those yet to come, will continue to depend on the habitability of the biosphere in order to exist with dignity. Likewise for the other living beings that accompany us and without whose coexistence the habitability of the planet for our species would be a chimera.

The fourth key is political: human beings are increasingly interdependent. One of the paradoxes of the global market economy is its capacity to dissolve the fragile institutional mechanisms of governance associated with the UN, as well as to weaken democracies wherever they existed. This situation is now more evident. An Environmental Education that contributes to eco-social transition is also a political praxis; that is, it must be aimed at cultivating a conscious and proactive citizenry that fosters a democratic culture which is also a culture of sustainability, both local and global; that is, a pedagogy that contributes to respecting the commons and works for the common good according to criteria of equity and social justice.

The 19 educational experiences and initiatives that comprise this monograph are aligned with these considerations. They demonstrate that it is possible to deepen the incorporation of the climate crisis into educational praxis from a complex and socially committed perspective in diverse educational and social settings. We simply need to multiply their scale and adapt the strategies, designs, and educational resources presented here in order to accelerate the adjustment of educational praxis to the scale of the civilizational challenge posed by the climate crisis.

Finally, we would like to recall the words written by William Ruckelshaus (1989, p. 167), the first Administrator of the United States Environmental Protection Agency; he asked:

> Can we move nations and people in the direction of sustainability? Such a move would be a modification of society comparable in scale to only two other changes: the agricultural revolution of the late Neolithic and the Industrial Revolution of the past two centuries. Those revolutions were gradual, spontaneous and largely unconscious. This one will have to be a fully conscious operation, guided by the best foresight that science can provide-foresight pushed to its limit. If we actually do it, the undertaking will be absolutely unique in humanity's stay on the earth.

This challenge justifies the need to resize the mission and praxis of Environmental Education and, realistically, of Education without any qualifying adjective. The troubling fact is that these words were written in 1989, and we are now in 2025. Hence, we cannot avoid the question that troubles us but also mobilizes us: what are we still on time for?

References

Conde, J. J., García-Vinuesa, A., Bisquert, K. M., & Meira Cartea, P. Á. (2026). RESCLIMA: Two decades of research in educational and social responses to climate change. In W. Leal Filho, M. Sima, A. Lange Salvia, M. Kovaleva, & E. Manolas (Eds.), *University initiatives on climate change education and research* (pp. 2087–2118). Springer Nature. https://doi.org/10.1007/978-3-031-59731-2_109

Richardson, K., Steffen, W., Lucht, W., Bendtsen, J., Cornell, S. E., Donges, J. F., Drüke, M., Fetzer, I., Bala, G., Von Bloh, W., Feulner, G., Fiedler, S., Gerten, D., Gleeson, T., Hofmann,

M., Huiskamp, W., Kummu, M., Mohan, C., Nogués-Bravo, D., et al. (2023). Earth beyond six of nine planetary boundaries. *Science Advances, 9*(37), eadh245. https://doi.org/10.1126/sciadv.adh2458

Ruckelshaus, W. D. (1989). Toward a sustainable world. *Scientific American, 261*(3), 166–175. https://www.jstor.org/stable/24987408

MIX
Papier aus verantwortungsvollen Quellen
Paper from responsible sources
FSC® C105338

If you have any concerns about our products,
you can contact us on
ProductSafety@springernature.com

In case Publisher is established outside the EU,
the EU authorized representative is:
Springer Nature Customer Service Center GmbH
Europaplatz 3, 69115 Heidelberg, Germany

Printed by Libri Plureos GmbH
in Hamburg, Germany